Sustainable Diets

Sustainable Diets
Linking Nutrition and Food Systems

Edited by

Barbara Burlingame

Massey University, Wellington, New Zealand

and

Sandro Dernini

Forum on Mediterranean Food Cultures
Rome, Italy

CABI

CABI is a trading name of CAB International

CABI
Nosworthy Way
Wallingford
Oxfordshire OX10 8DE
UK

Tel: +44 (0)1491 832111
Fax: +44 (0)1491 833508
E-mail: info@cabi.org
Website: www.cabi.org

CABI
745 Atlantic Avenue
8th Floor
Boston, MA 02111
USA

Tel: +1 (617)682-9015
E-mail: cabi-nao@cabi.org

A catalogue record for this book is available from the British Library, London, UK.

Library of Congress Cataloging-in-Publication Data

Names: Burlingame, B. A. (Barbara A.), editor. | Dernini, Sandro, 1949- editor. | C.A.B. International, issuing body.
Title: Sustainable diets / edited by Barbara Burlingame and Sandro Dernini.
Other titles: Sustainable diets (Burlingame)
Description: Wallingford, Oxfordshire ; Boston, MA : CABI, [2019] | Includes bibliographical references.
Identifiers: LCCN 2018033249 | ISBN 9781786392848 (hb) | ISBN 9781786392862 (ePub) | ISBN 9781786392855 (ePDF)
Subjects: | MESH: Food Supply | Diet | Nutritive Value | Nutrition Policy | Environmental Policy | Conservation of Natural Resources
Classification: LCC RA784 | NLM WA 695 | DDC 613.2/5--dc23 LC record available at https://lccn.loc.gov/2018033249

ISBN-13: 978 1 786392848 (hbk)
9781786392862 (EPub)
9781786392855 (ePDF)

Commissioning Editors: Rachael Russell and Rebecca Stubbs
Editorial Assistants: Tabitha Jay and Alexandra Lainsbury
Production editor: James Bishop

Cover image by Sebastian Comelli
Typeset by SPi, Pondicherry, India
Printed and bound in the UK by Severn, Gloucester.

Contents

Preface

On planet Earth, sustainability is the catch-phrase on everyone's agenda, whether referring to agriculture, the environment or health. The concept of sustainable diets brings these sectors together to simultaneously address major global priorities. Today, in spite of many efforts, the food security and sustainability challenges are escalating within the context of a shrinking natural resources base and climate change, highlighting the inadequacy of present unsustainable food systems and unhealthy dietary patterns. This book positions sustainable diets as central to the Earth's future to tackle the pressing challenges that have taken us to the edge of, and beyond, the planet's resources. It urges transformational changes in policies and actions towards more sustainable diets and sustainable food systems, to better deliver food security and nutrition for all. The case is strongly made that these actions must be transdisciplinary to be successful.

Improving food systems for sustainable diets requires an intersectoral effort to reverse the simplification of diets, the degradation of ecosystems, and the erosion of biodiversity. The book takes a transdisciplinary approach and considers multisectoral actions, integrating health, agriculture, environment, economy, and socio-cultural issues, to comprehensively explore the topic of sustainable diets. It presents the latest findings and the associated challenges, arguments, perspectives, dilemmas, actions, policies and solutions. Consideration is given to the multi-dimensional nature of diets and food systems, and explores the challenging issues connecting food security and nutrition to sustainability, culture, tradition, and a broader range of scientific topics.

This book is the continuation of collaborative efforts over several years with many of the chapter authors. It is a follow-up to our work on *Sustainable Diets and Biodiversity: Directions and Solutions for Policy, Research and Action*, in which was published the consensus definition of sustainable diets, cited in many chapters. This definition, now broadly internationally accepted, acknowledges the interdependencies of food production and consumption with food requirements and nutrient recommendations, and at the same time, reaffirms the notion that the health of humans cannot be isolated from the health of ecosystems:

> *Sustainable diets are those diets with low environmental impacts which contribute to food and nutrition security and to healthy life for present and future generations. Sustainable diets are protective and respectful of biodiversity and ecosystems, culturally acceptable, accessible, economically fair and affordable; nutritionally adequate, safe and healthy; while optimizing natural and human resources.*
>
> (FAO, 2012)

As practised today, there is very little about diets that is sustainable. Intensive production systems, relying on ever-increasing inputs of agricultural chemicals, are progressively degrading ecosystems and pushing devastating losses of biodiversity. Consumption, with imbalanced and excessive intakes from these same productions systems, coupled with losses and waste along food chains and within food systems, is not sustainable. Obesity and diet-related chronic diseases are the expectation and norm for developed and developing countries alike.

This realization is not new. Human and planetary health have been the topics of research, policies and interventions for decades. Despite many efforts over time, the downward trajectory of food security and sustainability continues. The anthropometry of malnutrition has changed in many parts of the world, shifting from high prevalence of underweight to overweight and obesity. By measuring the quantity of food, or dietary energy supply and availability, food security has improved. But by measures of quality of foods, represented by nutrient adequacy, and micronutrients in particular, malnutrition rates are alarming. Combined with population growth, urbanization, degraded and diminishing natural resources and climate change, we have a diet crisis.

Sustainable Diets: Linking Nutrition and Food Systems, presents relevant global topics that must be properly understood in order to be effectively addressed. The transition to sustainable diets is explored within the context of sustainable food systems and the right to food, overcoming the divide between disciplines and linking food security and nutrition to sustainability. The content is provocative, exercising thought-leadership, and is not reticent in conveying the urgency of action. In the end, it aims to provide a way forward for achieving relevant goals, targets and commitments, both global and local, by highlighting how diets, food consumption and production are interconnected and ecosystem dependent. Although the evidence base must be improved, the chapter authors are unanimous in their view that immediate action is warranted to promote sustainable diets, linking nutrition to sustainable food systems, connecting the wellbeing of the individual and the community to the wellbeing of the planet.

Barbara Burlingame and Sandro Dernini

Contributors

Alicia Aguilar, Faculty of Health Sciences, Universitat Oberta de Catalunya, Barcelona, Spain. E-mail: aaguilarmart@uoc.edu

Marie-Josèphe Amiot-Carlin, Département Alimentation Humaine INRA MOISA, Université Montpellier, CIRAD, CIHEAM-IAAM, INRA, SUPAGRO, Montpellier, France. E-mail: marie-josephe.amiot-carlin@univ-amu.fr

Roberto Azofeifa Rodríguez, Sustainable Production Department, Ministry of Agriculture and Livestock (MAG), Costa Rica. E-mail: razof@mag.go.cr

Phillip Baker, Institute for Physical Activity and Nutrition (IPAN), School of Exercise and Nutrition Sciences, Deakin University, Geelong, Australia. E-mail: phil.baker@deakin.edu.au

Javier Aranceta Bartrina, Spanish Academy of Nutrition and Food Sciences, Barcelona; Spanish Society of Community Nutrition; University of Navarra, Pamplona, Spain. E-mail: jaranceta@unav.es

Caterina Batello, Plant Production and Protection Division, Food and Agriculture Organization of the United Nations, Rome, Italy. E-mail: Caterina.batello@fao.org

Jane Battersby, African Centre for Cities, University of Cape Town, Cape Town, South Africa. E-mail: jane.battersby.lennard@gmail.com

Elliot M. Berry, Braun School of Public Health, Hebrew University-Hadassah Medical School, Jerusalem, Israel; CIISCAM-International Inter-University Center for Mediterranean Food Culture Studies. E-mail: elliotb@ekmd.huji.ac.il

Marina Bortoletti, Environment Programme - Economy Division, United Nations, Paris, France. E-mail: marina.bortoletti@un.org

Gianluca Brunori, Department of Agriculture, Food and Environment, University of Pisa, Pisa, Italy. E-mail: gianluca.brunori@unipi.it

Camelia Bucatariu, Nutrition and Food Systems Division, Food and Agriculture Organization of the United Nations, Rome, Italy. E-mail: Camelia.Bucatariu@fao.org

Susanne Bügel, Department of Nutrition, Exercise and Sports, University of Copenhagen, Frederiksberg Campus, Denmark. E-mail: shb@nexs.ku.dk

Barbara Burlingame, Professor of Nutrition and Food Systems, College of Health Sciences, School of Public Health, Massey University, Wellington, New Zealand. E-mail: Barbara.burlingame@gmail.com

Roberto Capone, CIHEAM (International Center for Advanced Mediterranean Agronomic Studies), Bari, Italy. E-mail: capone@iamb.it

Sandro Dernini, Forum on Mediterranean Food Cultures; IFMeD-International Foundation of Mediterranean Diet, London, UK. E-mail: s.dernini@tiscali.it

Lorenzo Donini, Experimental Medicine Department, Medical Pathophysiology, Food Science and Endocrinology Section, Food Science and Human Nutrition Research Unit, Sapienza University, Rome; Scientific Council Member of CIISCAM-International Inter-University Center for Mediterranean Food Culture Studies, Rome, Italy. E-mail: lorenzomaria.donini@uniroma1.it
Corné van Dooren, Netherlands Nutrition Centre (Voedingscentrum), The Hague, Netherlands. E-mail: dooren@voedingscentrum.nl
Adam Drewnowski, Center for Public Health Nutrition, University of Washington, Seattle, WA, USA. E-mail: adamdrew@u.washington.edu
Marcela Dumani Echandi, Nutritionist, Professor of Food and Nutrition Security, Public Nutrition Section of School of Nutrition, University of Costa Rica, San José, Costa Rica. E-mail: marcela.ucr@gmail.com
Paul Eme, School of Public Health, Massey University, Wellington, New Zealand. E-mail: p.eme@massey.ac.nz
Jessica Fanzo, Berman Institute of Bioethics, School of Advanced International Studies and the Bloomberg School of Public Health, Johns Hopkins University, Baltimore, Maryland, USA. E-mail: jfanzo1@jhu.edu
Yon Fernandez de Larrinoa, Indigenous Peoples Team Leader, Food and Agriculture Organization of the United Nations, Rome, Italy. E-mail: yon.fernandezlarrinoa@fao.org
Jorge M. Fonseca, Food Systems Strategic Programme, Food and Agriculture Organization of the United Nations, Rome, Italy. E-mail: jorge.fonseca@fao.org
Emile Frison, Member of the International Panel of Experts on Sustainable Food Systems (IPES-Food) and Former Director General of Bioversity International, Rome, Italy. E-mail: e.frison@cgiar.org
Silvia Gaiani, Nutrition and Food Systems Division, Food and Agriculture Organization of the United Nations, Rome, Italy. E-mail: Silvia.Gaiani@fao.org; silviagaiani@yahoo.it
Andrea Polo Galante, Nutrition and Food Systems Division, Food and Agriculture Organization of the United Nations, Rome, Italy. E-mail: Andrea.PoloGalante@fao.org
Francesca Galli, Department of Agriculture, Food and Environment, University of Pisa, Pisa, Italy. E-mail: francesca.galli@unipi.it
Randall García Víquez, National Institute of Biodiversity, University of Costa Rica, San José, Costa Rica. E-mail: rgarcia@inbio.ac.cr
Kakoli Ghosh, Sustainable Agriculture Strategic Programme, Food and Agriculture Organization of the United Nations, Rome, Italy. E-mail: Kakoli.ghosh@fao.org
Vincent Gitz, CGIAR Research Program on Forests, Trees and Agroforestry, Center for International Forestry Research (CIFOR), Bogor, Indonesia. E-mail: v.gitz@cgiar.org
Esther González-Padilla, Research Institute of Biomedical and Health Sciences (IUIBS), Department of Clinical Sciences, University of Las Palmas de Gran Canaria, Las Palmas, Gran Canaria; Ciber Obn Physiopathology of Obesity and Nutrition, Institute of Health 'Carlos III', Madrid, Spain. E-mail: estherglzpadilla@gmail.com
Meredith Harper, School of Plant Sciences and Food Security, Faculty of Life Sciences, Tel Aviv University, Tel Aviv, Israel. E-mail: meredithmharper@gmail.com
Luis Antonio T. Hualda, School of Management, University of the Philippines Mindanao, Davao City, Philippines. E-mail: lthualda@up.edu.ph
Danny Hunter, Healthy Diets for Sustainable Food Systems Initiative, Biodiversity International, Rome, Italy. E-mail: d.hunter@cgiar.org
Massimo Iannetta, Biotechnologies and Agroindustry Division, ENEA - Italian National Agency for New Technologies, Energy and Sustainable Economic Development, Rome, Italy. E-mail: massimo.iannetta@enea.it
Nick Jacobs, Coordinator of the IPES-Food Panel, Brussels, Belgium. E-mail: nick.jacobs@ipes-food.org
Rebekah Jones, Department of Nutrition and Dietetics, Massey University, Auckland, New Zealand. E-mail: rebekah.airlie.jones@hotmail.com

Johannes Kahl, Section of Organic Food Quality and Food Culture, Kassel University, Witzenhausen, Germany. E-mail: kahl@uni-kassel.de

Rachel Bezner Kerr, Department of Development Sociology, College of Agriculture and Life Sciences, Cornell University, Ithaca, New York, USA. E-mail: rbeznerkerr@cornell.edu

Parviz Koohafkan, World Agricultural Heritage Foundation, Rome Italy. E-mail: parvizkoohafkan@gmail.com

Harriet Kuhnlein, Centre for Indigenous Peoples' Nutrition and Environment (CINE), McGill University, Montreal, Canada. E-mail: harriet.kuhnlein@mcgill.ca

Denis Lairon, Research Center for Cardiovascular Diseases and Nutrition (C2VN), Aix-Marseille University, Marseille, France. E-mail: denis.lairon@orange.fr

Tim Lang, Professor of Food Policy, City University of London, London, UK. E-mail: t.lang@city.ac.uk

Mark Lawrence, Institute for Physical Activity and Nutrition (IPAN), School of Exercise and Nutrition Sciences, Deakin University, Geelong, Australia. E-mail: mark.lawrence@deakin.edu.au

Rebecca Lindberg, Institute for Physical Activity and Nutrition (IPAN), School of Exercise and Nutrition Sciences, Deakin University, Geelong, Australia. E-mail: r.lindberg@deakin.edu.au

Allison Loconto, Institut National de la Recherche Agronomique (INRA), Laboratoire Interdisciplinaire Sciences Innovations et Sociétés (LISIS), Université Paris-Est Marne-la-Vallée, France; Plant Production and Protection Division, Food and Agriculture Organization of the United Nations, Italy. E-mail: allison-marie.loconto@inra.fr

James Lomax, Environment Programme, UN Environment's Division of Technology, Industry and Economics Paris, France. E-mail: james.lomax@un.org

Patricia Sedó Masís, Nutritionist, Professor of Culture and Food Heritage and Traditional Gastronomy, Public Nutrition Section of School of Nutrition, University of Costa Rica, San José, Costa Rica. E-mail: patricia.sedo@gmail.com

Pamela Mason, Consultant Public Health Nutritionist, Monmouthshire, UK. E-mail: pamelamason99@icloud.com

Dalia Mattioni, Department of Agriculture, Food and Environment, University of Pisa, Pisa, Italy; Nutrition and Food Systems Division, Food and Agriculture Organization of the United Nations, Rome, Italy. E-mail: daliamattioni@hotmail.com

F. Xavier Medina, UNESCO Chair on Food, Culture and Development, Faculty of Health Sciences, Universitat Oberta de Catalunya (UOC), Barcelona, Spain. E-mail: fxmedina@gmail.com

Alexandre Meybeck, Agriculture and Consumer Protection Department, Food and Agriculture Organization of the United Nations, Rome, Italy. E-mail: alexandre.meybeck@gmail.com

Patrick Mink, Federal Office for Agriculture, Switzerland, stationed in Rome, Italy. E-mail: patrick.mink@blw.admin.ch

Stineke Oenema, United Nations Systems Standing Committee on Nutrition, Rome, Italy. E-mail: stineke.oenema@fao.org

Nir Ohad, Director, Manna Center Program for Food Safety and Security, School of Plant Sciences and Food Security, Faculty of Life Sciences, Tel Aviv University, Tel Aviv, Israel. E-mail: nir.ohad@gmail.com

Adriana Ortiz-Andrellucchi, Research Institute of Biomedical and Health Sciences (IUIBS), Department of Clinical Sciences, University of Las Palmas de Gran Canaria, Las Palmas, Gran Canaria; Ciber Obn Physiopathology of Obesity and Nutrition, Institute of Health 'Carlos III', Madrid, Spain. E-mail: Adriana.ortiz@ulpgc.es

Ibukun Owoputi, Division of Nutritional Sciences, Cornell University, Ithaca, New York, USA. E-mail: io57@cornell.edu

Flavio Paoletti, Centre of Research on Food and Nutrition, Council for Agricultural Research and Economics (CREA), Rome, Italy. E-mail: flavio.paoletti@crea.gov.it

Suzanne Piscopo, Department of Health, Physical Education and Consumer Studies, Faculty of Education, University of Malta, Msida, Malta. E-mail: suzanne.piscopo@um.edu.mt

Maryam Rahmanian, Climate, Biodiversity, Land and Water Department (CBDD), Food and Agriculture Organization of the United Nations, Rome, Italy. E-mail: maryam.rahm@gmail.com

Rosa Rolle, Nutrition and Food Systems Division, Food and Agriculture Organization of the United Nations, Rome, Italy. E-mail: rosa.rolle@fao.org

Cristina Ruano-Rodriguez, Research Institute of Biomedical and Health Sciences (IUIBS), Department of Clinical Sciences, University of Las Palmas de Gran Canaria, Las Palmas, Gran Canaria; Ciber Obn Physiopathology of Obesity and Nutrition, Institute of Health 'Carlos III', Madrid, Spain. E-mail: Cristina.ruano@ulpgc.es

Pilar Santacoloma, Sub-Regional Office for Mesoamerica, Food and Agriculture Organization of the United Nations, Rome, Italy. E-mail: Pilar.santacoloma@fao.org

Lluis Serra-Majem, University of Las Palmas de Gran Canaria and CIBER OBN, Instituto de Salud Carlos III, Las Palmas, Spain; CIISCAM-International Inter-University Center for Mediterranean Food Culture Studies. E-mail: lluis.serra@ulpgc.es

Alon Shepon, Department of Plant and Environmental Sciences, Weizmann Institute of Science, Rehovot, Israel; Co-Founder and Chair of The Israeli Forum for Sustainable Nutrition, NGO, Tel Aviv, Israel. E-mail: alon.shepon@weizmann.ac.il

Michael Mulet Solon, World Wildlife Fund, Paris, France. E-mail: mmulet@wwf.fr

Andrea Sonnino, Biotechnologies and Agroindustry Division, ENEA - Italian National Agency for New Technologies, Energy and Sustainable Economic Development, Rome, Italy. E-mail: andrea.sonnino@enea.it

Milena Stefanova, Biotechnologies and Agroindustry Division, ENEA - Italian National Agency for New Technologies, Energy and Sustainable Economic Development, Rome, Italy. E-mail: milena.stefanova @enea.it

Carola Strassner, Department of Food – Nutrition – Facilities, FH Münster University of Applied Sciences, Correns Str. 25, D-48149 Münster, Germany. E-mail: carola.strassner@a-verdis.com

Haley Swartz, Linda Golodner Food and Nutrition Policy Fellow, National Consumers League, Washington, DC, USA. E-mail: hswartz5@jhu.edu

Luana F.J. Swensson, Nutrition and Food Systems Division, Food and Agriculture Organization of the United Nations, Rome, Italy. E-mail: Luana.Swensson@fao.org

Florence Tartanac, Nutrition and Food Systems Division, Food and Agriculture Organization of the United Nations, Rome, Italy. E-mail: Florence.tartanac@fao.org

Cristina Tirado, Institute of Environment and Sustainability, University of California Los Angeles, USA. E-mail: cristinatirado@ucla.edu

Antonia Trichopoulou, Hellenic Health Foundation, Athens, Greece. E-mail: atrichopoulou@hhf-greece.gr

Emilie Vandecandelaere, Nutrition and Food Systems Division and Investment Centre Division, Food and Agriculture Organization of the United Nations, Rome, Italy. E-mail: emilie.vandecandelaere@fao.org

Christopher Vogliano, School of Public Health, Massey University, Wellington, New Zealand. E-mail: c.vogliano@massey.ac.nz

Kate Wingrove, Institute for Physical Activity and Nutrition (IPAN), School of Exercise and Nutrition Sciences, Deakin University, Geelong, Australia. E-mail: k.wingrove@deakin.edu.au

Introduction

Barbara Burlingame and Sandro Dernini

Sustainable diets, as a concept, is not new. It was the reality for civilizations from time immemorial. For millennia, when diets and food systems were products of ecosystems, cultures and traditions, sustainable consumption and production were understood and practised.

This publication is a multi-authored monograph for scientists and practitioners, academics and students, policymakers and citizens of the global community. The subject matter is transdisciplinary in nature and thus, each chapter considers multisectoral actions, integrating health, agriculture, environment, economy and sociocultural issues to comprehensively explore for the first time the topic of sustainable diets within the broader context of the sustainable food systems. It has been prepared with the goal of bridging divides between and among disciplines and sectors in order to bring about the needed actions to redress the myriad of threats to the survival of people and planet.

The team of international authors informs readers with arguments, challenges, perspectives, policies, actions and solutions on global topics that must be properly understood in order to be effectively addressed. Each chapter is offered as an entry point to better understanding the interdependencies between nutrition and sustainable food systems.

Presenting the latest findings, they:

- Explore the transition to sustainable diets within the context of food systems, addressing the right to food, and linking food security and nutrition to sustainability.
- Convey the urgency of coordinated action, and consider how to engage multiple sectors in dialogue and joint research to tackle the pressing problems that have taken us to the edge, and beyond, of the planet's limits to growth.
- Review tools, methods and indicators for assessing sustainable diets.
- Describe lessons learned from case studies on both traditional food systems and current dietary challenges.

The 29 chapters in the book are classified under three main headings: *Grand Challenges*, *Qualitative and Quantitative Approaches* and *Moving Forward*. As is required for a complex topic of interconnected systems, '*sustainable diets*', both problems and solutions, is addressed through a variety of focal lenses.

Grand challenges

The first section, *Grand Challenges*, positions sustainable diets in the multi-perspective context of food systems. Within the current international debate, it introduces some overarching wicked problems, resistant to resolution in spite of the dire consequences of inaction. Nine chapters are presented on multisectoral policy, public health, sustainable food systems, climate change, biodiversity loss, agro-ecology, indigenous peoples, the role of cities, and food and waste.

In Chapter 1 (**Sustainable Diets: a Bundle of Problems (Not One) in Search of Answers**), Tim Lang and Pamela Mason reflect on the policy debates surrounding sustainable diets. They outline developments tried by a number of countries and actors at various policy levels, highlighting positions that have emerged through a process of democratic experimentation. Food's effect on ecosystems, health, the economy and society, they argue, has turned what could be positive into something starkly negative. Where reluctance once prevailed, these authors show that the direction pointed by sustainable diets now has a body of science behind it to justify strong multi-criteria policies and actions.

In Chapter 2 (**Sustainable Diets: the Public Health Perspective**), Mark Lawrence, Phillip Baker, Kate Wingrove and Rebecca Lindberg put forward the perspective that positions sustainable diets as a prerequisite for public health, directly through nutrition and indirectly through their environmental impacts. They review the literature that consistently identifies four key characteristics of sustainable diets to promote and protect public health: moderate consumption, shift to more plant-based diets, reduce consumption of ultra-processed foods, and reduce food waste. Priority activities for promoting sustainable diets for public health are presented.

Meredith Harper, Alon Shepon, Nir Ohad and Elliot Berry examine food system challenges in Chapter 3 (**The Challenges of Sustainable Food systems – Where Food Security Meets Sustainability – What are Countries Doing?**). Here they note the imperative to add sustainability to the existing four pillars of food security. As a practical exercise, they showcase eight different countries as examples of food systems. Country-level recommendations for policies and actions are described and proposed.

Presenting climate as a central challenge of our time, Cristina Tirado von der Pahlen describes in Chapter 4 (**Climate Change and Sustainable and Healthy Diets**) the impacts of climate change on our food systems and diets, and the role of food system practices and dietary patterns in contributing to climate change. She presents an analysis of the interconnections of sustainable dietary patterns, health and nutrition in a context of climate change mitigation. Co-benefits to health and climate from dietary change are discussed, including shifting away from the overconsumption of meat to more plant-based diets.

Along with climate change, biodiversity loss is among the key threats to sustainable diets. Emile Frison and Nick Jacobs present this subject in Chapter 5 (**Biodiversity Loss: We Need to Move from Uniformity to Diversity**). They describe the vicious cycles of the low-diversity industrial model and present options for fundamentally different models of agriculture providing a basis for secure farm livelihoods and diverse sustainable healthy diets.

One of those models is presented in the following Chapter 6 (**Agroecology and Nutrition: Transformative Possibilities and Challenges**), by Rachel Bezner Kerr, Maryam Rahmanian, Ibukun Owoputi and Caterina Batello. Agroecology, a holistic approach to agriculture takes into account the ecological, social, political and economic dimensions of producing food in order to build sustainable and resilient food systems that ensure food security and nutrition. While documenting the importance of agroecology and its resonance with sustainable diets, the authors recommend more research to better establish the relationship between agroecology and nutrition.

In Chapter 7 (**Indigenous Food Systems: Contributions to Sustainable Food Systems and Sustainable Diets**), Harriet Kuhnlein, Paul Eme and Yon Fernandez de Larrinoa examine the remarkable reservoirs of unique cultural knowledge grounded in historical legacy and spirituality, linking people with their sustainably managed resources. They cover the extreme disparities and environmental assaults upon Indigenous lands that contribute to the inability of many Indigenous Peoples to realize sustainable diets, and end with a plea to global leaders for recognition and protection of indigenous food systems.

Urban nutrition is the subject of Chapter 8 (**Can Cities – From the global South – be the Drivers of Sustainable Food Systems?**) by Jorge Fonseca, Jane Battersby and Luis Antonio Hualda.

Cities have been an easy target to promote non-sustainable consumption, due to a lifestyle that encourages it, and where 'convenience' is the driver. These authors describe the global context and identify current opportunities that cities can exercise to create sustainable food systems of the future. The question in the chapter title is answered using examples from Africa, with social and environmental inclusion in city-linked food systems.

The final chapter in Grand Challenges is contributed by Silvia Gaiani, Rosa Rolle and Camelia Bucatariu (Chapter 9, **Consumer Level Food Waste Prevention and Reduction toward Sustainable Diets**). They have identified six major challenges and present a matrix policy analysis based on a combination of initiatives as an approach to successfully address the problem. Their analysis is linked to Agenda 2030 for Sustainable Development, drawing particularly on Sustainable Development Goal 12 'to ensure sustainable consumption and production patterns', to embed prevention and reduction of food loss and waste in public and private sector strategies in order to contribute to more sustainable diets and consumption patterns.

Qualitative and Quantitative Approaches

The second section, *Qualitative and Quantitative Approaches*, contains eight chapters that discuss the topic from different cultural, sectoral and disciplinary angles. The issues are analysed with data and methods derived from social sciences, clinical sciences and experimental sciences. Perspectives and solutions, with evidence, are presented to underpin policies and interventions.

Jess Fanzo and Haley Swartz are the authors of this section's lead piece, Chapter 10 (**Attaining a Healthy and Sustainable Diet**). As the burden of diet-related chronic diseases escalates throughout the world, affecting billions of individuals and the countries in which they live, food systems come under close scrutiny. They describe both barriers and opportunities for achieving sustainable diets, for ourselves and for the planet, and solutions addressed to individuals, communities, and institutions.

Alexandre Meybeck and Vincent Gitz in Chapter 11 show that **Highlighting Interlinkages between Sustainable Diets and Sustainable Food Systems**, helps to orient action towards the eradication of hunger and malnutrition and the fulfilment of sustainable development goals. They describe diets as both the result and the driver of food systems, and emphasise that by linking the two different perspectives of sustainable diets – a nutrition perspective, person focused, and a global sustainability perspective -- more effective incentives and policies can be designed.

In Chapter 12 (**Understanding the Food Environment: the Role of Practice Theory and Policy implications**), Dalia Mattioni, Francesca Galli and Gianluca Brunori explore the linkages between diet quality and the underlying food systems through the intermediation of the food environment. They use social practice theory to contribute to a better understanding of the food environment. Through analysis of a number of studies, they show that to be effective, policies need to be consistent and coherent, and directed toward changing the material aspects of the food environment, as well as improving awareness of people to make the better, sustainable, choice for their diets.

F. Xavier Medina and Alicia Aguilar, in Chapter 13 take on the often-neglected dimension of **Sustainable Diets: Social and Cultural Perspectives.** They describe how anthropological concerns with food and nutrition have increased greatly in the last five decades. By highlighting the intrinsic relationship of diets, territories and sustainability, they show that multiple goals for human nutrition and environmental sustainability can be simultaneously achieved.

A large, expert team of researchers, Lorenzo Donini *et al.*, prepared Chapter 14 (**Nutritional Indicators to Assess the Sustainability of the Mediterranean Diet**). They show that the Mediterranean diet can guide innovative inter-sectoral efforts to counteract the degradation of ecosystems, loss of biodiversity and homogeneity of diets due to globalization, through the improvement of sustainable healthy dietary patterns. Their group presents a consensus position for a suite of nutrition and health indicators for assessing the sustainability of the Mediterranean diet, and for characterizing sustainable diets around the world.

In Chapter 15 (**Assessing the Environmental Impact of Diets**), Corné van Dooren addresses the planetary boundaries approach to prioritize the most pressing issues related to the agri-food system as a driver. Several quantitative methodologies are presented including Life Cycle Assessment with eleven pressure indicators, Nutrient Density Unit, and Sustainable Nutrient Rich Foods index. He concludes by proposing that an index such as SNRF on food product labels could assist consumers in making better informed food choices.

Rebekah Jones, Chris Vogliano and Barbara Burlingame present Chapter 16 (**Sustainable Diets and Food-based Dietary Guidelines**). They review guidelines historically based on diet-related morbidity and mortality, and put forward the case for inclusion of elements of environmental sustainability including biodiversity, sustainable fish consumption, meat and dairy consumption and production, water use, seasonality and local production, and waste. Examples of challenges and failures are discussed, along with recommendations for developing country-specific *sustainable* food-based dietary guidelines.

In Chapter 17 (**Costs and Benefits of Sustainable Diets: Impacts for the Environment, Society and Public Health Nutrition**), Adam Drewnowski presents and analyses the multiple and sometimes contradictory demands that sustainable diets require. Cost-benefit analyses based on multiple inputs, and diet quality measured through a variety of indices reveal the challenges for multi-sector engagement policy development.

Moving forward

The last section, *Moving Forward*, presents 12 chapters on selected innovations, initiatives, projects, case studies and programmes enhancing sustainable diets by linking nutrition to food systems. Although independent, there is mutual support and recognition among the chapters, providing the overarching goals and aspirations for moving forward.

This final section starts with Chapter 18 (**The One Planet Sustainable Food Systems (SFS) Programme as a Multi-stakeholder Platform for a Systemic Approach**), prepared by the team of Michael Mulet Solon, Patrick Mink, Sandro Dernini, Marina Bortoletti and James Lomax. They show that modern-day food production and consumption has a failing performance record in its delivery of sustainable diets, in terms of food security, nutrition, health, equality, environmental protection and climate change mitigation. The programme's efforts to accelerate the shift to sustainable food systems, in support of the implementation of the Agenda 2030 is explained, with success requiring a holistic, food systems approach with multi-stakeholder commitment.

In the next chapter, Sandro Dernini *et al.*, revisits the Mediterranean diet in a new iteration: Chapter 19 **The Med Diet 4.0: a Multidimensional Driver for Revitalizing the Mediterranean Diet (MD) as a Sustainable Diet Model to Drive Current Dietary Shifts towards more Sustainable Food Systems in the Mediterranean Region.** The erosion of the MD is presented, along with efforts to revitalize its practice in the countries of the Mediterranean basin. By connecting food consumption to production in the context of the improvement of the sustainability of food systems, the Med Diet 4.0 is shown as a sustainable diet model and provides useful insights and potential action points for policy, practice and education within an interconnected, globalized food system.

That model reappears in the next chapter by Antonia Trichopoulou (Chapter 20, **Traditional Foods at the Epicenter of Sustainable Food Systems**). She describes the concept of traditional foods that includes the preservation of traditional farming knowledge, local crop varieties and animal breeds, and native forms of socio-cultural organization. Local, traditional foods are highlighted as important components of a sustainable diet, and consequently of a sustainable food system. In addition to being vehicles for culture, they contribute to better nutrition with a corresponding diversity of crop varieties and animal breeds and associated lower risks for diet-related chronic diseases. The Mediterranean diet offers a clear example, partly attributable to its traditional foods.

Parviz Koohafkan presents Chapter 21 (**Globally Important Agricultural Heritage Systems (GIAHS): a Legacy for Food and Nutrition Security**). The unique systems testify to the inventiveness and ingenuity of farmers in their use and management of natural resources, biodiversity and inter-species dynamics, and the physical attributes of the landscape. International recognition, conservation and adaptive management of these systems, including support to local and indigenous communities, is laid out for safeguarding GIAHS as an important contribution to sustainable diets and for its role in improving efficiency and productivity within food systems.

Allison Marie Loconto and a team of authors contributed Chapter 22 (**Sustainability along All Value Chains: Exploring Value Chain Interactions in Sustainable Food Systems**). They delve into the recent advances in value chain theories, identifying innovations that bring new values such as environmental sustainability, into food systems. Their analytical lens shifts the focus away from specific commodities and towards new forms of organization – such as short supply chains, circular economies, gastronomy and geographical indications and how these contribute to sustainable diets.

Local food systems get attention again in Chapter 23 (**Sustainable and Healthy Gastronomy in Costa Rica: Betting on Sustainable Diets**). Authors Marcela Dumani Echandi, Patricia Sedó Masís, Randall García Víquez and Roberto Azofeifa Rodríguez present Costa Rica's 'National Plan on Healthy and Sustainable Gastronomy', as a successful example of a multi-stakeholder initiative to reverse increasing trends of unsustainable and unhealthy consumption patterns. Acknowledging that unhealthy diets are a major reason for health problems, environmental degradation and biodiversity loss in many parts of the world, the National Plan is offered as a new paradigm of sustainable development, based on agroecology and the efficiency of agri-food systems.

In Chapter 24 (**How Organic Food Systems Support Sustainability of Diets**), Johannes Kahl, Carola Strassner, Susanne Bügel, Denis Lairon and Flavio Paoletti present strong arguments for adopting organic as a model for transforming food systems. They show sustainability as an inherent property of a healthy food system and identify 'enabling mechanisms' from the organic food system actors' perspective to provide insights to drivers and factors shaping food systems. Organic farming is shown as contributing to sustainable diets in theory and in practice by providing a range of ecosystem services and allowing values-based ethical and personal responsibilities in food choices.

Institutional Food Procurement for Promoting Sustainable Diets is the topic of chapter 25 by Florence Tartanac, Luana Swensson, Andrea Polo Galante and Danny Hunter. They argue that institutional food procurement programmes (IFPP) hold considerable potential to influence both food consumption and food production and to deliver multiple social, economic, environmental, nutritional and health benefits that will contribute to sustainable diets. Examples of good practices from the Brazilian Food Procurement Programme, Cape Verde National School Feeding Programme and the municipality of Rome are presented.

Kakoli Ghosh authors Chapter 26 (**Renewing Partnerships with Non-state Actors for Sustainable Diets through Sustainable Agriculture**). Optimizing both natural and human resources, a key element of the definition of sustainable diets, requires strong partnerships amongst stakeholders engaged in production, delivery and disposal of food. She presents ways and means to strengthen sustainable diets by increasing collaborations among governments and non-state actors such as the civil society, farmers' organizations, private sector, academia and research institutions. Examples show that coordination and strengthening of strategic partnerships enhance knowledge and resource sharing, and develop capacities among countries in support of the sustainable development goals.

Lluís Serra-Majem and co-authors present Chapter 27 (**Decalogue of Gran Canaria for Sustainable Food and Nutrition in the Community**) for sustainable food and nutrition in the community. The aim of the declaration is to improve food sustainability across the globe. The science-based development and implementation of its 10 key elements for a healthier life and world requires commitments and accountability from citizens and governments alike if the virtuous circle between sustainable development and nutrition is to be fully realized.

Chapter 28 (**Ten Years to Achieve Transformational Change: the United Nations Decade of Action on Nutrition 2016–2025**) is presented by Stineke Oenema. The decade, proclaimed the United Nations, maps a 10-year window of opportunity to intensify policies, programming and actions to improve nutrition, one requirement of which is to transform food systems. Sustainable diets are an entry point which serve to promote people's health, promote the demand for sustainably produced food, and reduce the demand for products that have a high environmental footprint.

In the final chapter (Chapter 29, **Towards a Code of Conduct for Sustainable Diets**), Barbara Burlingame outlines the proposal to develop a document of principles involving multisectoral stakeholder groups deemed necessary for guiding the transition to sustainable diets. Progress to date is reviewed, establishing a rationale along with a draft code or set of guidelines for sustainable diets. Alignment with five global policy instruments is presented. Regardless of the mechanism for reaching the goal of sustainable diets, the need for urgent action is expressed.

Throughout this book, the authors of the various chapters position the issues of sustainable diets as central to the Earth's future. It is heartening to see the many global and local efforts around the world, directly and indirectly related to sustainable diets, but more is required, particularly in the form of political will and commitment. In our world with a plethora of goals and targets, declarations, calls for action, and universally-agreed commitments, results are the needed currency, and these have been unconscionably slow-moving, and even regressive on some issues. The utility of this book is that it reviews a number of options and opportunities for addressing sustainable diets by linking nutrition to food systems and aligning different sectors and disciplines and viewing the problems and solutions through a number of focal lenses. The overarching goal is to inform, inspire, and motivate to bring about fundamental changes. Survival of people and planet is at stake.

1 Sustainable Diets: a Bundle of Problems (Not One) in Search of Answers

Tim Lang and Pamela Mason

Abstract

This chapter reflects on the status of policy debate about sustainable diets. That the scientific case for shifting the population's diet into a more sustainable direction is now as certain as science can be. The effect of food on ecosystems, health, the economy and society has turned what could be positive into too negative effects. Yet a policy approach to the food system has remained largely in place, which perpetuates these impacts, seemingly unaffected by the evidence. The policy approach to food centres on output, maximizing consumer choice and cheaper prices. A gap has thus been created between what the evidence suggests needs to be addressed and what society actually delivers, eats and aspires to. The chapter uses the Nuffield Council of Bio-Ethics' Ladder of Interventions to gauge why action on sustainable diets is relatively so weak. The ladder posits that the lowest rung one is minimal intervention, and rises higher to invoke tough measures such as fiscal and legal action, and at the top of the ladder on rung eight, choice is totally reframed. The chapter argues that attention needs to be given to how to move up the ladder, so that policy on sustainable diets encourages the radical change suggested by the evidence. Attempts to create international and national policy frameworks for sustainable diets have been few. The reluctance even to step onto the ladder's first rung is remarkable. While the majority of politicians and food system actors seem reluctant to change, the chapter outlines developments tried by a number of countries and actors at various policy levels. These suggest that the 'business-as-usual' policy framework may be fraying at the edges. The chapter concludes by outlining policy arguments that have emerged in what it describes as a process of democratic experimentation, and proposes that policymakers should adopt multicriteria approaches to sustainable diets.

Introduction: the Philosophical Challenges of (Un)Sustainable Diets

Compared to ten or twenty years ago, there is growing realization today that sustainable diet is a problem desperately in need of policy solutions. But what exactly is the problem with which policymakers should engage? That is the question this chapter explores.

As we and others argue, the term 'sustainable diet' can appear deceptively simple (Nelson *et al.*, 2016; Mason and Lang, 2017). Two benign words jointly indicate a bundle of problems! Different perspectives can be taken. Some argue that the problem is best summarized as a matter of carbon plus calories, the solution to which is to pursue de-carbonization with reduced calorie intake throughout the food system (Cabinet Office Strategy Unit, 2008). Others argue – ourselves included – that sustainability of diets requires a broader perspective on the impact of diet on health (Mason and Lang, 2017). It needs to expand beyond just nutrients to the full gamut of health effects – safety, equity, culture, and so on. And it needs similarly to expand what is focussed on within environmental factors. Thus carbon, nutrients and safety need to be dovetailed with biodiversity, soil, water, land use and, yes, carbon. Indeed, we have gone further, arguing that a more complex approach to both health and environment inevitably requires the insertion of other factors such as social, economic, ethical and governance elements of diet as well. Sustainable diets, we argue, inevitably have to adopt a 'multicriteria' framework of thinking, or it falsifies reality (Lang and Mason, 2017).

This growth of perspective can make policymakers tired! It requires too much, they cry. Keep it simple, and we can try to do something about it. Broaden it, and people (usually they mean politicians) get lost or lose interest! One can have sympathy for this reaction, yet the enormity of the impacts now known to be driven by diet means one cannot rationally take the 'keep it simple, take it slowly' approach (Garnett, 2014). Unless the data are completely wrong, the case for recalibrating what is meant by a good diet is one of the most pressing challenges facing public policy today. And that is what the notion of sustainable diets is all about. The benign phrase 'sustainable diets' – who could be against it? – in fact carries a searing critique. So many patterns of diet at the population level are not sustainable. Indeed, one could argue that 20th century food progress has been about systematically ensuring that they are not sustainable. Progress has been driven by consumerism, excess, waste and over-consumption. The fetish of consumer choice denies that there are limits. Consumer freedom trumps both planetary and population health.

This combination of intellectual, scientific, practical and political challenges is what makes the 'sustainable diet' topic of this book so exciting. Research in the area has grown exponentially. What seems solid one moment can easily become fluid the next. One moment carbon is the most important criterion being urged onto policymakers, then others argue: what about water or biodiversity? And others say: it is all a matter of trade-offs, so surely, realpolitik must kick in. Then along comes the Paris Climate Change Accord in 2015, and seemingly overnight even previously resistant big food businesses want to rally round carbon as *the* goal for restructuring the food system. But unless food systems and the ceaseless pressure from unsustainable consumption alter, the chances of achieving the two degree CO_2 growth limit are slight. In academia, meanwhile, scientists have realized that no-one's speciality is more important that the others. The impact of diets on people and the planet necessitates a complex perspective. Multicriteria thinking may be brain numbing, but it is correct, nonetheless. The problem becomes political: how to achieve leverage? How to translate this complexity into terms that policymakers can engage with and deliver change on.

Unpicking (Un)Sustainable Diets

The phrase sustainable diet yokes two already charged notions – diet and sustainability – and raises many questions for scientific inquiry and cultural change. If diets are said to be unsustainable, what is it about diets that makes them so? And how rigorous is the notion of sustainability in the first place?

And how can sense be made of the weight and range of evidence about dietary unsustainability? Is it just about science? Can policymakers hope to do anything about it? Moral, social and political judgements are almost inevitable over sustainable diets, as over much to do with food matters. Food is the intimate commodity. People eat food; they do not eat iron or gold or bitcoins, other tradeable commodities. Diet and food are big business, with a biological materiality, yet are highly infused with values, ethics and meanings. Those values infuse the eating experience. It is also why food companies spend so many billions trying to mould consumers' minds. One soft drink company alone spends on marketing twice the whole World Health Organization's annual budget. They know that food and drink are culture, not just carbon.

Little wonder, one might think, that there is some reluctance by policymakers so far to engage with sustainable diets. There's big money at stake. Moreover, the relationship between policy, evidence, practice and impact is famously tricky in the case of food. Only the naïve believe that the goal of 'evidence-based policy' is easy to achieve or logical in application. There can be policies with partial or out-of-date evidence and, perish the thought, there can be policies that deny or distort the evidence. Despite this, the pressure to address sustainable diets grows. Whether policymakers or particular industry sectors resist or not, pressure from outside is building up to do something. One should note the political furore in the USA in 2015 when the Secretaries of State for Health and Human Services and for Agriculture rejected the scientific advice from its Dietary Guidelines for Americans (DGA) committee, which had proposed the new DGA should have an environmental dimension (US Dietary Guidelines Advisory Committee, 2015). Unprecedented waves of reaction from US consumers flooded into Washington. And this under President Obama, pre-President Trump! The reaction was noisy but the policy block repeated what had happened in Australia, Sweden and the UK earlier (Lang and Mason, 2017).

It seems clear that, partly, the issue of (un) sustainable diets is a political not just a philosophical problem. The scientific community therefore needs to ask itself: how can we help unlock this policy lock-in? What needs to happen with policy processes and institutions to steer the food system in a safer direction?

Old doubts about the food system have resurfaced under the sustainable diet umbrella. The astonishing growth of food output in the 20th century is no longer the 100% success it once seemed. The improvement of public health by setting out to raise global food production and to make food cheaper and thus affordable and available to the poor is not the perfect solution it first appeared, either. Dietary improvement in lands of hunger still means making food more available but across vast swathes of the globe that policy equation no longer holds. Simply eating more food, or more of scientifically approved intake does not tick all the boxes, any more than tackling on-farm waste has stopped food waste. The rich world now wastes less before or just after the food leaves the farm than in the 1930s, but consumers have learned how to waste food on unprecedented scales, encouraged by cheaper prices and more plentiful food (Gustavsson *et al.*, 2011).

So, what are the messages for the general eating public from the sustainable diet discourse? Eat less but better food? That message certainly is meaningful in the rich West but it is trickier to apply for undernourished or malnourished populations. It may tame their aspirations to overeat or mal-consume as rich societies do, perhaps, but does not recognize the sensitivities of income differentials let alone market realities. The 21st century requires a new model of dietary progress in which scientifically sound multicriteria are applied to eating. This means re-engineering the food supply to provide the means for those better consumption patterns.

Faced with the complexity, some industry thinkers in private will admit that they can reformulate and change products, or whole meals, by reformulating this or that, and by altering product sizes, for example. But there comes a point when the consumers also have to change. They have to 'choose' to eat differently. This consumer change policy route is the politicians' nightmare. It means going up the Nuffield Ladder of Interventions (Nuffield Council on Bioethics, 2007). The lowest of the eight rungs the Nuffield Council described on its theoretical policy 'ladder' is to do nothing. The next is to provide information. Then to enable choice, then to guide choice by changing the default policy, then to guide choice by adding positive incentives, then to do that but with disincentives, then to restrict choice, and finally to eliminate choice. On sustainable diets, at present, policy hovers across rung one – do nothing – and rung two – provide information. These are the weakest policy actions. The onus is on the consumer. The power lies with those who frame the situation. Table 1.1 applies the Nuffield Ladder of Interventions to sustainable diets and indicates some possible actions.

One might think, with the rhetoric of consumer sovereignty, that policy makers would want to be seen to help 'empower' consumers, moving from rung one to at least rung five, incentivized choice. Often, in fact, they do not want to be the ones breaking policy ranks. There appears to be a self-imposed ceiling on how far up the ladder of intervention they are prepared to go. In part, they are held back by the argument that intervention opens them up to the criticism of being a 'nanny state', a status demonized by neoliberal or post Washington Consensus politics. This tends to lionize a reduction in state role and responsibility being ceded to market forces (Williamson, 2004). In fact, of course, the room for manoeuvre that individual consumers have is limited. They cannot possibly exert the kind of power of food supply that companies or governments do, unless they act in concert, but by definition in markets, they make individualized choices; hence the need for support and guidance. Good consumption requires an infrastructure and channels of

Table 1.1. Applying the Nuffield Ladder of Policy Intervention to sustainable diets.

Rung	Policy option	Level of intervention	Description	Comment in relation to sustainable diets
8	Eliminate choice	***'HARD'***	Channel actions only to the desired end and isolate inappropriate actions	Implies a zero negative impact food system
7	Restrict choice	↑	Remove inappropriate choice options	Implies rationing of choices on a considerable scale
6	Guide choices through disincentives		Apply taxes or charges	Some interventions are emerging e.g. carbon and sugar taxes
5	Guide choices through incentives		Use regulations or financial incentives	Requires active Government and a willing consuming public
4	Guide choice by changing default policy		Provide 'better' options	Hard to do in a supermarket context but is possible in food service by menu planning
3	Enable choice		Enable individuals to change behaviour	The market economics position, currently manifest via logos and branding appeals
2	Provide information	↓	Inform or educate the public	There are many appeals to eat differently, led by non-governmental organizations, brands and some commercial interests
1	Do nothing	***'SOFT'***	No action or only monitor situation	An all too common government position at present

Source: Authors

support if it is to become the default behaviour pattern. Yet from a scientific perspective, the build-up of evidence as to the unsustainability of current dietary patterns amplifies pressure for public policy to move up the policy ladder.

An impasse appears to have emerged from the juncture of: (i) pressure from evidence to change default diets; and (ii) resistance from political structures and cultures. Perhaps surprisingly, there is some interest from some (not all) food companies to take a longer view and to break the impasse. Companies want to be operating a decade or more ahead. Most politicians and political parties' attention is on the next election. And food companies are increasingly aware of the looming sustainability crisis, hence the strong pressure they exerted on resistant governments at the Paris Climate Change negotiations in 2015 to agree on the Paris Accord. Progressive food companies, reading the climate change writing on the wall, needed to recalibrate the baseline for decarbonization; no company could go it alone. This required systems change. There are many corporate initiatives aiming to make food products more sustainable including: product reformulation; size change; packaging change; in-factory efficiency modernization; new management structures and responsibilities (such as creating sustainability managers); and producing externally verified annual sustainability audits. But food companies also know that there comes a point where changing the food *before* consumers get to choose meets its limitations. Consumption patterns themselves must change.

The impasse we identified above can be addressed but has limits. This is why many academics and scientists now agree that everyday cultural norms and assumptions require more policy attention. Default behaviour at a population level needs to be 're-booted'. Sustainable diets – whatever form they take, in different parts of the globe – need to be normalized much as unsustainable diets have been normalized in the 20th century. The clocks are ticking, not just in relation to climate change but across the multicriteria field of sustainable diets. Almost wherever one starts in the literature, the weight of evidence points to the case for a fundamental shift in the medium term. The scale and pace of biodiversity loss, land-use change, obesity and other diet-related healthcare costs, the impact of the growth of animal and meat-based diets, the social inequalities from mal-distribution of food and the cultural power of irresponsible corporate marketing, all contribute to a policy lock-in.

No wonder the notion of sustainable diet can be so threatening. It seems to point to a system change rather than a minor readjustment here or there. One area where this itself is being questioned is the issue of meat. Rising meat consumption and the resources used by meat – land, water, feed, labour, capital – have all pointed to it being a test case for whether the world will take sustainable diets seriously (Smil, 2013; Bailey *et al.*, 2014; Green *et al.*, 2015; Garnett, 2016). Yet as those data consolidate, investment in technical options has accelerated, producing new meat substitutes such as lab-based meat (BBC News, 2013; Singer, 2013), industrial production of insects (ICIPE, 2011; van Huis *et al.*, 2013), plant-based substitutes (Beyond Meat, 2017), animal feed alternatives (Forum for the Future, Protein Challenge Partnership, 2018), and new generations of synthetic biology, genetic and nutrigenomics (German *et al.*, 2011).

These initiatives are receiving serious US capital investment, some of it from software finance looking to diversify (Bradshaw, 2014).

Sustainable Diets: a New Old Problem

Can we relax, therefore, and allow new market forces to exert their dynamic effects? Some perspective can be given by looking back while looking ahead. Arguably, the entire discourse on sustainable diets can be traced at least to Malthus' 1798 *Essay on the Principle of Population* (Malthus, 1798). In this Malthus made his dire assessment that population growth would outstrip the capacity of agriculture to increase output. He focussed on Britain but, ever the intellectual imperialist, he proposed his analysis as a universal truth. Humanity could not resolve the impasse, he argued. Yet the rest of the 19th century, and particularly the 20th century, proved him wrong by unleashing chemical, technical, transport and land-use changes on an unprecedented scale. But the potential for demographic and political disequilibrium from food supply had been logged. More importantly, before we relax and say 'well, this can be resolved once more', the problems the food system today faces are heavily framed by the 'solutions' that emerged in the 19th and 20th centuries: oil-dependency in the form of fertilizers, profligate destruction of biodiversity, plant and animal breeding for output, de-forestation to create new cropland, and so on.

In the health sphere, the late 19th and early 20th centuries saw modern nutrition exploring the variability and limits of human physiological need for food. From the 1890s, W.O. Atwater in the USA began to calculate for the US government the nutritional needs for different modes of labour (Atwater, 1891, 1894, 1895). These set benchmarks used by Seebohm Rowntree, the philanthropic son of a giant chocolate manufacturing family, at the start of his 50 year exploration of the (in)adequacy of his workforce's dietary intake (Rowntree, 1901, 1913, 1941). This baseline work was used by policymakers in World War I and more extensively and systematically in World War II. It informed how rationing was set by occupation, gender and life stage (Minns, 1980; Zweiniger-Bargielowska, 2000). While adopting a positive and optimistic belief that more people could be fed well, these reformers also realized that the answers to the Malthusian problem were not simply technical. It required social actors – particularly the state – to ensure fair distribution of food resources (Vernon, 2007).

In this expanding discourse, the notion of the 'food environment' also subtly shifted from the immediate social environment (determined by factors such as wealth, income, occupation and geographical location) to what today we call the ecosystems environment, a shift from proximal to distal shaping factors. Some old concerns resurfaced in this shift, such as energy, soil quality, the nitrogen and phosphorus cycles, but the urgency and scale of the challenge was both new with regard to scale and sobering beyond Malthus' fears. That food is a driver of basic infrastructures in planetary existence became clear, with the ecological sciences expressing serious concerns about how profligate food methods threatened water, biodiversity, land use and climate change stability.

Even back in the 1970s, the range of this new holistic discourse was emerging. It was daringly articulated from outside science in Frances Moore Lappé's (1971) best-selling *Diet for a Small Planet*. That book tapped a US West Coast zeitgeist of living simply with a more vegetarian and less meat-oriented diet but it offered a pro-planet food *cultural* outlook. This sociocultural dimension was important but edged out by scientific and business policy reports such as the Club of Rome *Limits to Growth* in 1972 (Meadows *et al.*, 1972), and demographic arguments such as from Ehrlich (1971) and Commoner (1972). One could argue that these were restating the Malthusian arguments, but there was a new appeal to cultural change.

Today, most researchers on sustainable diets begin their literature reviews with the short 1986 paper 'Dietary Guidelines for Sustainability' by US nutritionists Joan Gussow and Kate Clancy (Gussow and Clancey, 1986). They proposed that the problem of unsustainable diets needed to be tackled by revising national dietary guidelines – such as the USA's DGA – to include sustainability. Environmental criteria should accompany food-based or nutrient-based foci, was their argument.

Since the 1980s, numerous reports and papers across a wide range of scientific disciplines and foci have simply amplified their argument, showing that the current dietary transitions have serious impacts on ecosystems (UNEP *et al.*, 2009), public health (Popkin, 2002, 2009; Monteiro *et al.*, 2013) and healthcare costs (WHO, 2015), climate change (Watts *et al.*, 2015), biodiversity (Burlingame and Dernini, 2012; Lawrence *et al.*, 2015) and land use (Smith, 2012; Tilman and Clark, 2014). As this literature expanded, consortia of scientists proposed that planetary boundaries – such as nitrogen, phosphorus, carbon, rates of biodiversity extinction – were in danger of exceeding safe limits (Rockström *et al.*, 2009; Steffen *et al.*, 2015). Food, again, was implicated and affected.

By the 2000s, this intellectual pressure was building up once more on policymakers. In 2010 the Food and Agriculture Organization (FAO) and Bioversity International hosted a large scientific conference (FAO, Bioversity International, 2010). In 2012, the United Nations Environment Programme argued that sustainable diets needed support to prevent future famines (UNEP, 2012). In 2014, the Food Climate Research Network published a review (Garnett, 2014). In 2015, the United Nations-sponsored Paris Climate Change Accord was agreed (UN Framework Convention on Climate Change, 2015), as were seventeen new Sustainable Development Goals (SDGs), with many targets all pointing to the need to tackle food. Policymakers were under pressure to act on the unsustainability of current dietary trends. Is the discourse thus back to the impasse: solid evidence meeting resistance to cultural change? Time will tell. Already, however, it is clear that single-issue approaches to sustainability of diets is inadequate. In the next section, we explore the advantages of multicriteria thinking for policymakers. They and we must get used to accepting that the world of food and diet is complex. Reductionist thinking is unlikely to help. It certainly muddles policy processes with false promises.

Multicriteria Versus Single-focus Policy Approaches

We see an important policy clash between viewing sustainable diets through the lens of complexity or simplicity. Our argument is that even if one desires a simple, unifactorial approach for pragmatic reasons, one is inevitably led into a multicriteria complex food world. Factors in food dynamics lead us to complexity. To the 1970s generation, the dietary problem was clear: people had to change how and what they consumed. It was a cultural problem, not in a reductionist sense but as total living (Lappé, 1971). It could be chosen by informed consumers. That generation was reacting to and questioning the earlier post-World War II approach to diet which posited that the food problem was mostly one of under production. If only more food could be produced, more mouths could be fed, prices would therefore drop and diet-related ill health would improve with affordability (Boyd Orr, 1943; 1966; Boyd Orr and Lubbock, 1953; Lang and Heasman, 2015).

The analysis was that the salient dietary problems of hunger and stunting were due to mal availability and under availability. Therefore, policy should focus on producing more food. This 'productionist' food policy was a social policy, too, reflecting the 1930s/1940s scientific consensus. But today, this no longer fits the evidence about how food systems have generated hidden externalized costs (TEEB, 2015), or have created massive impacts on environment and health through over-consumption. In the 1940s, the world of today with its global obesity epidemic and the spread of diet-related non-communicable disease with the nutrition transition was literally unthinkable. Today, simply to produce more food is unlikely to resolve the complex pattern of problems the science indicates (Gladek *et al.*, 2016).

Whether Simple or Complex, who is Going to Take a Lead?

A process of multilevel democratic policy experimentation is under way (Lang and Mason, 2017). At the global level, the SDGs put great emphasis on requiring the food system in general and nutrition and consumption in particular to change. The SDGs and the Paris Climate Change Accord, both passed in 2015, set clear targets for which dietary change is essential. Importantly, in 2010 the FAO hosted the first official global attempt to define and characterize sustainable diets (admirably led by the editors of this volume). Box 1.1 gives the definition (Burlingame, 2012). At the national policy level, however, there has been less policy engagement. No trade agreements, no intergovernmental accords (e.g. the European Union (EU), Mercosur or Asia-Pacific Economic Cooperation) are yet in play. The EU began with interest in this policy terrain at the 1992 UN Conference on Environment and Development, offered to lead on sustainable consumption and production work thereafter, with Sweden delegated as the EU internal lead – hence its shock at the 2010 rebuff over formal sustainable dietary advice. One should perhaps not be surprised at the slow policy development at the national level. It is where tensions can most excite. It should be noted, too, that the one country simply to produce and announce sustainable dietary guidelines is rather more authoritarian a political state than many. Qatar issued clear guidelines in 2014 (Qatar Supreme Council of Health, 2014a, 2014b; Seed, 2014), fully aware not just of how the nutrition transition was wreaking havoc on its public health, but also that its geography was in the front line of climate change, and its political neighbours were not insignificant actors in the CO_2 oil economy.

> **Box 1.1.** The FAO-Bioversity International 2010 definition of sustainable diets.
>
> Sustainable Diets are those diets with low environmental impacts which contribute to food and nutrition security and to healthy life for present and future generations. Sustainable diets are protective and respectful of biodiversity and ecosystems, culturally acceptable, accessible, economically fair and affordable; nutritionally adequate, safe and healthy; while optimizing natural and human resources.
>
> Source: FAO (2010)

At the local level, however, there is a welcome flowering of activity with campaigns, dietary diversity and cultural pitching, all leading to local political interest, for example through the sustainable food cities movement (Sustainable Food Cities, 2014). Over 140 cities worldwide signed up to the Milan Urban Food Policy Pact in 2014, which committed mayors to nurture sustainable diets (Pact MUFP, 2015).

This democratic growth suggests that policy interest in sustainable diets has already left the conventions of recent politics, that is, matters can be left to market dynamics or to individual consumer choice (i.e. the lower rungs in the Nuffield Ladder of Intervention summarized in Table 1.1). Something larger, perhaps messier but more inclusive appears to be emerging, in which popular citizens' action is noticeable. But, a policy gap has also emerged between high level global commitments or analyses and the local and sub-national.

At the national level, there is a small but useful experience of governmental engagement, such as by the Netherlands, which has slowly built up evidence-led policy (Health Council of the Netherlands, 2011; Netherlands Scientific Council for Government Policy, 2015; Netherlands Nutrition Centre, 2016), Sweden which did the right thing straight away and then had to regroup after a rebuff (National Food Administration,

2009; Livsmedelsverket, 2015), and Brazil which tailored its sustainable dietary advice through cultural language, after a patient and exhaustive national consultative process (Ministry of Health (Brazil), 2014a, 2014b). These all took different trajectories, but united on the need for significant dietary change. We draw hope from these countries but note they are in a tiny minority, which is why lessons from their experience must remain tentative pending wider experience. Meanwhile, other attempts to provide national sustainable dietary guidelines met with defeat (in the USA, Australia and the UK, again all for differing reasons and via different routes), dismissal (USA), sectoral power (Australia) and erosion and abandonment (UK) (Lang and Mason, 2017).

Lessons must be learned from blocks on attempts to turning formal national dietary guidelines into sustainable dietary guidelines. Guidelines to consumers in cultural terms, on the other hand, appear to survive lobby pressure. Brazil's official dietary guidelines (Box 1.2) were proofed for environmental impact yet were promulgated in the form of cultural advice rather than as sustainable diet. These guidelines have survived tumultuous political change, from left- to right-wing governments – a test of robust policy effectiveness (Rutter *et al.*, 2011). By contrast, when Sweden issued the first official national proposals for sustainable dietary guidelines in 2009 (National Food Administration, 2009), it withdrew them a year later following complaints by one other EU government (Dahlbacka *et al.*, 2010). The precise details of this *volte-face* remain unclear, but it was widely perceived to have been triggered by opposition to the recommendation that Swedish consumers should eat more locally and seasonally, and to eat less meat. Some say this offended the EU promotion of the single market; others that it offended US meat interests represented within Europe (Boyle, 2012). Happily, the Swedish authorities did not give up. A few years later, they issue not dissimilar sustainable dietary advice but entirely framed as cultural advice (Livsmedelsverket, 2015). Perhaps the cultural route to sustainable diet will be quicker and more consensual than hard environmental advice (Lang and Mason, 2017).

Box 1.2. The ten main recommendations in the Brazilian 2014 nutrition guidelines.

1. Prepare meals from staple and fresh foods.
2. Use oils, fats, sugar and salt in moderation.
3. Limit consumption of ready-to-consume food and drink products.
4. Eat regular meals, paying attention, and in appropriate environments.
5. Eat in company whenever possible.
6. Buy food at places that offer varieties of fresh foods. Avoid those that mainly sell products ready for consumption.
7. Develop, practice, share and enjoy your skills in food preparation and cooking.
8. Plan your time to give meals and eating proper time and space.
9. When you eat out, choose restaurants that serve freshly made dishes and meals. Avoid fast food chains.
10. Be critical of the commercial advertisement of food products.

Source: Ministry of Health of Brazil (2014a)

Conclusion: the Limits to Individual Choice as a Motor of Change

The problem of (un)sustainable diets is complex. It almost certainly requires multicriteria, multilevel, multisector action. Appeals to individual change are unlikely to be of sufficient policy weight to be effective. The policy task ahead is how to shift mass trends within the wider challenge of pressures on ecological public health. Policy makers must help re-shape conditions and must aim to normalize sustainable diets. This is too important to leave to the vagaries of choice. The alternative is to allow continued drift or partial action, insufficient to make a difference.

Since the millennium, there has been some useful policy formulation and experimentation. Thus far, it lacks viable multi-institutional frameworks and clarity of political leadership. Recent global commitments such as the Paris Climate Change Accord and the SDGs begin to fill that gap, but lack specificity and uptake particularly at the national level. Targets exist but require actors to deliver them. Different styles and forms of policy engagement can be noted. These range from 'soft' to 'hard' approaches, from reliance on 'choice-editing' before consumers have the chance to select foods, to new cultural 'rules' for eating such as have been offered by Brazil and

Sweden. Meanwhile, the evidence grows on the urgency of taming runaway dietary change.

Some interest within commerce is real and to be welcomed but an element of corporate action is driven by anticipating reputational risk. Food consumption remains a policy 'elephant in the room' in a commercial world of ruthless competition, cost shaving and jostling for shelf space. Policymakers are reluctant to work with the consuming public to deliver mass change. Yet, as one industry spokesperson commented to one of the present authors (T.L.) a few years ago, 'we can shave carbon and other effects out of food beneath the radar up to a point, but ultimately consumers will have to change'.

Academics and scientists have a useful role in debating, researching and promoting the specifics of (un)sustainable diets. They have contributed important, detailed work that has raised the issue's profile. Better composite indicators are probably needed to help policymakers link the different aspects of sustainability. Those indicators must help answer some old, basic, post-Malthusian questions. What is a good diet? Where lies food progress? How much must be left to consumer engagement and how much follows simply by creating new norms and recalibrating default actions? In a world where food is a hugely significant driver of ecosystems and health damage, consumers and the food industries alike deserve better policy frameworks. There is much work to do to make policy structures effective in this herculean task. But the mapping has begun.

References

Atwater, W.O. (1891) *Investigations on the Chemistry and Economy of Food*. US Department of Agriculture Bulletin 21. Department of Agriculture, Washington DC, USA.

Atwater, W.O. (1894) *Foods, Nutritive Value and Cost*. US Department of Agriculture, Farmers Bulletin 23. US Department of Agriculture, Washington DC, USA.

Atwater, W.O. (1895) *Methods and Results of Investigations on the Chemistry and Economy of Food*. US Department of Agriculture Bulletin 21. Department of Agriculture, Washington DC, USA.

Bailey, R., Froggatt, A., Wellesley, L. (2014) *Livestock – Climate Change's Forgotten Sector: Global Public Opinion on Meat and Dairy Consumption*. Royal Institution of International Affairs, London, UK.

BBC News (2013) First lab grown burger eaten. Available at http://www.bbc.co.uk/news/av/science-environment-23582684/first-lab-grown-burger-cooked-and-eaten-in-london (accessed 27 June 2018).

Beyond Meat (2017) Available at http://beyondmeat.com/ (accessed 24 January 2018).

Boyd Orr, J. (1943) *Food and the People*. Target for Tomorrow No 3. Pilot Press, London, UK.

Boyd Orr, J. (1966) *As I Recall: The 1880's to the 1960's*. MacGibbon and Kee, London, UK.

Boyd Orr, J. and Lubbock, D. (1953) *The White Man's Dilemma*. George Allen and Unwin, London, UK.

Boyle, E. (2012) *High Steaks: Why and How to Eat Less Meat*. New Society Publishers, Gabriola Island, Canada.

Bradshaw, T. (2014) Food 2.0: the future of what we eat. Financial Times. Available at http://www.ft.com/cms/s/2/bfa6fca0-5fbb-11e4-8c27-00144feabdc0.html (accessed 11 January 2015).

Burlingame, B. Dernini, S. (Eds) (2012) Sustainable Diets and Biodiversity: Directions and Solutions for Policy, research and action. *Proceedings of the International Scientific Symposium Biodiversity and Sustainable Diets United against Hunger*, 3–5 November 2010, FAO Headquarters, Rome. http://www.fao.org/docrep/016/i3004e/i3004e.pdf (accessed 28 August 2018).

Cabinet Office Strategy Unit (2008) *Food Matters: Towards a Strategy for the 21st Century*. Cabinet Office Strategy Unit, London, UK.

Commoner, B. (1972) *The Closing Circle: Confronting the Environmental Crisis*. Cape, London, UK.

Dahlbacka, B. and Spencer, P. (2010) *Sweden Withdraws Proposal on Climate Friendly Food Choices*. December 2, 2010. GAIN Report SW1007. USDA Foreign Agricultural Service Global Agricultural Information, Stockholm, Sweden.

Ehrlich, P.R. (1971) *The Population Bomb*. Pan Books, London, UK.

FAO, Bioversity International (2010) Final Document: International Scientific Symposium: Biodiversity and Sustainable Diets – United against Hunger, 3-5 November 2010, FAO Headquarters, Rome, Italy. Available at http://www.eurofir.net/sites/default/files/9th%20IFDC/FAO_Symposium_final_121110.pdf (accessed 24 January 2018).

FAO (2010) Definition of sustainable diets. Final Document – International Scientific Symposium: Biodiversity and Sustainable Diets, 3–5 November 2010, FAO Headquarters, Rome. http://www.fao.org/ag/humannutrition/23781-0e8d8dc364ee46865d5841c48976e9980.pdf (accessed 24 June 2018).
Forum for the Future, Protein Challenge Partnership (2018) The feed behind our food: time to act. Available at https://www.forumforthefuture.org/sites/default/files/files/feed%20behind%20our%20food_artwork_lr_compressed.pdf (accessed 24 June 2018).
Garnett, T. (2014) *What is a Sustainable Diet? A Discussion Paper*. Food and Climate Research Network, Oxford Martin School, University of Oxford, Oxford, UK, p. 31.
Garnett, T. (2016) Plating up solutions: Can eating patterns be both healthier and more sustainable? *Science* 353(6305), 1202–1204.
German, J.B., Zivkovic, A.M., Dallas, D.C. and Smilowitz, J.T. (2011) Nutrigenomics and personalized diets: what will they mean for food? *Annual Review of Food Science and Technology* 2, 97–123. DOI: 10.1146/annurev.food.102308.124147
Gladek, E., Fraser, M., Roemers, G., Munoz, O.S., Hirsch, P. *et al.* (2016) *The Global Food System: An Analysis – Report to WWF*. WWF Netherlands, Amsterdam, Netherlands, p. 188.
Green, R., Milner, J., Dangour, A.D., Haines, A., Chalabi, Z., *et al.* (2015) The potential to reduce greenhouse gas emissions in the UK through healthy and realistic dietary change. *Climatic Change* 129(1–2), 253–265.
Gussow, J.D. and Clancy, K.L. (1986) Dietary guidelines for sustainability. *Journal of Nutrition Education* 18(1), 1–5.
Gustavsson, J., Cederberg, C., Sonnesson, U., van Otterdijk, R., Meybeck, A., *et al.* (2011) *Global Food Losses and Food Waste: Extent, Causes and Prevention*. Food and Agriculture Organization of the United Nations, Rome, Italy.
Health Council of the Netherlands (2011) *Guidelines for a Healthy Diet: The Ecological Perspective*. Health Council of the Netherlands, The Hague, Netherlands.
ICIPE (2011) *Insects and Africa's Health: 40 years of ICEPE*. International Centre of Insect Physiology and Ecology (ICIPE), Nairobi, Kenya, p. 64.
Lang, T. and Heasman, M. (2015) *Food Wars: The Global Battle for Mouths, Minds and Markets,* 2e. Routledge Earthscan, Abingdon, UK.
Lang, T. and Mason P. (2017) Sustainable diet policy development: implications of multi-criteria and other approaches, 2008–2017. *Proceedings of the Nutrition Society* 4, 1-16. DOI: 10.1017/S0029665117004074
Lappé, F.M. (1971) *Diet for a Small Planet*. Ballantine Books, New York, USA.
Lawrence, M., Burlingame, B., Caraher, M., Holdsworth, M., Neff, R. and Timotijevic, L. (2015) Public health nutrition and sustainability. *Public Health Nutrition* 18(13), 2287–2292. DOI: 10.1017/S1368980015002402
Livsmedelsverket, National Food Administration (Sweden) (2015) *Find your Way to Eat Greener, Not Too Much and be Active*. Livsmedelsverket / National Food Administration, Stockholm, Sweden, p. 26.
Malthus, T.R. (1798) *An Essay on the Principle of Population, as it Affects the Future Improvement of Society with Remarks on the Speculations of Mr. Godwin, M. Condorcet and Other Writers*. Printed for J. Johnson, London, UK.
Mason, P. and Lang, T. (2017) *Sustainable Diets: How Ecological Nutrition can Transform Consumption and the Food System*. Routledge Earthscan, Abingdon, UK.
Meadows, D.H., Meadows, D.L., Randers, J., Behrens, W. (1972) *The Limits to Growth: A Report for the Club of Rome's Project on the Predicament of Mankind*. Universe Books, New York, USA.
Ministry of Health (Brazil) (2014a) *Guia Alimentar para a População Brasileira.* Ministério da Saúde, Brasilia, Brazil.
Ministry of Health (Brazil) (2014b) Dietary Guidelines for the Brazilian Population. Available at http://189.28.128.100/dab/docs/portaldab/publicacoes/guia_alimentar_populacao_ingles.pdf (accessed 28 June 2018).
Minns, R. (1980) *Bombers and Mash: The Domestic Front 1939-1945*. Virago, London, UK.
Monteiro, C.A., Moubarac, J., Cannon, G., Ng, S.W. and Popkin, B. (2013) Ultra-processed products are becoming dominant in the global food system. *Obesity Reviews* 14(S2), 21–28. DOI: 10.1111/obr.12107
National Food Administration, Sweden's Environmental Protection Agency (2009) Environmentally effective food choices: Proposal notified to the EU, 15 May 2009. National Food Administration and Swedish Environmental Protection Agency, Stockholm, Sweden.
Nelson, M.E., Hamm, M.W., Hu, F.B., Abrams, S.A. and Griffin, T.S. (2016) Alignment of healthy dietary patterns and environmental sustainability: A systematic review. *Advances in Nutrition* 7, 1005–1025. DOI: 10.3945/an.116.012567
Netherlands Nutrition Centre (Voedingscentrum) (2016) Netherlands Nutrition Centre guidelines wheel of five (Appendix 13 on sustainability issues). Available at http://www.voedingscentrum.nl/Assets/

Uploads/voedingscentrum/Documents/Professionals/Schijf%20van%20Vijf/Voedingscentrum%20 Richtlijnen%20Schijf%20van%20Vijf%202016%204.pdf (accessed 28 June 2018).

Netherlands Scientific Council for Government Policy (WRR) (2015) Towards a Food Policy – Synopsis of Naar een Voedselbeleid. Netherlands Scientific Council for Government Policy (WRR), The Hague, and Amsterdam University Press, Amsterdam, Netherlands, p. 17.

Nuffield Council on Bioethics (2007) Public health: ethical issues. Available at http://nuffieldbioethics.org/project/public-health/ (accessed 24 June 2018).

Pact MUFP (2015) Milan Urban Food Policy Pact signed by 100 cities October 15 2015. Available at http://www.milanurbanfoodpolicypact.org/wp-content/uploads/2016/06/Milan-Urban-Food-Policy-Pact-EN.pdf (accessed 24 June 2018).

Popkin, B.M. (2002) An overview on the nutrition transition and its health implications: the Bellagio meeting. *Public Health Nutrition* 5(1A), 93–103.

Popkin, B.M. (2009) Reducing meat consumption has multiple benefits for the world's health. *Archives of International Medicine* 169(6), 543–545.

Qatar Supreme Council of Health (2014a) *Diet and Nutrition Profile for Qatar National Dietary Guidelines*. Supreme Council of Health, Doha, Qatar.

Qatar Supreme Council of Health (2014b) *Qatar Dietary Guidelines Evidence Base*. Supreme Council of Health, Doha, Qatar.

Rockström, J., Steffen, W., Noone, K., Persson, Å., Chapin, III, F.S., *et al.* (2009) Planetary boundaries: exploring the safe operating space for humanity. *Ecology and Society* 14(2), 32. DOI: 10.5751/ES-03180-140232

Rowntree, B.S. (1901) *Poverty: A Study of Town Life*. Macmillan & Co., London, UK.

Rowntree, B.S. (1913) *How the Labourer Lives*. Thomas Nelson & Sons, London, UK.

Rowntree, B.S. (1941) *Poverty and Progress*. Longmans, London, UK.

Rutter, J., Marshall, E. and Sims, S. (2011) The "S" Factors: Lessons from IFG's Policy Success Reunions. Institute for Government, London, p. 124.

Seed, B. (2014) Sustainability in the Qatar national dietary guidelines, among the first to incorporate sustainability principles. *Public Health Nutrition* 18(13), 2303–2310. DOI: 10.1017/S1368980014002110

Singer P. (2013) The world's first cruelty-free hamburger. *The Guardian*, 5 August. Available at https://www.theguardian.com/commentisfree/2013/aug/05/worlds-first-cruelty-free-hamburger (accessed 24 June 2018).

Smil, V. (2013) *Should We Eat Meat? Evolution and Consequences of Modern Carnivory*. Wiley-Blackwell, Chichester, UK.

Smith, P. (2012) Delivering food security without increasing pressure on land. *Global Food Security* 2(1), 18–23. DOI: 10.1016/j.gfs.2012.11.008

Steffen, W., Richardson, K., Rockström J, Cornell, S.E., Fetzer, I., *et al.* (2015) Planetary boundaries: Guiding human development on a changing planet. *Science* 347(6223), 1259855.

Sustainable Food Cities (2014) Sustainable food cities network. Available at http://sustainablefoodcities.org/ (accessed 24 June 2018).

TEEB (2015) *TEEB for Agriculture and Food: An Interim Report*. United Nations Environment Programme, Geneva, Switzerland, p. 124.

Tilman, D. and Clark, M. (2014) Global diets link environmental sustainability and human health. *Nature* 515(7528), 518–522. DOI: 10.1038/nature13959

UN Framework Convention on Climate Change (2015) Paris Convention of the Parties (COP21). Available at http://unfccc.int/2860.php (accessed 24 June 2018).

UNEP (2012) *Avoiding Future Famines: Strengthening the Ecological Basis of Food Security through Sustainable Food Systems*. United Nations Environment Programme, Nairobi, Kenya.

UNEP, Nellemann, C., MacDevette, M., Eickhout, B., Svihus, B., *et al.* (2009) *The Environmental Food Crisis: The Environment's Role in Averting Future Food Crises*. A UNEP rapid response assessment. United Nations Environment Programme / GRID-Arendal, Arendal, Norway.

US Dietary Guidelines Advisory Committee (2015) Scientific Report of the 2015 Dietary Guidelines Advisory Committee to the Secretaries of the U.S. Department of Health and Human Services and the U.S. Department of Agriculture. US Department of Health and Human Services, Washington DC, USA.

van Huis, A., Van Itterbeeck, J., Klunder, H., Mertens, E., Halloran, A., *et al.* (2013) *Edible Insects: Future Prospects for Food and Feed Security*. FAO Forestry Paper 71. Food and Agriculture Organization of the United Nations, Rome, Italy.

Vernon, J. (2007) *Hunger: A Modern History*. Harvard University Press, Cambridge, Massachusetts, USA.

Watts, N., Adger, W.N., Agnolucci, P., Blackstock, J., Byass, P., *et al.* (2015) Health and climate change: policy responses to protect public health. *The Lancet* 386(10006), 1861–1914. DOI: 10.1016/S0140-6736(15)60854-6.

WHO (2015) *WHO Estimates of the Global Burden of foodborne Diseases: Foodborne Diseases Burden Epidemiology Reference Group 2007–2015*. World Health Organization, Geneva, Switzerland, p. 225.

Williamson, J. (2004) A short history of the Washington Consensus. From the Washington Consensus towards a new global governance [hosted by Fundación CIDOB]. Institute for International Economics, Barcelona, Spain.

Zweiniger-Bargielowska, I. (2000) *Austerity in Britain: Rationing, Controls, and Consumption, 1939–1955*. Oxford University Press, Oxford, UK.

2 Sustainable Diets: the Public Health Perspective

Mark Lawrence, Phillip Baker, Kate Wingrove and Rebecca Lindberg

Abstract

Sustainable diets are a prerequisite for public health directly through their impact on nutrition and indirectly through their impact on the environment. Dietary patterns have implications for the use of finite resources, biodiversity and the production of waste including greenhouse gas emissions. In turn, these environmental implications affect the quantity, quality, safety and diversity of the food supply, food and nutrition security and, ultimately, public health. In this chapter we present a review that: conceptualizes the relationship between sustainable diets and public health; describes current dietary patterns and their impacts on the environment and nutrition; explains the characteristics of sustainable diets for protecting public health; and provides policy and practice suggestions for promoting sustainable diets. Current diets have been shaped by transitions in the supply of and demand for food driven by economic, agricultural and food policies, combining with technological innovations and the interests of powerful transnational corporations. The diets are characterized by over consumption and, in particular, a relatively high consumption of animal sourced foods, vegetable oils, caloric sweeteners and ultra-processed food products. These food supply transitions are both a cause and effect of once food literate citizens who were actively engaged with food supply chains becoming progressively passive food consumers whose food demands are mediated via external influencers. Consequently, current diets are having adverse impacts on the environment and nutrition. They are non-sustainable and the leading contributors to the global burden of disease. The literature consistently identifies four key characteristics of sustainable diets to promote public health: moderate consumption; shift current dietary patterns to more plant-based diets; reduce consumption of ultra-processed food products; and reduce food waste. Priority activities for promoting sustainable diets for public health are: policies to promote sustainable diets; empowering people to consume sustainable diets; and research to better understand and promote sustainable diets.

Introduction

Public health is concerned with the health of populations. Diets are a fundamental determinant of public health directly through their impact on nutrition and indirectly through their impact on the environment. From a nutrition perspective, diets consisting of a variety of nutritious and safe foods consumed in balanced and moderate proportions help protect people against all forms of malnutrition, including undernutrition, overweight, obesity and diet-related non-communicable diseases (WHO, 2015). Conversely, dietary inadequacies, excesses and imbalances are responsible for risk factors that are the leading contributors to the global burden of disease (GBD 2016 Risk Factors Collaborators, 2017) and this disease burden is not spread equally within and across populations. There are inequities in food access, affordability and availability and hence there exists a social gradient in nutrition-related public health problems. The prevalence and severity of all forms of malnutrition being faced by the world has been described as 'a global nutrition crisis' (UNSCN, 2017a: p. 9).

From an environment perspective, diets are comprised of combinations of foods that are made available within the physical constraints of an environment containing finite resources and the conditions necessary to support biological systems. Diets that place excessive demands on environmental resources, narrow biodiversity and/or generate unnecessary greenhouse gas emissions, can disrupt environmental systems. In turn, such disruption can contribute to diminishing environmental capacity to provide for food and nutrition security (Lawrence *et al.*, 2015). Consequently, many nutritionists are extending the scope of the diet and public health agenda from its conventional biological foundations to one that is inclusive of social and environmental dimensions (Cannon and Leitzmann, 2005). The nutrition agenda is being reframed using terms such as 'ecological public health' (Lang and Rayner, 2012) and 'environmental nutrition' (Sabaté *et al.*, 2016).

A sustainable diet is the common thread for tackling the dietary risk factors for the nutrition- and environment-oriented problems confronting public health (Tilman and Clark, 2014). In this context we adopt the following definition of 'sustainable diets':

> those diets with low environmental impacts which contribute to food and nutrition security and to healthy life for present and future generations. Sustainable diets are protective and respectful of biodiversity and ecosystems, culturally acceptable, accessible, economically fair and affordable; nutritionally adequate, safe and healthy; while optimizing natural and human resources.
>
> (Burlingame & Dernini, 2012: p. 7)

Critical in this regard is the commitment by governments to the United Nations Sustainable Development Goals (UNDP, 2017) and its Decade of Action on Nutrition, 2016–2025 which identifies as the first of its six cross-cutting action areas, 'Sustainable, resilient food systems for healthy diets' (UNSCN, 2017b: p. 4). In this chapter we present a review that: conceptualizes the relationship between sustainable diets and public health; describes current dietary patterns and their impacts on the environment and nutrition; explains the characteristics of sustainable diets for protecting public health; and provides policy and practice suggestions for promoting sustainable diets.

The Conceptual Basis to the Relationship Between Sustainable Diets and Public Health

Humans (and their nutritional requirements) have evolved over millions of years consuming diets that have been sustainable in the sense that their impact on public health has been compatible with environmental continuity and human survival, at least to reproductive age. This evolutionary process has taken place within an ecological system that has nurtured human populations by providing a sufficient quantity, quality and variety of safe food to meet physiological requirements for nutrients and energy (Eaton *et al.*, 1996).

The relationship between sustainable diets and public health and how it is embedded within the ecological system is depicted in Fig. 2.1. Sustainable diets are the outcome of the combination of the food supply and demand for food. Sustainable diets then determine nutrition outputs, which in turn influence public health. In addition, sustainable diets share a cyclic relationship with the environment. The environment affects the capacity of the food supply to provide for food and nutrition-secure diets, and dietary characteristics then have impacts on the amount and quality of environmental resources, biodiversity and climatic conditions (Turner *et al.*, 2017), and the cycle continues.

Contemporary Changes in Food Supply and Demand and Dietary Impacts

Recently, by evolutionary standards, human populations have undergone a series of social, economic and technological transitions. Approximately 10,000 years ago many populations began transitioning from hunter–gatherer lifestyles to living in agricultural settlements. Then approximately 200 years ago the advent of the industrial revolution witnessed the beginning of

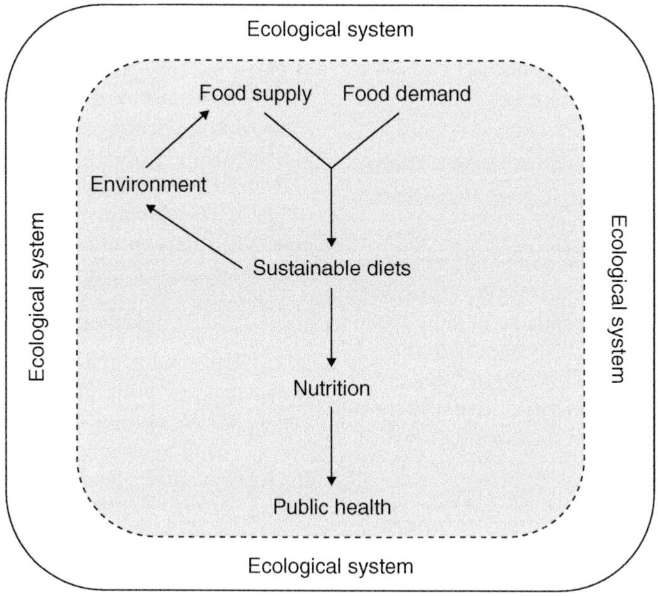

Fig. 2.1. Relationship between sustainable diets and public health.

large-scale migration from agricultural settlements to large towns and cities. By mid-2009, for the first time in recorded history, more people across the world were living in urban areas than rural areas (UN DESA, 2010).

More recently, accelerated urbanization, income growth, labour market transformations and technological development have been powerful drivers of food demand and dietary changes. On the supply side, the liberalization of food trade and investment, the growth and global expansion of trans-national food and beverage corporations, and the growing ubiquity and intensity of food marketing have transformed food systems at all levels. These changes have been enabled by economic, agricultural and trade policies that prioritize the making of as much food as possible, achieving low food prices through mass production, and the globalization of markets (IPES-Food, 2016). These transitions are both a cause and effect of the once high level of food literacy among populations diminishing over time with an increasing proportion of modern day consumers' food demands being mediated via external influences.

The operation of modern (industrialized) food systems is increasingly dependent on chemical inputs. These inputs are associated with improved efficiency, but there is evidence of adverse effects on public health (IPES-Food, 2017). These effects can occur through direct exposure or as a result of environmental contamination. For example, chronic exposure to chemicals used in food production, processing and packaging is associated with increased risk of endocrine system disruption. There is emerging evidence that exposure to these chemicals increases the risk of adverse reproductive outcomes and affects thyroid function, brain function, metabolism, and insulin and glucose homeostasis. In addition to the risks associated with chronic exposure to the endocrine disrupting chemicals used in industrial agriculture, farm workers can be exposed to the risk of acute pesticide poisoning (IPES-Food, 2017).

Inevitably, changes in food demand and supply have resulted in significant shifts in diets around the world away from those followed for the vast majority of human existence. The pace and extent of the dietary change has outstripped the adaptive capacity of biological systems, including the human genome. There is, however, no single nutrition transition across populations. Although there may be a global convergence in the consumption of a limited number of foods (i.e. a 'universal' nutrition transition), there are also wide divergences in consumption patterns resulting from demographic, cultural, socioeconomic and market factors that shape food systems, dietary preferences and consumer demand at regional and local levels.

The Impacts of Contemporary Diets on the Environment and Nutrition

In evolutionary terms the scale and pace of the food supply and demand transitions have been more substantial and rapid than the adaptive capabilities of the ecological system (Eaton *et al.*, 1996). This disruption has manifested in terms of adverse impacts on the environment and nutrition and with subsequent risks to public health.

Dietary impacts on the environment

Diets impact on the environment through the signals they send to the production, processing and distribution steps in the food supply chain. Dietary patterns place demands on land, soil, water and nutrient resources, have implications for biodiversity, and generate significant waste as a by-product at each step of the food supply chain. For instance, modern food production practices are characterized by profligate land clearing and degradation of natural habitats and consequent losses of biodiversity as well as the over-exploitation of aquifers (IPES-Food, 2016). Collectively, it is estimated that the global food supply chain is responsible for 19–29% of all human-induced greenhouse gas emissions, up to 70% of fresh water use and 60% of terrestrial biodiversity loss (UNSCN, 2017c).

Subsequently, these dietary impacts on the environment are feeding back to affect the food supply. For example, climate change alone has been associated with reducing food production yields (Porter *et al.*, 2014), altering the macronutrient and micronutrient composition of certain foods (Fernando *et al.*, 2012), increased food safety risk (McMichael, 2006) and adversely affecting food quality (Porter *et al.*, 2014).

Dietary impacts on nutrition

The food systems and dietary changes described in previous sections have had significant impacts on the world's nutritional status, most dramatically in low- and middle-income countries (Global Panel on Agriculture and Food Systems for Nutrition, 2016). A 'double burden' of malnutrition is now widespread; of 129 countries with available data, 57 (44.2%) have high-levels of both undernutrition and adult overweight/obesity (IFPRI, 2016). Malnutrition is a leading contributor to the global burden of disease. Overweight and obesity and child and maternal undernutrition alone account for 4.9% and 7.0% of global disability-adjusted life years, respectively (GBD 2013 Risk Factor Collaborators, 2015). The global scale of malnutrition is immense: 794 million people cannot meet their daily dietary energy requirements; 2 billion are micronutrient deficient; and 1.9 billion overweight or obese (IFPRI, 2016). Within countries around the world there is a gradient in the burden of malnutrition and associated risk factors among population groups differentiated by income, gender, education and rural/urban status. Although global food production has kept pace with growing human population size, the distribution and availability of food is highly inequitable within and between countries (IFPRI, 2016).

The characteristics of sustainable diets for protecting public health

The priority response to the current dietary impact on the environment and nutrition and subsequent poor public health outcomes is to shift current diets to sustainable diets. And just as current diets have 'co-risks' for the environment and nutrition components of public health, the literature often identifies that shifting to sustainable diets will have 'co-benefits' for the environment and nutrition and hence public health (Tirado, 2015). As one systematic review has reported, a change from current to sustainable diets could have diet-related greenhouse gas emissions and land use (Hallström *et al.*, 2015). Sustainable diets delivering this level of environmental benefit have also been estimated to achieve a 6–10% reduction in nutrition-related global mortality (Springmann *et al.*, 2016).

A first step in shifting current diets is to identify the characteristics of sustainable diets. Although there is no one sustainable diet as such because diets are complex combinations of food and beverage types, amounts and variety, the literature reports a number of complementary dietary characteristics that are associated with relatively low adverse environment impacts. The key characteristics of sustainable diets along with their co-benefits for nutrition and the environment are: (i) moderate consumption; (ii) shift current dietary patterns to more plant-based diets; (iii) reduce consumption of ultra-processed food products; and (iv) reduce food waste (Friel *et al.*, 2013; Jones *et al.*, 2016). These characteristics are described below.

Moderate consumption

Moderate dietary consumption entails eating to satisfy and not exceed energy and nutrient requirements for growth, activity and repair. The nutrition benefit of moderate consumption is that a healthy weight is most likely to be achieved and maintained, thus avoiding health risks associated with overweight and obesity. The environmental benefit of moderate consumption is that relative to the current dominant dietary pattern of consuming beyond nutrition requirements, there is less demand for finite resources to produce, process and distribute extra food (Friel *et al.*, 2013).

Shift current dietary patterns to more plant-based diets

Plant-based diets contain mostly fruit, vegetables, whole grains and pulses, and modest amounts of animal products. Relative to animal-based diets that predominate in many parts of the world, plant-based diets are associated with beneficial health outcomes of themselves (Hu, 2003) and because they moderate excessive consumption of red meat (Popkin, 2009). Environmental benefits would be gained by shifting to plant-based diets because they would effectively be displacing excessive red meat consumption which generally is associated with relatively high levels of greenhouse gas emissions (Gerber *et al.*, 2013) and threats to biodiversity (Machovina *et al.*, 2015).

Reduce consumption of ultra-processed food products

Diets consisting predominantly of a balanced variety of whole or minimally processed foods consumed in moderate amounts are associated with multiple nutrition benefits and are the basis of national dietary guideline recommendations around the world (FAO, 2017). Conversely, basing a diet on ultra-processed food products that tend to be high in energy and low in nutrients is likely to result in dietary excesses and imbalances. Environmentally, there are benefits in replacing ultra-processed food products with whole foods because ultra-processed food products are associated with a relatively higher number of energy- and water-intensive processing steps.

Reduce food waste

Worldwide an estimated one-third of food is wasted from the paddock to the plate (FAO, 2011). There are benefits for food and nutrition security in minimizing this wastage so that more food is available. Environmentally, there are direct benefits to be gained from diets that comprise minimal food waste because less of the finite resources that are required in producing, processing and delivering the food to the plate are wasted. It is also indirectly beneficial because there is less organic food that is sent to landfill to degrade and release methane, a potent greenhouse gas (HLPE, 2014).

There is a need to avoid over-generalizing these key dietary characteristics as the context and circumstances in which foods and beverages are produced, processed and distributed can have a significant influence on their environmental impact. For example, although animal-based diets typically are more resource demanding and greenhouse gas emission-intensive than plant-based diets, the assessment depends on how the animals and plants are produced. Smaller scale livestock production may have benefits for the environment as regards manure production that can substitute for synthetic fertilizer inputs in mixed crop-livestock systems to support nitrogen cycling and biodiversity (Garnett *et al.*, 2017). Moreover, vegetables and fruits can be resource intensive with regard to agrochemical use and energy consumption when produced in greenhouses and/or when requiring excessive packaging materials (UNSCN, 2017a).

Priority Activities for Promoting Sustainable Diets for Public Health

Policies to promote sustainable diets

Policies to promote sustainable diets need to be formulated using a food systems approach for three particular reasons. First, there are multiple threats to and opportunities for promoting sustainable diets and although they might appear to be discrete and unrelated, when viewed through a food-systems lens their interrelationships become more visible and can be more efficiently addressed (IPES-Food, 2017). Second, the urgent and substantial nature of the task to promote sustainable diets means that policies need to deliver transformative change, and this requires acting on the economic, social and political drivers of diets (HLPE, 2017), a level of engagement that can only be achieved at the systems level. Third, the holistic orientation of a food systems perspective enables the contexts and circumstances in which foods are produced, processed and distributed to be more accurately and comprehensively taken into account in planning policy activities.

Polices are essential to effectively create healthier food systems and food environments that enable behaviour changes conducive to sustainable diets. Dietary guidelines are core reference standards for food and nutrition policies and as such, it is critical that sustainable diet principles are integrated into their recommendations (Gonzalez Fischer and Garnett, 2016). National agricultural policies should enable the production of diverse foods that support good nutrition, biodiversity and the sustainable use of finite environmental resources (FAO and WHO, 2014).

Empowering people to consume sustainable diets

The capacity for individuals and populations to consume sustainable diets is influenced by a diversity of factors including food availability, accessibility, convenience, taste, social desirability, food-related knowledge and skills, and habituation (FAO and WHO, 2014). A mix of social marketing campaigns, education programmes and food labelling initiatives have been adopted by governments, citizens' groups and market actors to raise awareness and help guide food choices (Meybeck and Gitz, 2017).

Empowering people to consume sustainable diets requires not only interventions to increase awareness and knowledge, but also a focus on strengthening food literacy to empower people to be more in control of the many factors influencing their diet. Food literacy is based on building an appreciation and understanding of the social, cultural, and environmental dimensions of food alongside practical food skills to build resilience and skills in critical analysis and problem solving (Azevedo Perry *et al.*, 2017).

Research to better understand and promote sustainable diets

According to McMichael (2005: p. 706), 'The essential challenge for nutrition science is to develop new understanding and strategies to enable a balance between promoting, equitably, the health of humans while sustaining the long-term health of the biosphere'. This requires rethinking how evidence is generated, synthesized and translated. In the context of contemporary nutrition and environmental problems, 'conventional reductionist, nutrient-oriented approaches for generating and translating nutrition evidence into public health practice are no longer sufficient' (Lawrence, 2017: p. 62). These approaches broadly reflect the principles of evidence-based medicine, which is appropriate in a clinical nutrition context, but tend to be inappropriate in a public health and food systems context.

Alternative approaches for generating, synthesizing and translating evidence into policy and practice exist but currently are underutilized. Dietary pattern research involves the collection and analysis of data describing the quality, quantity and combination of foods frequently consumed by populations and the impact this has on public health (Cespedes and Hu, 2015). Research into sustainable diets extends this thinking by considering the social, cultural, economic and environmental impacts associated with dietary patterns. Food systems research is transdisciplinary and can be used to explore the health, environmental and socioeconomic impacts associated with dietary patterns. This approach can account for other determinants of food choices including food availability driven by food production and trade (Turner *et al.*, 2017).

Conclusion

Sustainable diets are a fundamental prerequisite for public health. However, current diets and their nutrition and environment outputs are incompatible with the parameters of the ecological system within which humans have evolved and food systems operate. Significant diet-related public health problems exist, and the burden of these problems is inequitably spread within and across populations. Just as there are co-risks for nutrition and the environment from current non-sustainable diets, there are co-benefits to be gained from shifting to sustainable diets that moderate consumption, are plant-based, substitute whole foods for ultra-processed food products, and minimize food waste. Promoting sustainable diets is a fundamental public health challenge and opportunity for the twenty first century.

References

Azevedo Perry, E., Thomas, H., Samra, H.R., Edmonstone, S., Davidson, L., *et al.* (2017) Identifying attributes of food literacy: A scoping review. *Public Health Nutrition* 20, 2406–2415. DOI: 10.1017/S1368980017001276

Burlingame, B. and Dernini, S. (Eds) (2012). Sustainable Diets and Biodiversity: Directions and solutions for policy, research and action. *Proceedings of the International Scientific Symposium Biodiversity and Sustainable Diets United against Hunger*, 3–5 November 2010, FAO Headquarters, Rome. Available at http://www.fao.org/docrep/016/i3004e/i3004e.pdf (accessed 28 August 2018).

Cannon, G. and Leitzmann, C. (2005) The new nutrition science project. *Public Health Nutrition* 8, 673–694. DOI: 10.1079/phn2005819

Cespedes, E.M. and Hu, F.B. (2015) Dietary patterns: From nutritional epidemiologic analysis to national guidelines. *American Journal of Clinical Nutrition* 101, 899–900. DOI: 10.3945/ajcn.115.110213

Eaton, S.B., Eaton, S.B., Konner, M.J. and Shostak, M. (1996). An evolutionary perspective enhances understanding of human nutritional requirements. *Journal of Nutrition* 176, 1732–1740.

FAO (2011) Global food losses and food waste. Extent, causes and prevention. Food and Agriculture Organization of the United Nations, Rome, Italy. Available at http://www.fao.org/docrep/014/mb060e/mb060e00.pdf (accessed 4 October 2017).

FAO (2017) Food-based dietary guidelines. Available at http://www.fao.org/nutrition/nutrition-education/food-dietary-guidelines/en/ (accessed 6 June 2017).

FAO and WHO (2014) Second International Conference on Nutrition: Rome declaration on nutrition. Food and Agriculture Organization of the United Nations, World Health Organization, Rome, Italy. Available at http://www.fao.org/3/a-ml542e.pdf (accessed 12 August 2017).

Fernando, N., Panozzo, J., Tausz, M., Norton, R., Fitzgerald, G., *et al.* (2012). Rising atmospheric CO2 concentration affects mineral nutrient and protein concentration of wheat grain. *Food Chemistry* 133, 1307–1311. DOI: 10.1016/j.foodchem.2012.01.105

Friel, S., Barosh, L.J. and Lawrence, M. (2013) Towards healthy and sustainable food consumption: an Australian case study. *Public Health Nutrition* 17, 1156–1166. DOI: 10.1017/S1368980013001523

Garnett, T., Godde, C., Muller, A., Röös, E., Smith, P., *et al.* (2017) Grazed and confused? Ruminating on cattle, grazing systems, methane, nitrous oxide, the soil carbon sequestration question – and what it all means for greenhouse gas emissions. Food Climate Research Network, Oxford, UK. Available at http://www.fcrn.org.uk/sites/default/files/project-files/fcrn_gnc_report.pdf (accessed 9 October 2017).

GBD 2013 Risk Factor Collaborators (2015) Global, regional and national comparative risk assessment of 79 behavioural, environmental and occupational, and metabolic risks or clusters of risks in 188 countries, 1990–2013: A systematic analysis for the Global Burden of Disease Study 2013. *The Lancet* 386, 2287–2323. DOI: 10.1016/S0140-6736(15)00128-2

GBD 2016 Risk Factors Collaborators (2017) Global, regional, and national comparative risk assessment of 84 behavioural, environmental and occupational, and metabolic risks or clusters of risks, 1990–2016: A systematic analysis for the Global Burden of Disease Study 2016. *The Lancet* 390, 1345–1422. DOI: 10.1016/S0140-6736(17)32366-8

Gerber, P.J., Steinfeld, H., Henderson, B., Mottet, A., Opio, C., *et al.* (2013). *Tackling climate change through livestock: A global assessment of emissions and mitigation opportunities*. Food and Agriculture Organisation of the United Nations, Rome, Italy. Available at http://www.fao.org/3/a-i3437e.pdf (accessed 11 October 2017).

Global Panel on Agriculture and Food Systems for Nutrition (2016) Food systems and diets: Facing the challenges of the 21st century. Global Panel on Agriculture and Food Systems for Nutrition, London, UK. Availaible at http://glopan.org/sites/default/files/ForesightReportDec2016.pdf (accessed 2 December 2016).

Gonzalez Fischer, C. and Garnett, T. (2016) Plates, pyramids, planet. Developments in national healthy and sustainable dietary guidelines: A state of play assessment. Food Climate Research Network, Oxford, UK. Available at http://www.fcrn.org.uk/sites/default/files/ppp_final_10-5-2016.pdf (accessed 15 August 2017).

Hallström, E., Carlsson-Kanyama, A. and Börjesson, P. (2015) Environmental impact of dietary change: A systematic review. *Journal of Cleaner Production* 91, 1–11. DOI: 10.1016/j.jclepro.2014.12.008

HLPE (2014) Food losses and food waste in the context of sustainable food systems. A report by the High Level Panel of Experts on food security and nutrition. Committee on World Food Security, Rome, Italy. Available at http://www.fao.org/fileadmin/user_upload/hlpe/hlpe_documents/HLPE_Reports/HLPE-Report-8_EN.pdf (accessed 4 October 2017).

HLPE (2017) Nutrition and food systems. A report by the High Level Panel of Experts on food security and nutrition. Committee on World Food Security, Rome, Italy. Available at http://www.fao.org/fileadmin/user_upload/hlpe/hlpe_documents/HLPE Reports/HLPE-Report-12_EN.pdf (accessed 6 October 2017).

Hu, F.B. (2003) Plant-based foods and prevention of cardiovascular disease: An overview. *American Journal of Clinical Nutrition* 78, 544S–551S.

IFPRI (2016) Global nutrition report 2016. From promise to impact: ending malnutrition by 2030. International Food Policy Research Institute, Washington, DC. Available at http://ebrary.ifpri.org/utils/getfile/collection/p15738coll2/id/130354/filename/130565.pdf (accessed 12 August 2017).

IPES-Food (2016) From uniformity to diversity: A paradigm shift from industrial agriculture to diversified agroecological systems. International Panel of Experts on Sustainable Food Systems. Available at http://www.ipes-food.org/images/Reports/IPES_ExSummary02_1606_BRweb_pages_br.pdf (accessed 14 August 2017).

IPES-Food (2017) Unravelling the Food–Health Nexus: Addressing practices, political economy, and power relations to build healthier food systems. Global Alliance for the Future of Food and the International Panel of Experts on Sustainable Food Systems. Available at http://www.ipes-food.org/images/Reports/Health_FullReport.pdf (accessed 8 October 2017).

Jones, A.D., Hoey, L., Blesh, J., Miller, L., Green, A., *et al.* (2016) A systematic review of the measurement of sustainable diets. *Advances in Nutrition* 7, 641–664. DOI: 10.3945/an.115.011015

Lang, T. and Rayner, G. (2012) Ecological public health: the 21st century's big idea? An essay by Tim Lang and Geof Rayner. *BMJ* 345, e5466. DOI: 10.1136/bmj.e5466.

Lawrence, M. (2017) Rethinking the translation of nutrition evidence into public health practice. *Journal of Nutrition and Intermediary Metabolism* 8, 62. DOI: 10.1016/j.jnim.2017.04.006

Lawrence, M., Burlingame, B., Caraher, M., Holdsworth, M., Neff, R., *et al.* (2015) Public health nutrition and sustainability. *Public Health Nutrition* 18, 2287–2292. DOI: 10.1017/S1368980015002402

Machovina, B., Feeley, K.J. and Ripple, W.J. (2015) Biodiversity conservation: The key is reducing meat consumption. *Science of the Total Environment* 536, 419–431. DOI: 10.1016/j.scitotenv.2015.07.022

McMichael, A.J. (2005) Integrating nutrition with ecology: Balancing the health of humans and biosphere. *Public Health Nutrition* 8, 706–715. DOI: 10.1079/PHN2005769

McMichael, A.J. (2006) Climate change and human health: Present and future risks. *The Lancet* 367, 859–869. DOI: 10.1016/S0140-6736(06)68079-3

Meybeck, A. and Gitz, V. (2017) Sustainable diets within sustainable food systems. *Proceedings of the Nutrition Society* 76, 1–11. DOI: 10.1017/S0029665116000653

Popkin, B.M. (2009) Reducing meat consumption has multiple benefits for the world's health. *Journal of the American Medical Association* 169, 543–545.

Porter, J.R., Xie, L., Challinor, A.J., Cochrane, K., Howden, S.M., *et al.* (2014) Food security and food production systems. In: *Climate change 2014: Impacts, adaptation and vulnerability. Part A: Global and sectoral aspects. Contribution of the Working Group II to the Fifth Assessment Report of the Intergovernmental Panel on Climate Change*. Cambridge University Press, Cambridge, UK and New York, Available at http://www.ipcc.ch/report/ar5/wg2/ (accessed 25 March 2016).

Sabaté, J., Harwatt, H. and Soret, S. (2016) Environmental nutrition: a new frontier for public health. *American Journal of Public Health* 106, 815–821. DOI: 10.2105/AJPH.2016.303046

Springmann, M., Godfray, H.C.J., Rayner, M. and Scarborough, P. (2016) Analysis and valuation of the health and climate change cobenefits of dietary change. *Proceedings of the National Academy of Sciences of the United States of America* 113, 4164–4151. DOI: 10.5287/bodleian:XObxm2ebO

Tilman, D. and Clark, M. (2014) Global diets link environmental sustainability and human health. *Nature* 515, 518–522. DOI: 10.1038/nature13959

Tirado, C. (2015) *Sustainable food systems and health. The convenient truth of addressing climate change while promoting health.* United Nations Standing Committee on Nutrition Rome, Italy. Available at https://www.unscn.org/files/Announcements/EXE_2_Sustainable_Food_systems_and_health.pdf (accessed 28 September 2017).

Turner, G.M., Larsen, K.A., Candy, S., Ogilvy, S., Ananthapavan, J., *et al.* (2017) Squandering Australia's food security: The environmental and economic costs of our unhealthy diet and the policy path we're on. *Journal of Cleaner Production*, 195, 1581–1599. DOI 10.1016/j.jclepro.2017.07.072

UN DESA (2010) Urban and rural areas 2009. Available at http://www.un.org/en/development/desa/population/publications/pdf/urbanization/urbanization-wallchart2009.pdf (accessed 28 September 2017).

UNDP (2017) Sustainable Development Goals. Available at http://www.undp.org/content/undp/en/home/sustainable-development-goals.html (accessed 9 October 2017).

UNSCN (2017a) UNSCN News 42: A spotlight on the Nutrition Decade. United Nations Standing Committee on Nutrition, Rome, Italy. Available at https://www.unscn.org/en/Unscn-news?idnews=1682 (accessed 5 July 2017).

UNSCN (2017b) United Nations decade of action on nutrition (2016-2025): Work programme. United Nations Standing Committee on Nutrition, Rome, Italy. Available at https://www.unscn.org/uploads/web/news/Work-Programme_UN-Decade-of-Action-on-Nutrition-20170517.pdf (accessed 26 September 2017).

UNSCN (2017c) Sustainable diets for healthy people and a healthy planet. United Nations Standing Committee on Nutrition, Rome, Italy. Available at https://www.unscn.org/uploads/web/news/document/Climate-Nutrition-Paper-EN-WEB.pdf (accessed 13 September 2017).

WHO (2015) Healthy diet fact sheet no. 394. Available at http://www.who.int/mediacentre/factsheets/fs394/en/ (accessed 8 September 2017).

3 The Challenges of Sustainable Food Systems Where Food Security Meets Sustainability – What are Countries Doing?

Meredith Harper, Alon Shepon, Nir Ohad and Elliot M. Berry

Abstract

The evolutionary history of the concepts of food security (FS) and sustainability have run in parallel for many years. After the food crisis of 2008, stability was added to definition of FS as a short-term time dimension to express the ability to withstand shocks to the food system caused by natural or man-made disasters. We have proposed that sustainability be added as a fifth long-term time dimension, thus bringing together FS and sustainability. In 2015, the United Nations described the seventeen sustainable development goals. We believe that FS involves *all* the goals to a greater or lesser extent. The challenge ahead is to build and integrate FS on the sustainability agenda and vice versa. The final common pathway for all these efforts is for countries to develop their most appropriate sustainable food systems. As a practical exercise towards this aim, we have reviewed what eight different countries (United States, Brazil, France, Greece, Spain, the Netherlands, United Kingdom, and the Scandinavian nations) are doing regarding their food systems. We have compared their programmes according to an operational template for recommendations for Israel based on eight consensus criteria.

Introduction

According to Professor Klaus Bosselmann in the Foreword of *Sustainability: A Cultural History*, 'at its core, sustainability relates to the basic human need to maintain and to nurture the conditions on which life depends' (Grober, 2012). Sustainable food systems (SFS) take into account the environmental needs along the entire food system chain, from production to consumption. It also incorporates social, health and economic concerns. The goal of SFS is a world where the earth can produce enough nutritious, safe, affordable food to feed the growing population while preserving the biodiversity and ecological needs of the planet.

Food Security and Sustainability

The development of the concepts of FS and of sustainability have run in parallel over the past 40 years or so. Food security originally had three elements – availability, accessibility and utilization. It was only after the food crisis of 2007–2008 that a fourth component was added – that of stability. This introduced a time dimension with the ability to withstand shocks to the food system caused by natural or man-made disasters.

Sustainable development also originally involved three aspects – environment, economic and social. It is a systems approach to harness capital (natural, produced and social) for the welfare of present and future generations and incorporates issues such as land and natural resource conservation and development as well as the wider concerns of human development, including public health, nutrition education and standard of living.

From Pillars to Pathways

We have proposed to change the notion of the four FS dimensions from a portrayal as independent pillars of equal importance to a pathway of interacting elements with a time dimension (see Fig. 3.1). Further, in the pathway model we suggest that there is a *feedback loop* (not shown) whereby, if there is FS at the individual level, then human capital is made available for maintaining the upstream components of *both* FS and sustainability. ('A well-fed nation is a healthy nation is a productive and sustainable nation'.) Food security and sustainability have overlapping and mutual concerns regarding sustainable agriculture, biodiversity and renewable natural resources. Thus, the two concepts of FS and sustainability come together naturally, with the latter being considered as the long-term time (fifth) dimension of the former (Berry *et al.*, 2015). Together, FS and sustainability will promote resilience of our planet following the statement of Aiken that:

> sustainability is not a static notion but a moving target which should be understood as a challenge to preserve the adaptability and resilience of the natural (biotic and abiotic) systems that form the basis of economic and social development.
>
> (Aiking, 2014)

To emphasize the inter-relations between FS and sustainability, Fig. 3.2 is an attempt to position all the seventeen sustainable development goals (SDGs) within the framework of FS. All SDGs have been classified within the framework of SFS. The SDGs have been aligned to the dimension of sustainability (Environment) and FS.

1. Environmental aspect at regional level (SDGs 7, 13, 14 and 15).
2. Availability aspect at national level (SDGs 9 and 11).
3. Accessibility aspect at household level (SDGs 1, 4, 5, 8 and 10).
4. Utilization aspect at individual level (SDGs 2, 3 and 6).

Three SDGs (12, 16, 17) are necessary for the overall functioning of the whole sustainable food system.

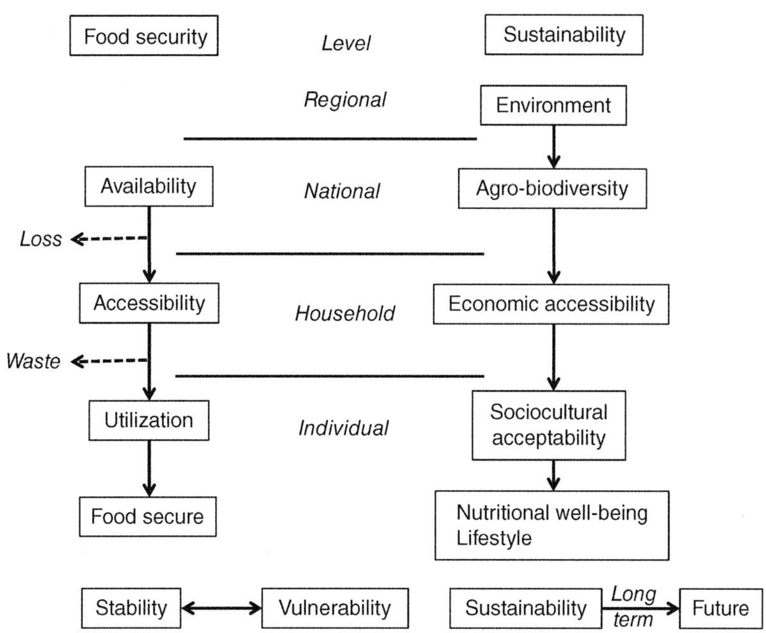

Fig. 3.1. The multi-level interactions between food security and sustainability.

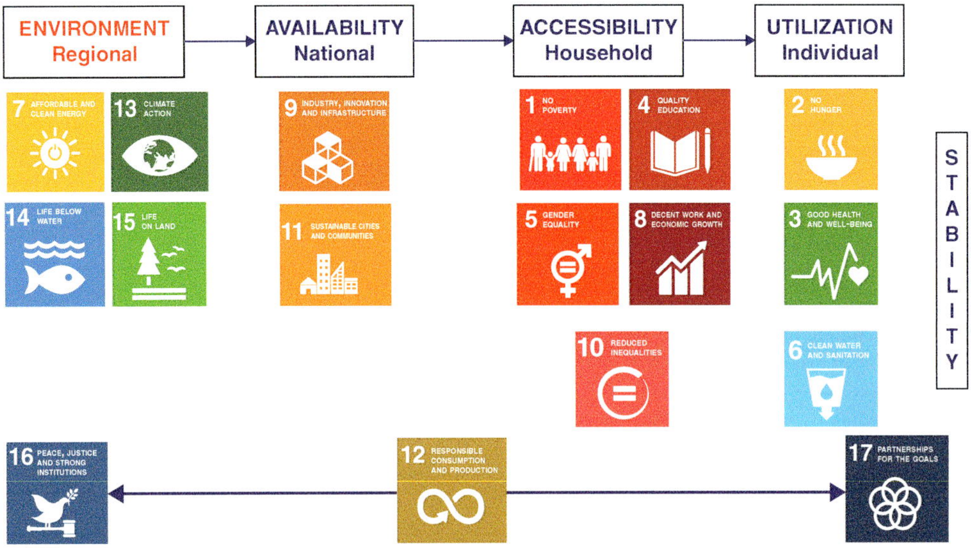

Fig. 3.2. Sustainable food (secure) systems and sustainable development goals (SDGs). An exercise in positioning all 17 SDGs within the framework of sustainable food security.

In June 2016, an international conference entitled 'Sustainable Food Systems: Agriculture, Environment, and Nutrition' was convened in Tel Aviv, Israel. One of the outcomes of this meeting was an eight-point call to action to delineate the steps that should be taken towards securing a sustainable food system for Israel, and as possible guidelines for other countries (Shepon *et al.*, 2017). This call to action was a consensus statement based on discussions during and after the conference, involving academics, policymakers, representatives from the food industry and civil society. We have used this as a template to study and compare what different countries – United States, Brazil, France, Greece, Spain, the Netherlands, United Kingdom and the Scandinavian nations – are already doing to promote food sustainability and security, according to the eight-point call described below, which also covers all the relevant seventeen SDG goals.

Action point 1: Make food systems sustainable along the entire food chain – from production to consumption, protecting resources such as soil, air and water in the light of climate change challenges – and reduce food losses and waste (SDGs 7, 12, 13, 14 and 15)

Of the countries reviewed, France probably goes the furthest to making a national effort towards SFS. France has created a Ministry for Sustainable Development and has added a Charter for the Environment to its constitution. France's National Strategy for Sustainable Development 2010–2013 included several agricultural goals: organic agriculture on 20% of French farmland by 2020; public hotels, restaurants and institutions must contain 20% organic products by 2012; 50% of French farms to have organic certification by 2012; 30% of farms to have low energy dependence by 2013; and reduce pesticide use by 50% by 2018 (Henard, 2012).

Both Brazil and the Scandinavian nations include sustainability in their national dietary guidelines. Brazil's dietary guidelines state:

> Dietary recommendations need to take into account the impact of the means of production and distribution of food on social justice and environmental integrity. Thus, these Guidelines consider the means by which food is produced, distributed, and sold, favoring those methods which are socially and environmentally sustainable.
> (Ministry of Health of Brazil, 2014)

The Nordic Nutrition Recommendations devote an entire chapter to sustainability (Nordic Council of Ministers, 2012). While this information is certainly important, it would be interesting to see what percentage of consumers are familiar with their countries' dietary guidelines and what effect they have on consumer choices.

The federal governments in both the United States and Britain have devoted funds towards research in sustainable agriculture. A grant programme through the United States Department of Agriculture (USDA) Sustainable Agriculture Research and Education (SARE) promotes innovations in sustainable agriculture (USDA, 2017a). Also, in the US, the National Institute of Food and Agriculture (NIFA) offers funding to projects that address national agriculture priorities, including sustainability and FS (USDA, 2017b). In Britain, the Department for Environmental, Food and Rural Affairs has launched a £4.5 million research project to investigate ways to increase farm productivity while decreasing negative environmental effects (Case, 2014).

Food loss and waste is becoming an increasingly relevant topic; however, from reviewing the literature, it appears that addressing this problem is primarily from initiatives of private organizations rather than governments. Some notable organizations devoted to reducing food waste include From Waste to Taste (Finland), the Real Junk Food Project (UK), InStock (Netherlands), Espigoladores (Spain), Bouroume (Greece) and Leket (Israel).

The US government has set its first national food waste reduction goal of 50% by the year 2030 (USDA, 2015). However, it is difficult to determine what exactly is being done to meet this goal. The city of Oslo has taken it upon itself to implement a new food waste bag system to recycle household food waste (Holmertz, 2015). A new law in France made big news in 2015: supermarkets are now forbidden from throwing away good food. Instead, the food must be donated to food banks and charities (Chrisafis, 2016).

Action point 2: Strengthen the current agricultural basis, recognize its vital importance in providing local food and strive to align it towards the best practices in sustainable agriculture (SDGs 13, 14 and 15)

The first question that must be asked when it comes to this point is *what is local?* Most food movements define 'local' as food that is grown/produced within 100 miles of purchase and consumption. According to the United States 2008 Farm Bill, food can be advertised as local if it is transported less than 400 miles from its point of origin (Johnson, 2016). For Israel, this would often mean that food produced outside of the country would be considered 'local'.

Regardless of the definition, the idea of eating local foods is becoming more acceptable in communities across the globe. People often believe that local food is of higher quality, tastes better, may be more environmentally sustainable and supports the local economy. In the US, the USDA Local Food Promotion Program offers grants to local and regional agriculture and business enterprise development and expansion (USDA, 2017c). In Brazil, a 2009 law requires that 30% of federal school lunch funds goes towards food produced by local family farms. In turn, this promotes the local economy, rural development and small farmers (Huber, 2016). Urban, non-commercial farming is also becoming increasingly popular. These, again, appear to be primarily the work of local initiatives (individual and non-profit organizations) rather than government. The largest European urban farm is found in the Netherlands. It resides in a seven-storey former telecommunications building (Stadslandbouw, 2017). Of course, just because food is locally produced does not automatically ensure that it is grown sustainably or is of good nutritional quality. It will be important to promote local food initiatives that work within a sustainable food system framework.

Action point 3: Protect the fundamental human right of all residents to sustainable, healthful, nutritious and adequate food. A well-fed nation is a healthy nation is a sustainable, productive nation. This means ensuring food security and nutrition for all ('no-one goes to bed hungry') at present and in the future (SDGs 1, 2, 3, 5, 6, 12, 16 and 17)

Of all of the countries reviewed for this article, only Brazil includes the right to food in its constitution. Brazil's Zero Hunger Program is often lauded as a major success story in terms of beating hunger. The Zero Hunger Program is a national programme that provides financial support to family farmers and others, ensures free school lunches and even has developed subsidized restaurants. Since its inception, malnutrition in children under 2 years old decreased from 12.7% to 3.5%, and there was a 47% drop in infant mortality. Stunting fell from 13.5% to 6.8%. Additionally, malnutrition in the poorest parts of the country fell from 17.9% to 6.6% (WFP, 2010).

Action point 4: Make a nation's food of high nutritional value with a nationwide programme for enriching and monitoring nutrient values of the food supply through a national data base monitor, making sure it is safe, free of pathogens and environmentally friendly (SDGs 9, 10 and 12)

The fourth point in the call to action requires 'high nutritional value' be further defined prior to implementation. Each country reviewed here already has some sort of food safety regulatory body in place at the national level, such as the Hellenic Food Authority in Greece, the Food Standards Agency in Britain, and the Food and Drug Administration (FDA) and the USDA in the US. The European Union has such a regulatory organization for European nations. However, foods and food additives that may be categorized as safe in one country may be considered unsafe in another. Additionally, there are conflicting policies between nations regarding the regulation of nutrition supplements. Ensuring that food is grown and distributed in an environmentally friendly manner, however, does not appear to play a role in governmental food safety agencies. The topic of genetically modified foods is very relevant here (but beyond the scope of this chapter) and highlights the complex interactions between nutrition, economics, science and pseudo-science.

Several countries use standardized food fortification to promote the health and nutrition of their citizens. The United States boasts iodized salt, vitamin D in milk, calcium in juice and cereals, and flour enriched with B vitamins and iron (De Lourdes *et al.*, 2012). Salt is also iodized in Sweden. In the UK, fat spreads (such as margarine) are regularly fortified with vitamins A and D (Food Standards Agency, 2010). The Netherlands fortifies margarine with vitamin D, and Finland allows vitamin D fortification of both milk and margarine. Some food companies choose to further fortify their foods in the hope of boosting sales; for example, yogurts with fibre or eggs with omega-3 fatty acids. Bioavailability of the added nutrients, cost, and preservation of quality must be considered prior to mandating national food fortification programmes (Fletcher *et al.*, 2004).

Action point 5: Direct the food industry, through legislation, to produce healthy, nutritious (minimally processed) foods in a sustainable manner with less sugars, salt and additives that may adversely affect health. Production and marketing must be honest and transparent. Institute legible, intelligible food labelling. Restrict the marketing of junk food and sweet beverages especially to children (SDGs 4 and 9)

In order to promote healthy, nutritious foods and restrict junk food, these terms require definitions as well as those for processed and ultra-processed foods. Currently, there are advertising laws that affect food in the Scandinavian countries, the UK and the Netherlands. As of 2007, the UK restricts advertising of 'junk food' to children (BBC News, 2008). All forms of food advertising are banned for children under age 13 in the Netherlands (Harrison-Dunn, 2015). France requires warnings on advertisements for unhealthy foods. For example, one such warning is: 'for your health, avoid eating too much fat, too much sugar, too much salt' (NBC News, 2007). Brazil attempted to enact anti-junk food advertising laws; however, they failed due to industry opposition (Huber, 2016). Instead, they include warnings in their dietary guidelines:

> Be wary of food advertising and marketing. The purpose of advertising is to increase product sales, and not to inform or educate people. Be critical and teach children to be critical of all forms of food advertising and marketing.
>
> (Ministry of Health of Brazil, 2014)

The dietary guidelines go on to encourage citizens to talk to their congressional representatives about the need for bills to protect people from food advertising exposure. Industry advertising is self-regulated in Spain, per a voluntary code (Aranceta, 2009). Advertising to children in the United States also follows self-regulation of the food industry (FTC, 2006).

Food labels and their contents are a source of debate for the industry, consumers and legislators. The US recently updated food labels to include added sugars and serving sizes based on amounts that people typically consumed rather than recommended serving sizes (FDA, 2016). The UK makes use of some front of package traffic light labelling; however, this has been opposed by other European Union nations prior to the UK's Brexit vote. The opposition comes from such concerns that olive oil would be given a red label due to its high fat content, despite scientific consensus that olive oil is integral to the Mediterranean diet (which has the best evidence base for being healthy and also sustainable), and that this labelling may negatively impact sales (Dernini *et al.*, 2016; Robinson, 2014). Allergen labelling on packaged foods tends to be required throughout the countries reviewed for the protection of individuals with food allergies.

In the US, legislation enforcing food labelling has demonstrated some success with encouraging the food industry to develop healthier products rather than face the stigma associated with certain labels. This has best been demonstrated with trans-fat labelling laws, which went into effect in 2006. This resulted in a significant decrease of trans-fats in foods, though a food could still legally claim to be 'trans-fat free' if it contained less than 0.5 g of trans-fat per serving. Consumption of trans-fat among American consumers fell 78% from 2003 to 2012. Recently, in 2016, the FDA officially declared trans-fat as not GRAS (generally recognized as safe) and food companies now have three years to remove trans-fat from their products (Christensen, 2015).

Unfortunately, some legislation to restrict unhealthy food faces difficulty as junk food and soda companies are known to fund health groups in the US. Soft drink companies alone have been found to fund 96 health groups (Sifferlin, 2016). One example of the effect this may have is Save the Children, which previously supported a soda tax but stopped its support after receiving $5 million from Coke and Pepsi in 2009. The Academy of Nutrition and Dietetics (AND) is also known to receive funding from the soda industry, and they released a statement saying it did not support New York legislation limiting soda portions. AND was also paid by Kraft to use their 'Kids Eat Right' seal on their 'cheese' singles in 2015. This resulted in a national outcry in print media, online and television news (Nestle, 2015).

Action point 6: Train nutritionist staff to educate the children in our schools (from kindergarten onwards) how to lead a healthy, sustainable lifestyle that includes good nutrition (Mediterranean diet pattern, cooking skills, eating behaviour, sustainability, growing vegetables, fruits and herbs) and regular physical activity (SDGs 3 and 4)

The Mediterranean diet pattern primarily emphasizes plant-based foods such as fruits, vegetables and legumes, as well as healthy fats such as nuts and olive oil. It has been shown to not only have numerous health benefits, but also to be favourable in terms of sustainability (Dernini and Berry, 2015). A study in Spain noted that following a Mediterranean diet results in a 72% decrease in greenhouse gas emissions, 58% decrease in land use, 52% reduction in energy usage and a 33% decline in water consumption (Saez-Alemendros, 2013).

Nutrition education appears to be taught primarily by teachers rather than registered dietitians. In the US, it is estimated that 50% of schools have state or district requirements for nutrition education. It is also estimated that 86% of nutrition education materials are provided by the food industry (National Center for Education Statistics, 1996). Nutrition education in Britain is required as part of core competencies for students (GOV.UK, 2015). Oftentimes, nutrition education is provided as part of biology or health courses, but in Finland, cooking and nutrition are taught in home economics classes, which are required for both boys and girls (Sarlio-Lahteenkorva and Manninen, 2010).

Action point 7: Promote access of healthy, sustainable foods in restaurants, work-place cafeterias, vending machines, medical facilities (including hospitals), sports arenas, public spaces, schools and day-care centres. Remove junk food chains from hospitals, health clinics and educational places like schools and academia (SDGs 4, 5, 8, 10 and 11)

In the US, former First Lady Michelle Obama's Let's Move Initiative focused on promoting healthy lifestyles and reducing obesity, particularly in children. The campaign included many strategies, such as updating school meal nutrition standards, ensuring that any food or beverage marketed to children at school meets certain nutrition standards, improve nutritional quality of meals and snacks in childcare, and teamed up with Sesame Workshop to market fresh fruit and vegetable consumption to children. Let's Move Cities, Towns, and Counties have committed to implementing strategies to improve the health of their citizens, and 225 corporations committed to help make healthy choices an easy choice.

Most nations reviewed have nutrition requirements in place for their school meals. For example, in Britain, schools must provide fruits and vegetables, chicken or oily fish, high-quality meat and cereals. No crisps, chocolates, sweets or drinks with added sugar are allowed in school meals or vending machines, and fried food is limited to twice per week (GOV.UK, 2017).

In 2015, the Physician's Committee for Responsible Medicine in the US found that of 208 hospitals surveyed, 20% had fast food chains. A number of these hospitals had additional agreements with fast food chains, such as allowing advertising throughout the hospital or even delivery from the fast food establishment to patients' rooms (Schumaker, 2015). Meanwhile, the British government has proposed banning sugar-sweetened beverages from hospital vending machines and restaurants (Boseley, 2016).

Action point 8: Remember that eating is not a list of do's and don'ts but should be a pleasurable and tasty experience. Meal times are important opportunities for socializing and building relationships. Traditional and cultural preferences in food choices should be respected (SDGs 4, 5 and 10)

Brazil's food guide includes a section on the importance of enjoying mealtimes and traditional cultural foods. Otherwise, this aspect of the proposed call to action appears largely overlooked by governments around the world.

Summary and Conclusions

Summarizing the action points, we find that eight SDG goals (1. End Poverty; 2. End Hunger; 6. Clean Water; 7. Affordable Clean Energy; 8. Decent Work and Economic Growth; 11. Sustainable Cities/Communities; 16. Peace, Justice and Strong Institutions; and 17. Global Partnerships) are represented once; five (3. Health and Well-Being; 9. Industry, Innovation, Infrastructure; 13. Climate Action; 14. Life Below Water; and 15. Life on Land) are represented twice; three (5. Gender Equality; 10. Reduced Inequalities; and 12. Responsible Consumption and Production) are represented three times; and one (4. Quality Education) is represented four times. From Fig. 3.2 the representation of SDGs in the action points for food security is Environment (7); Availability (3); Accessibility (12) and Utilization (4).

This exercise may give some insights as to the prioritization, relevance and inter-relations between SDGs and FS and SFS.

Many of the countries surveyed above are aware of the challenges (sometimes competing) of ensuring FS and sustainability. Public and political pressure will be essential to ensure the sustainability of initiatives to promote action by the food industry and encourage public awareness, especially in reducing food waste.

From all of the above, we would like to highlight the following actions that countries might take to strengthen the ties between FS and sustainability:

1. Sustainable food systems should be on the agenda of any government plans involving food consumption and production.
2. Work with industry to reduce sugar, sodium and fat in processed foods; increase fortification.
3. Ban 'junk food' adverts for kids or require them to feature health messages/warnings, following the legislation set by France.
4. Grants/tax incentives for sustainable agriculture practices.
5. Make it illegal for supermarkets to waste food as in France.
6. Work with registered dietitians/nutritionists to develop mandatory nutrition and lifestyle (activity) curricula for schools.
7. Make food more affordable – subsidize healthy foods.
8. Pair local farms with schools, and so on, to increase healthy food in schools while helping local farmers economically – require a certain percentage of foods come from local farmers.
9. Develop and implement easily legible food labels for nutrition, environmental sustainability.
10. Encourage the socio-cultural aspects of experiencing different cuisines.

Challenges for the future

Achieving SFS is an enormous global challenge, and the era of climate change and its potential to increase food insecurity is a tremendous challenge. Apart from Brazil, this review has covered only industrialized countries from the United States and Europe, and thus is not representative of the world picture. However, by reviewing what these countries have accomplished, other countries should be able to move forward with their own action plans. Sustainable food systems result in long-term benefits for the planet, people, and the economy. We must continue to organize, promote, and implement strategies for such sustainability.

The real test is political rather than nutritional, since the benefits of sustainability are long-term and of little immediate electoral advantage. The concept also needs to be made more concrete and person-centred in order to be more acceptable and understandable. As Isatou Jallow has stated: 'Political will plus people's will equal sustainable will' (Jallow, personal communication). The task ahead is to build and integrate FS on the sustainability agenda and vice versa. The final common pathway for all these efforts is towards SFS and nutrition. Not all food secure diets are sustainable, but all sustainable diets are food secure.

Acknowledgements

We thank Dorit Adler, Aron Troen and Hagit Ulanovsky for constructive discussions about some of the issues raised in this chapter.

References

Aiking, H. (2014) Protein production: planet, profit, plus people? *American Journal of Clinical Nutrition* 100(1), 483S–489S.

Aranceta, J., Lobo, F., Viedma, P., Salvador-Castell, G., Martínez De Victoria, E., *et al.* (2009) Community nutrition in Spain: advances and drawbacks. *Nutrition Reviews* 67(1), S135–S139. DOI: 10.1111/j.1753-4887.2009.00176.x

BBC News. (2008) Ban on junk food ads introduced. Available at http://news.bbc.co.uk/2/hi/health/7166510.stm (accessed 28 September 2017).

Berry, E.M., Dernini, S., Burlingame, B., Meybeck, A. and Conforti, P. (2015) Food security and sustainability: can one exist without the other? *Public Health Nutrition* 16, 1–10.

Boseley, S. (2016) NHS England proposes hospital ban on sugar-sweetened drinks. *The Guardian*, 9 November. Available at https://www.theguardian.com/society/2016/nov/09/nhs-england-chief-proposal-sugar-ban-drinks-hospitals (accessed 28 September 2017).

Case, P. (2014) Defra launches 4.5 million sustainable intensification project. *Farmers Weekly* 5 December. Available at http://www.fwi.co.uk/news/defra-launches-sustainable-intensification-project.htm (accessed 28 September 2017).

Chrisafis, A. (2016) French law forbids food waste by supermarkets. *The Guardian* 4 February. Available at https://www.theguardian.com/world/2016/feb/04/french-law-forbids-food-waste-by-supermarkets (accessed 28 September 2017).

Christensen, J. (2015) FDA orders food manufacturers to stop using trans-fat within three years. Available at http://edition.cnn.com/2015/06/16/health/fda-trans-fat/index.html (accessed 28 September 2017).

De Lourdes, S.V., Alonso-Aperte, E. and Varela-Moreiras, G. (2012) Vitamin food fortification today. *Food and Nutrition Research* 56. DOI: 10.3402/fnr.v56i0.5459

Dernini, S. and Berry, E.M. (2015) Mediterranean diet: from a healthy diet to a sustainable dietary pattern. *Frontiers in Nutrition* 2, 15. DOI: 10.3389/fnut.2015.00015

Dernini, S., Berry, E.M., Serra-Majem, L., La Vecchia, C., Capone, R., *et al.* (2016) Med Diet 4.0: the Mediterranean diet with four sustainable benefits. *Public Health Nutrition* 20(7), 1322–1330. DOI: 10.1017/S1368980016003177

FDA (2016) Changes to the nutrition facts label. Available at http://www.fda.gov/Food/GuidanceRegulation/GuidanceDocumentsRegulatoryInformation/LabelingNutrition/ucm385663.htm (accessed 28 September 2017).

Fletcher, R.J., Bell, I.P. and Lambert, J.P. (2004) Public health aspects of food fortification: a question of balance. *Proceedings of the Nutrition Society* 63, 605–614.

Food Standards Agency (2010) Guidance on legislation for spreadable fats and other yellow fat spreads. Available at https://www.food.gov.uk/sites/default/files/multimedia/pdfs/yellowfatguidance0610.pdf (accessed 28 September 2017).

FTC (2006) Perspectives on marketing, self-regulation, and childhood obesity. *A Report on a Joint Workshop of the Federal Trade Commission and the Department of Health and Human Services*. Available at https://www.ftc.gov/sites/default/files/documents/reports/perspectives-marketing-self-regulation-childhood-obesity-report-joint-workshop-federal-trade/perspectivesonmarketingself-regulation childhoodobesityftcandhhsreportonjointworkshop.pdf (accessed 28 September 2017).

GOV.UK. (2017) School meals – healthy eating standards. Available at https://www.gov.uk/school-meals-healthy-eating-standards (accessed 28 September 2017).

Grober, U. (2012) *Sustainability: A Cultural History.* UIT Cambridge Ltd, Cambridge, p. 7.

Harrison-Dunn, A.R. (2015) Netherlands bans food ads to children under 13. *Food Navigator* 7 January. Available at http://www.foodnavigator.com/Policy/Netherlands-bans-food-ads-to-children-under-13 (accessed 28 September 2017).

Henard, M.C. (2012) GAIN Report: France's Sustainable Agriculture Initiative. Available at http://sustainable-agriculture.org/wp-content/uploads/2012/11/France-Actions-Towards-a-More-Sustainable-Agriculture_Paris_France_10-23-2012.pdf (accessed 28 September 2017).

Holmertz, S. (2015) Oslo's colourful solution to waste management. *Waste Management World,* 25 June. Available at https://waste-management-world.com/a/oslos-colourful-solution-to-waste-management (accessed 28 September 2017).

Huber, B. (2016) Welcome to Brazil, where a food revolution is changing the way people eat. *The Nation*, 28 July. Available at https://www.thenation.com/article/slow-food-nation-2/ (accessed 28 September 2017).

Johnson, R. (2016) The role of local and regional food systems in US farm policy. Congressional Research Service, 18 February. Available at https://fas.org/sgp/crs/misc/R44390.pdf (accessed 28 September 2017).

Ministry of Health of Brazil (2014) Dietary guidelines for the Brazilian population. Available at http://189.28.128.100/dab/docs/portaldab/publicacoes/guia_alimentar_populacao_ingles.pdf (accessed 28 September 2017).

National Center for Education Statistics (1996) Nutrition education in public elementary and secondary schools. Available at https://nces.ed.gov/pubs/96852.pdf (accessed 28 September 2017).

NBC News (2007) French cracking down on junk food ads. Available at http://www.nbcnews.com/id/17402161/ns/health-diet_and_nutrition/t/french-cracking-down-junk-food-ads/#.WNkQSHR95mA (accessed 28 September 2017).

Nestle, M. (2015) Dietitians in turmoil over conflicts of interest: it's about time. *Food Politics* blog, 18 March. Available at https://www.foodpolitics.com/2015/03/dietitians-in-turmoil-over-conflicts-of-interest-its-about-time/ (accessed 28 September 2017).

Nordic Council of Ministers (2012) Nordic nutrition recommendations 2012. Available at http://norden.diva-portal.org/smash/get/diva2:704251/FULLTEXT01.pdf (accessed 28 September 2017).

Robinson, N. (2014) EU challenges UK traffic light scheme. *Food Manufacture*, 9 July. Available at http://www.foodmanufacture.co.uk/Packaging/Europe-investigating-UK-food-labelling (accessed 28 September 2017).

Saez-Alemendros, S., Obrador, B., Bach-Faig, A. and Serra-Majem, L. (2013) Environmental footprints of Mediterranean versus Western dietary patterns: beyond the health benefits of the Mediterranean diet. *Environmental Health* 12, 118.

Sarlio-Lahteenkorva, S. and Manninen, M. (2010) School meals and nutrition education in Finland. *Nutrition Bulletin* 35(2), 172–174.

Schumaker, E. (2015) Fast food chains are allowed inside hospitals in these 15 states. *The Huffington Post*, 16 April. Available at http://www.huffingtonpost.com/2015/04/10/fast-food-hospitals_n_7033300.html (accessed 28 September 2017).

Shepon, A., Harper, M., Troen, A. and Dernini, S. (2017) Call for action for food security and sustainability in Mediterranean countries. In: Proceedings of a Technical Workshop: Development of Voluntary Guidelines for the Sustainability of the Mediterranean Diet in the Mediterranean Region. FAO/CIHEAM, 2017, pp 97–100. Available at http://www.fao.org/3/a-i7557e.pdf (accessed 28 September).

Sifferlin, A. (2016) Soda companies fund 96 health groups in the US. *Time Magazine*, 10 October. Available at http://time.com/4522940/soda-pepsi-coke-health-obesity (accessed 28 September 2017).

Stadslanbouw den Haag (2017) The new farm: vertical farming. Europe's biggest vertical urban farming. Available at projecthttp://stadslandbouwdenhaag.nl/index.php/de-schilde-vertical-farming/ (accessed 28 September 2017).

USDA (2015) US Food Challenge FAQs. Available at https://www.usda.gov/oce/foodwaste/faqs.htm (accessed 28 September 2017).

USDA (2017a) SARE (Sustainable Agriculture Research and Education). Available at http://www.sare.org/ (accessed 28 September 2017).

USDA (2017b) NIFA (National Institute of Food and Agriculture): About NIFA. Available at https://nifa.usda.gov/about-nifa (accessed 28 September 2017).

USDA (2017c) USDA Local Food Promotion Program. Available at https://www.ams.usda.gov/services/grants/lfpp (accessed 28 September 2017).

WFP (2010) Brazil shows world how to beat hunger, says WFP. Available at https://www.wfp.org/stories/brazil-shows-world-how-beat-hunger-says-wfp-head (accessed 28 September 2017).

4 Climate Change and Sustainable and Healthy Diets

Cristina Tirado von der Pahlen

Abstract

Promoting good nutrition, health and sustainable food systems in the context of population growth, dietary transition and a changing climate is a central challenge of our time. While climate change has an impact on our food systems and diets, our food systems and dietary patterns also affect climate change. This chapter presents an analysis of the interconnections of sustainable dietary patterns, health and nutrition in a context of climate-change mitigation. It outlines the global frameworks and agreements on climate change, food and nutrition, exploring the many, complex ways in which diet affects climate change, and vice versa. It looks at diets that boost health and are environmentally sustainable, as well as the measures needed to steer food production and consumption in that direction. The chapter identifies policies based on co-benefits to health and climate of dietary change and opportunities for joint action on nutrition, health, and climate policy. There are co-benefits of measures that reduce climate-altering emissions and, at the same time, improve health by shifting away from the overconsumption of meat from ruminant sources in high-meat-consuming societies. A general transition to more plant-based diets could lead to lower climate-altering emissions and likely reductions in diet-related non-communicable diseases. In this context, it is critical to promote demand-side climate mitigation options for the agriculture and food sector, such as changes in dietary patterns towards less emissions-intensive, healthier, more plant-based diets. From the health perspective, transitioning towards more plant-based diets in line with World Health Organization and other international dietary guidelines could decrease global mortality, shrink the global food gap and substantially reduce diet-related climate-altering emissions. The chapter concludes with recommendations to integrate food consumption and nutrition considerations into climate adaptation and mitigation planning and financing, emphasizing the importance of concerted and coherent policymaking to develop sustainable food systems and diets, while safeguarding the planet.

Introduction

Promoting good nutrition, health and sustainable food systems in the context of population growth, dietary transition and a changing climate is a central challenge of our time. While climate change has an impact on our food systems and diets, our food systems and dietary patterns also affect climate change (UNSCN, 2017a). Food production and consumption are responsible for 19–29% of the human-induced greenhouse gas (GHG) emissions, 60% of the terrestrial biodiversity loss and 70% of freshwater use. Animal-based foods are the main culprit (Steinfeld *et al.*, 2006; Vermeulen *et al.*, 2012; Tubiello *et al.*, 2014; CBD, 2015), with livestock accounting for an estimated 14.5% of GHG emissions (FAO, 2013a). By 2050, GHG emissions from food and agriculture could rise by as much as 80% due to the increased consumption of

 Sustainable Diets: Linking Nutrition and Food Systems (eds B. Burlingame and S. Dernini)

animal products (Popp *et al.*, 2010; Hedenus *et al.*, 2014; Springmann *et al.*, 2016; Tilman and Clark, 2014). Indeed, food-related GHG emissions could account for half of all emissions allowed for keeping the global rise in temperature to less than 2°C by the middle of the century and could exceed total permissible levels by 2070 (Hedenus *et al.*, 2014; Springmann *et al.*, 2016).

Diets, meanwhile, have deteriorated globally (GLOPAN, 2016), leading to an increase in non-communicable diseases (NDCs), particularly type II diabetes, coronary heart disease and some cancers (Lim *et al.*, 2010; Tilman and Clarke, 2014; Sabate and Soret, 2014).

Economic development, globalization, urbanization and lifestyle changes have caused major shifts towards poor diet, excessive caloric intake and low levels of exercise. The result has been a rapid rise in obesity and non-communicable diseases. The alarming pace of climate and environmental change and its effects on food systems, nutrition and health require a major rethink of how food is produced and consumed (UNSCN, 2017a).

This chapter presents a descriptive analysis of the interconnections of sustainable food systems, dietary patterns, health and nutrition, and climate-change mitigation. The chapter presents a framework for integrated policy development based on co-benefits to health and climate mitigation of dietary change and identifies policy opportunities for joint action on nutrition, health and climate change by promoting sustainable and healthy food systems and diets. It outlines the global frameworks and agreements on climate change, food and nutrition. It looks at diets that boost health and are environmentally sustainable, as well as the measures needed to steer food production and consumption in that direction, emphasizing the importance of concerted and coherent policymaking to develop sustainable food systems and diets, while safeguarding the planet (UNSCN, 2017a).

Climate Policy Framework

The United Nations Framework Convention on Climate Change (UNFCCC) is the primary international, intergovernmental forum for negotiating the global response to climate change and provides a series of technical and financial support mechanisms to build national capacity for a more comprehensive and systematic response to climate change. Despite the opportunities for integrating health, nutrition and diets into the UNFCCC work (Tirado *et al.*, 2013; IPCC, 2014; WHO, 2014), there remains vast potential to expand the work to address nutrition-related issues in climate adaptation and mitigation measures. The Paris Climate Agreement, adopted in 2015, ushered in a new era in the global response to climate change. The Agreement aims to keep the global temperature rise this century well below 2°C above pre-industrial levels and to pursue efforts to limit the increase even further, to 1.5°C. The agreement states that the right to health will be central to national climate action and recognizes the social, economic and environmental value of voluntary mitigation actions and their co-benefits for adaptation, health and sustainable development. In this context, the promotion of sustainable food systems and healthy dietary patterns is critical to reducing emissions and meeting climate mitigation, nutrition and health goals (WHO, 2016). The Paris Agreement is the first international climate change agreement to prioritize food security. In 2016, parties to the UNFCCC ratified the request for further health-related work under the Nairobi Work Programme (NWP) on the effects of and vulnerability and adaptation to climate change, including malnutrition, presenting a further opportunity to promote nutrition and healthy diets.

At the core of the Paris Agreement are the Nationally Determined Contributions (NDCs). The NDCs lay out national plans to reduce GHG emissions and improve countries' resilience to climate change. The development of guidance on and the periodic revision of NDCs offers an opportunity for the health and nutrition communities to work on strengthening the commitments made in the NDCs, with an eye to integrating food security, nutrition and the promotion of sustainable and healthy dietary patterns into climate change action plans, both from an adaptation and a mitigation point of view.

Under the UNFCCC, countries have also been developing cross-sectoral National Adaptation Programmes of Action (NAPAs) and, more recently, National Adaptation Plans (NAPs), which give countries a process by which to identify priority actions in response to their urgent need to adapt to climate change. These NAPAs and NAPs usually identify health, agriculture and food security as priority sectors, but frequently do not consider the nutritional aspects. Similarly, the Nationally Appropriate Mitigation Actions (NAMAs) to date have not explored demand-side mitigation options, such as changes in dietary patterns with a view to less GHG-intensive diets, or other changes in lifestyle, such as active transportation (UNFCCC, 2015). The link between climate change and diet has not received adequate attention from the international community.

Interdependence of Climate Change, Food Systems, Diets, Nutrition and Health

Conceptual framework of the nexus between climate change, food systems and diet

Food systems and dietary patterns are key determinants of nutrition and health. At the same time, they play a significant role in environmental degradation and climate change. The global food system, spanning food production, consumption and waste, accounts for a substantial portion of the GHG emissions that are leading to climate and environmental change. Simultaneously, climate change is influencing the food system, food and bioenergy production, the food environment and socioeconomic conditions and affecting dietary quality and malnutrition. The dynamics between climate change, health and nutrition are diverse and complex. Climate change influences the key determinants of malnutrition, for example, food access, maternal and child care, access to health services, and environmental health. These determinants of malnutrition are shaped, in turn, by other socioeconomic factors that are also affected by climate change. These include income, wealth, education, social safety nets, food aid, institutional inequities, trade, economic, infrastructure, resources, political structures and the full realization of human rights (UNSCN, 2017a).

Climate-related extreme weather events can have a negative impact on diets, too. Undernutrition, meanwhile, weakens the resilience to climatic shocks and the coping strategies of vulnerable populations, reducing their capacity to resist and adapt to the consequences of climate change. The destitute and marginalized are even more affected and therefore the priority should be given explicitly and systematically to those social groups (UNSCN, 2017a).

A combination of climate adaptation and mitigation measures and disaster risk reduction and management could lessen the threats to nutrition from climate change. Climate change adaptation is key to managing the impact of climate change on the food system, food environment, health and nutrition. Early intervention is important, as options for successful adaptation diminish and the associated costs increase as climate change intensifies. Mitigation strategies to reduce food-related GHG emissions from the agricultural sectors and food system – such as sustainable food production, healthy dietary patterns and reducing food waste and loss – have co-benefits on climate, nutrition, human health and the environment (UNSCN, 2017a).

Global food demand and dietary patterns: impacts on climate change and health

Global diets are the tie that binds environmental sustainability and human health. Foods differ substantially, though, when it comes to the quantity of land, water and energy needed per unit of energy and protein consumed, as well as the amount of GHG generated (UNSCN, 2017a).

A 70% increase in total food demand is expected between 2005–2007 and 2050 (FAO, 2013b). At the same time, dietary patterns are shifting, with more food of animal origin, including fish, being consumed (FAO, 2013b). According to the World Resources Institute (WRI) report 'Shifting Diets for a Sustainable Food Future: Creating a Sustainable Food Future', global demand for beef is likely to increase by 95%, and for animal-based foods generally by 80%, between 2006 and 2050. The growth is likely to be concentrated in urban areas of emerging economies, particularly China and India (WRI, 2016).

The growing global meat consumption is expected to boost food-related GHG emissions from 30% to 80% by 2050. Higher demand for meat products may also have profound, long-term effects on the availability and pricing of certain basic food commodities and on access to nutritionally diverse food sources (Friel *et al.*, 2009).

Predictive studies show that if global diets change in an income-dependent way (i.e. tend to contain more animal protein), global average per capita dietary GHG emissions from crop and livestock production could increase 32% between 2009 and 2050 (Tilman and Clarke, 2014). It is estimated that alternative balanced or healthier diets, such as Mediterranean, pescatarian or vegetarian, could reduce emissions from food production to below those of the projected 2050 income-dependent diet, with potential per capita reductions of 30%, 45% and 55%, respectively (Tilman and Clarke, 2014). These studies underline the need to move towards more sustainable and healthy food-consumption patterns in the coming decades (UNSCN, 2017a).

Red meat consumption has declined everywhere in recent years, except East Asia, where it has risen by nearly 40% (GLOPAN, 2016), suggesting that it is possible to reduce meat consumption if the appropriate drivers are in place. This may reflect a shift in dietary pattern as countries become wealthier and prefer 'healthier components' found in higher-quality diets. This could also reflect a substitution of red meat for other types of fresh meats, however, more in-depth analysis is required. Cutting the level of animal-sourced foods in the diets of high meat-consuming countries, for example, needs to become a key element of climate mitigation strategies (Hedenus *et al.*, 2014; Ripple *et al.*, 2014).[1]

Sustainable and Healthy Diets

Ensuring that the planet's 9 billion people will have access to a nutritious and healthy diet, produced in a sustainable manner, by 2050 is an immense global challenge. The Intergovernmental Panel on Climate Change (IPCC)[2] Fifth Assessment Report (AR5) highlighted the opportunities to achieve co-benefits from actions that reduce emissions and also improve health in high meat-consuming countries, by shifting consumption away from animal products, especially from ruminant sources, towards less emission-intensive diets (Smith *et al.*, 2014). In very-low-income settings, however, better access to animal protein can be essential to improving nutrition for groups lacking diverse food sources (UNSCN, 2017a).

Sustainable diets have been defined as those 'with low environmental impacts which contribute to food and nutrition security and to healthy life for present and future generations'. These diets are protective and respectful of biodiversity and ecosystems, culturally acceptable, accessible, economically fair and affordable, nutritionally adequate, and safe and healthy, while optimizing natural and human resources (Burlingame and Dernini, 2012). The more diverse the system, the higher its resilience in the face of climate change and other stressors (UNSCN, 2017a).

Co-benefits of sustainable and healthy diets

Diets can improve public health and nutritional outcomes, while also helping to reduce GHG emissions (Friel, 2009; HLPE, 2012; Tilman and Clark, 2014; Green *et al.*, 2015; Springmann *et al.*, 2016; Milner *et al.*, 2017; UNSCN, 2017a). A transition to more nutritious and diverse diets (with fewer processed foods and more fruit and vegetables) is frequently projected to result in reduced GHG emissions, as well as likely reductions in non-communicable diseases (Green *et al.*, 2015; Milner *et al.*, 2015). For example, if the average adult diet in the UK conformed to WHO recommendations, its associated GHG emissions would be reduced by 17% (Green *et al.*, 2015). Further emission cuts of around 40% could be achieved by making realistic modifications to diets so that they contained fewer animal products and processed snacks, and more fruit, vegetables and cereals (Green *et al.*, 2015). In India dietary changes in line with nutritional guidelines (lower amounts of wheat, dairy, and poultry, and increased amounts of legumes) could help to address projected reductions in the availability of freshwater for irrigation in 2050 and, simultaneously reduce diet-related greenhouse gas emissions and improve diet-related health outcomes (Milner *et al.*, 2017).

Globally, it is estimated that transitioning to more plant-based diets, in line with WHO recommendations on healthy eating (WHO, 2015) and guidelines on human energy requirements (WHO, 2004) and recommendations by the World Cancer Research Fund (WCRF/AICR, 2007), could reduce global mortality by 6–10% and food-related greenhouse gas emissions by 29–70% compared with a reference scenario for 2050 (Springmann *et al.*, 2016). Yet, less than half of all regions meet, or are projected to meet, dietary recommendations for the consumption of fruit, vegetables and red meat, while also exceeding the optimal total energy intake (UNSCN, 2017a), therefore significant changes in the global food system would be necessary for regional diets to match these global healthy dietary patterns (Springmann *et al.*, 2016).

Comparisons of omnivorous diets to more sustainable alternatives, such as Mediterranean, pescatarian and vegetarian diets, have shown the latter to reduce emissions from food production and decrease disease risk globally. The incidence rates of type II diabetes were reduced by 16–41% and of cancer by 7–13%, while relative mortality rates from coronary heart disease were 20–26% lower and overall mortality rates for all causes combined were cut by 0–18% (Tilman and Clark, 2014).

However, diets can have a low environmental impact and still be harmful to human health (UNSCN, 2017a). Ultra-processed items high in sugars, fats or salt can have lower GHG emissions, but be less healthy than the carbohydrate-rich staple foods they displace. Similarly, while there are synergies between healthy diets and reduced emissions, cardio-protective diets are not always environmentally sustainable. For example, the consumption of nuts and fish has been associated with a reduction in the risk of cardiovascular disease (Zhao *et al.*, 2015; Mozaffarian *et al.*, 2012). However, there appear to be trade-offs between the health benefits and environmental impact of increasing fish[3] and nut consumption, which tend to have large ecological and water footprints, respectively (Downs and Fanzo, 2015). We, therefore, need to look at the entire food system, especially food production, to ensure diversity increases across the board. This requires research and dedicated resources. In addition, consumers need to choose from the most sustainable alternatives (e.g. eating nuts with a lower water footprint, sustainably caught fish or underutilized species). Educational initiatives to increase consumer knowledge and informed decision making, as well as incentives to make these foods more affordable (Downs and Fanzo, 2015) and accessible will help (UNSCN, 2017a).

Shifting dietary patterns towards sustainable and healthy diets

Strategies, policies and measures to make diets healthier and sustainable include economic interventions, changes to the governance of production or consumption, as well as changes to the context, defaults and norms of production or consumption. In more detail this could be taxation of unhealthy food or subsidizing providing economic incentives for the consumption of healthier food options, promoting collaboration and shared agreements, and sustainability in dietary guidelines, public education campaigns, educational programmes in schools, and labelling, among others (UK Foresight, 2011; Garnett *et al.*, 2015). An example of this is the recent commitment on taxation and labelling made by the Government of Brazil to the Nutrition Decade (UNSCN, 2017a, 2017b).

On the production front, eliminating agricultural subsidies for commodities that adversely affect human health and encourage the local production of fruit and vegetables have the potential to make healthy foods more accessible to lower-income communities, as well as support environmental goals (UK Foresight, 2011; Jacoby *et al.*, 2014; UNSCN, 2017a). The promotion of healthy diets based on the local, seasonal production of agro-ecological foods, along with the promotion of short marketing circuits, have been proposed as opportunities to increase added value and forge closer ties between farmers, consumers and the land (Jacoby *et al.*, 2014). The development of local food chains, especially for healthy, fresh and perishable products, could facilitate the commercialization of less standardized products and reduce food waste from transport and consumption (HLPE, 2014).

On the consumer side, taxing food-related emissions and creating economic incentives could make diets more sustainable and healthier (Springmann *et al.*, 2017). A GHG emissions tax on foods (corresponding to their emissions intensities), if properly designed, could be a powerful health-promoting climate policy affecting health improvements worldwide. Sparing food groups known to be beneficial for health – such as fruits and vegetables – from taxation, selectively compensating for income losses associated with tax-related price increases, and using a portion of tax revenues for health promotion are potential policy options that could help avert most of the negative health impacts experienced by vulnerable groups, while still promoting changes towards diets that are more environmentally sustainable (Springmann *et al.*, 2017). Only governments have the necessary resources and legitimacy to establish a global regulatory and fiscal framework that puts diets on a more sustainable and plant-based track (Wellesley *et al.*, 2015).

Positive changes in the variety and quality of diets can be brought about by innovative education campaigns that target young consumers, in particular, as well as economic incentives that align the marketing practices of retailers and processors with public health and climate goals (Foresight, 2011). Public-sector incentives for food service companies, retailers and distributors are another potential way of promoting sustainable healthy eating patterns. Such incentives can encourage the development of healthier foods and food labelling (for nutritional content, carbon and water footprints, etc.) in a way that helps consumers achieve nutritional requirements while meeting environmental goals (UNSCN, 2017a).

Food-based dietary guidelines are a key means of encouraging healthy, sustainable and climate-friendly diets. To date, only a few countries (notably Brazil, Germany, Qatar and Sweden) have included sustainability criteria in their national dietary guidelines (FAO and FCRN, 2016). Broadly, the advice issued by these countries focuses on reducing meat consumption, choosing seafood from non-threatened stocks, eating more plants and plant-based products, reducing energy intake and reducing food waste. Sweden and its Nordic neighbours have emphasized the environmental impact of diet in their sustainability criteria. Brazil's guidelines also address the social and economic aspects of sustainability and urge people to avoid ultra-processed foods that damage traditional food cultures and health (UNSCN, 2017a). It is important to note that most of the advances in addressing the issue of environmental sustainability in dietary guidelines to date have been made in developed countries. This suggests that investment in interdisciplinary research and action to address the broader social and economic dimensions of sustainable diets is needed, especially in developing countries (FAO and FCRN, 2016).

Research and investment needs

The complexity of the determinants of sustainable diets, such as agriculture, health, socio-cultural, environmental and socioeconomic factors, frequently makes it challenging for policymakers to understand the benefits of such diets (Johnston *et al.*, 2014). In addition, there are technical and political challenges to developing effective metrics for sustainable diets, particularly in developing countries (UNSCN, 2017a).

Investment in research is essential to obtain the data and evidence needed to develop sustainable and healthy diets in different socioeconomic and cultural environments and to measure their contribution to health and climate goals. The sustainability of and trade-offs involved in diverse production approaches also need to be considered, in order to make sure that needs and the rights of the most marginalized people are prioritized while these choices are made. Assessing and monitoring sustainability and health outcomes requires a reliable global database of food-consumption patterns (national and regional), national health profiles, food composition, and a life-cycle analysis of global food supply, in particular, by geo-climatic region (Johnston *et al.*, 2014).

Metrics and indicators must be developed to assess the impact of the various determinants on the sustainability of a diet and the trade-offs associated with recommendations aimed at making dietary patterns more sustainable (Johnston *et al.*, 2014). This is crucial to providing data and evidence of the co-benefits to climate and health of sustainable and healthy diets to scientific bodies, such as the IPCC and the UNFCCC Subsidiary Body for Scientific and Technological Advice (SBSTA), and thus making the argument for it to be recognized as a priority for climate financing (UNSCN, 2017a).

Integrated and coherent policies

Policies on, but not limited to, agriculture, health, food and nutrition, dietary guidance, environment, water, food waste and bioenergy production, trade, transportation and economics need to be integrated via a multi-stakeholder process to promote sustainable and healthy food systems. This will spark numerous challenges, as will bringing about behavioural change, but we should draw on both the positive and negative experiences of trying to bring about behavioural change in other aspects of health promotion (e.g. focusing on children in terms of protecting them and as advocates for positive change) (UNSCN, 2017a).

Strategies for accelerating a shift towards less GHG emission-intensive, healthier diets should evolve from the socioeconomic and cultural context and conditions of the food system in question, and embrace government, consumers and producers. Government and policymakers are crucial to creating the regulatory framework and institutional capacity for shifting to healthier, more sustainable dietary patterns (Wellesley *et al.*, 2015). These initiatives need to be framed within the context of the climate change and SDG agendas (UNSCN, 2017a).

Conclusions

Developing a global food system to deliver healthy diets for a growing population, while reducing the environmental impact and climate change, is one of the greatest global challenges of our time (UNSCN, 2017a).

- *Sustainable and healthy diets can bring co benefits to the environment and to people's well-being and nutritional status.* Dietary and nutritional considerations should be integrated into the climate-change agenda. The IPCC has highlighted the co-benefits of measures that reduce climate-altering emissions and, at the same time, improve health, for example, a shift away from the overconsumption of meat from ruminant sources in high-meat-consuming societies. A general transition to more plant-based diets could lead to lower GHG emissions and likely reductions in diet-related non-communicable diseases. In this context, it is critical to promote demand-side climate mitigation options for the agriculture and food sector, such as changes in dietary patterns towards less GHG-intensive, healthier, more plant-based diets containing more fruit, vegetables, whole grains and pulses (UNSCN, 2017a).
- *Adopting food-based dietary guidelines that include sustainability criteria is crucial.* Food-based dietary guidelines that include sustainability criteria are key to changing dietary patterns towards more sustainable, healthier diets. Transitioning towards more plant-based diets in line with WHO and other international dietary guidelines could decrease global mortality, shrink the global food gap and substantially reduce diet-related GHG emissions. The inclusion of sustainability criteria in food-based dietary guidelines requires a methodology for developing context-specific, healthy and sustainable national dietary guidelines (UNSCN, 2017a).
- *The world needs to engage in climate actions that support nutrition.* The nutrition community should engage in multi-sectoral decision-making processes that support nutrition-sensitive climate adaptation, mitigation, disaster risk reduction and sustainable development initiatives that promote sustainable and healthy food systems and diets. Nutrition should be considered in national climate-action processes under the UNFCCC, such as NAPs, NDCs and NAMAs. The nutrition community should contribute to IPCC assessments and to the work on adaptation and health (including malnutrition) carried out via the NWP on the impact of and vulnerability and adaptation to climate change. Stakeholders involved in UNFCCC work on agriculture and food security should draw on support from the UNSCN and related international institutions, such as the Committee on World Food Security, to integrate the right to food and other human rights as guiding principles for climate action (UNSCN, 2017a).
- *Promoting sustainable and healthy diets and nutrition should be recognized as a priority for climate financing.* The transition towards low-emission, sustainable and healthy food systems and diets should be a priority for climate funding and should be carried out by adopting a human rights-based approach. The most vulnerable countries should receive help to develop strategies that facilitate access to climate-change finance to enhance nutrition and ensure sustainable and healthy food systems and diets (UNSCN, 2017a).
- *Investment and research are needed to bolster knowledge on sustainable and healthy diets and to spawn effective measures for shifting dietary patterns.* Investment in multidisciplinary research is needed to glean the evidence necessary to affect a shift towards sustainable and healthy diets in different socioeconomic and cultural environments, particularly in low-income countries. To make sustainable

diets a priority when it comes to climate funding, research efforts should support the development of metrics and indicators of the co-benefits to climate and health of sustainable and healthy diets for scientific bodies, such as the IPCC and the UNFCCC's SBSTA (UNSCN, 2017a).

- *The sustainable development goals provide crucial frameworks for joint action to nourish the world sustainably.* The delivery of the 2030 Agenda requires a reshaping of the global food system into one that is efficient, inclusive, climate-smart, sustainable, nutrition- and health-driven (IFPRI, 2016). Transformed food systems should leave no one behind. Starting from a human rights-based approach, and investing in longer term support to allow for the necessary transition to a more equitable food and nutrition security system, will help. Governments, business and civil society must collaborate across sectors to implement international targets that support a transition to more sustainable and healthy food systems and diets as part of SDG implementation. Further recognition and enforcement of rights-based principles of sustainability in promoting healthy diets is needed, because human rights should be the basis from which to decide the trade-offs between environment, health, economy and other sectors. Integrated policies implemented through collaborative action to reduce climate change and its consequences, while underwriting better outcomes in nutrition and health, are critical to the implementation of the SDGs and the 2030 Agenda (UNSCN, 2017a).

Notes

[1] It is important to highlight, however, that in regions affected by severe undernutrition, where people often rely on few staple crops and poor-quality diets, higher meat intake could be nutritionally beneficial.

[2] The Intergovernmental Panel on Climate Change (IPCC) is the international body for assessing the science related to climate change.

[3] Although fish does not have a significant carbon footprint, the ecological footprint associated with unsustainable fishing practices is relatively high (e.g. overfishing, the use of trawlers, etc.).

References

A/RES/70/1 (2015) *Transforming our World: the 2030 Agenda for Sustainable Development*. Resolution adopted by the 70th General Assembly 2015.

A/RES/70/259 (2016) *United Nations Decade of Action on Nutrition (2016–2025)*. Resolution adopted by the 70th General Assembly 2016.

Brown, M.E., Antle, J.M., Backlund, P., Carr, E.R., Easterling, W.E., *et al.* (2015) Climate Change, Global Food Security, and the U.S. Food System. Available at http://www.usda.gov/oce/climate_change/FoodSecurity2015Assessment/FullAssessment.pdf (accessed 1 October 2018).

Burlingame, B. and Dernini, S. (Eds) (2012) Sustainable Diets and Biodiversity: Directions and solutions for policy, research and action. *Proceedings of the International Scientific Symposium Biodiversity and Sustainable Diets United against Hunger*, 3–5 November 2010, FAO Headquarters, Rome. http://www.fao.org/docrep/016/i3004e/i3004e.pdf (accessed 28 August 2018).

Committee on World Food Security (2012) Policy recommendations: food security and climate change. Available at http://www.fao.org/3/a-me421e.pdf (accessed 1 October 2018).

Danysh, H.E., Gilman, R.H., Wells, J.C., Pan, W.K., Zaitchik, B., *et al.* (2014) El Niño Adversely Affected Childhood Stature and Lean Mass in Northern Peru. *Climate Change Responses* 1(1), 7.

Downs, M. and Fanzo, J. (2015) Is a Cardio-Protective Diet Sustainable? A Review of the Synergies and Tensions Between Foods that Promote the Health of the Heart and the Planet. *Current Nutrition Reports* 4, 313–322.

FAO (2013a) *Tackling Climate Change Through Livestock – A Global Assessment of Emissions and Mitigation Opportunities.* Food and Agriculture Organization of the United Nations, Rome, Italy.

FAO (2013b) *The State of Food and Agriculture*. Food and Agriculture Organization of the United Nations, Rome, Italy.

FAO (2016) 2015–2016 El Niño: early action and response for agriculture, food security, and nutrition. Working draft. Available at www.fao.org/fileadmin/user_upload/emergencies/docs/FAOEl%20Nino ReportMarch2016.pdf (accessed 1 October 2018).

FAO and FCRN (2016) Plates, pyramids and planets. Available at http://www.fao.org/3/a-i5640e.pdf (accessed 1 October 2018).

Foresight: The future of food and farming. Final Project Report. Futures. 2011, Government Office for Science, London.

Friel, S., Dangour, A.D., Garnett, T., Lock, K., Chalabi, Z., *et al.* (2009) Public health benefits of strategies to reduce greenhouse-gas emissions: food and agriculture. *Lancet* 374, 2016–2025.

Garnett, T., Mathewson, S., Angelides, P. and Borthwick, F. (2015) Policies and actions to shift eating patterns: What works? *Foresight* 515, 518–522.

GLOPAN (2016) *Food Systems and Diets: Facing the Challenges of the 21st Century.* GLOPAN, London, UK.

Green, R., Milner, J., Dangour, A.D., Haines, A., Chalabi, Z., *et al.* (2015) The potential to reduce greenhouse gas emissions in the UK through healthy and realistic dietary change. *Climate Change* 129, 253–265. DOI: 10.1007/s10584-015-1329-y

Hedenus, F., Wirsenius, S. and Johansson D.J.A. (2014) The importance of reduced meat and dairy consumption for meeting stringent climate change targets. *Climatic Change* 124, 79–91.

HLPE (2012) *Food security and climate change: A report by the High Level Panel of Experts on Food Security and Nutrition.* Committee on World Food Security, Rome, Italy.

HLPE (2014) *Food losses and waste in the context of sustainable food systems: A report by the High Level Panel of Experts on Food Security and Nutrition.* Committee on World Food Security, Rome, Italy.

IFPRI (2016) *Global Nutrition Report 2016: From Promise to Impact: Ending Malnutrition by 2030.* International Food Policy Research Institute, Washington DC, USA.

IPCC (2014) *Summary for policymakers.* In: *Climate Change 2014: Impacts, Adaptation, and Vulnerability. Part A: Global and Sectoral Aspects.* Contribution of Working Group II to the Fifth Assessment Report of the Intergovernmental Panel on Climate Change. Cambridge University Press, Cambridge, UK, pp. 1–32.

Jacoby, E., Tirado, C., Diaz, A., Pena, M., Sanches, A. and Coloma, M. (2014) Family farming, food security and public health in the Americas. *World Nutrition* 5(6), 537–551.

Johnston, J., Fanzo, J. and Cogill, B. (2014) Understanding sustainable diets: a descriptive analysis of the determinants and processes that influence diets and their impact on health, food security, and environmental sustainability. *Advances in Nutrition* 5, 418–429.

Lim, S.S., Vos, T., Flaxman, A.D., Danaei, G., Shibuya, K. and Adair-Rohani, H. (2010) A comparative risk assessment of burden of disease and injury attributable to 67 risk factors and risk factor clusters in 21 regions, 1990–2010: a systematic analysis for the Global Burden of Disease Study. *Lancet* 380(9859), 2224–2260.

Milner, J., Green, R. and Dangour, A.D. (2015) *Health effects of adopting low greenhouse gas emission diets in the UK. BMJ Open* 5, e007364. DOI: 10.1136/bmjopen–2014–007364

Milner, J., Joy, E.J., Green, R., Harris, F., Aleksandrowicz, L., *et al.* (2017) Projected health effects of realistic dietary changes to address freshwater constraints in India: a modelling study. *The Lancet Planetary Health* 1(1), e26–e32.

Mozaffarian, D., Micha, R. and Michas, G. (2012) Unprocessed red and processed meats and risk of coronary artery disease and type 2 diabetes – an updated review of the evidence. *Current Atherosclerosis Reports* 14(6), 515–524. DOI: 10.1007/s11883-012-0282-8

Popp, A., Lotze-Campen, H. and Bodirsky, B. (2010) Food consumption, diet shifts and associated non-CO_2 greenhouse gases from agricultural production. *Global Environmental Change* 20, 451–462.

Ripple, W.J., Smith, P., Haberl, H., Montzka, S.A., McAlpine, C. and Boucher, D.H. (2014) Ruminants, climate change and climate policy. *Nature Climate Change* 4(1), 2–5.

Rosegrant, M.W. (2008) *Biofuels and Grain Prices: Impacts and Policy Responses.* International Food Policy Research Institute, Washington DC, USA, p. 4.

Sabate, J. and Soret, S. (2014) *Sustainability of Plant-Based Diets: Back to the Future.* American Society for Nutrition, Rockville, Maryland, USA.

Smith, P., Bustamante, M., Ahammad, H., Clark, H., Dong, H., *et al.* (2014) Agriculture, Forestry and Other Land Use (AFOLU). In: Edenhofer, O., Pichs-Madruga, R., Sokona, Y., Farahani, E., Kadner, S., *et al.* (Eds) *Climate Change 2014: Mitigation of Climate Change.* Contribution of Working Group III to the

Fifth Assessment Report of the Intergovernmental Panel on Climate Change. Cambridge University Press, Cambridge, UK.

Springmann, M., Godfray, H.C.J., Rayner, M. and Scarborough, P. (2016) Analysis and valuation of the health and climate change co-benefits of dietary change. *Proceedings of the National Academy of Sciences of the United States* 113(15), 4146–4151.

Springmann, M., Mason-D'Croz, D. and Robinson, S. (2017) Mitigation potential and global health impacts from emissions pricing of food commodities. *Nature Climate Change* 7, 69–74.

Steinfeld, H., Gerber, P., Wassenaar, T., Castel, V. and de Haan, C. (2006) *Livestock's Long Shadow: Environmental Issues and Options*. Food and Agriculture Organization of the United Nations, Rome, Italy.

Tilman, D. and Clark, M. (2014) Global diets link environmental sustainability and human health. *Nature* 515(7528), 518–522. DOI: 10.1038/nature13959

Tirado, M.C., Crahay, P., Mahy, L., Zanev, C., Neira, M., *et al.* (2013) Climate change and nutrition: creating a climate for nutrition security. *Food and Nutrition Bulletin* 34(4), 533–547.

Tubiello, F.N., Salvatore, M., Ferrara, A.F., House, J., Federici, S., *et al.* (2015) The Contribution of Agriculture, Forestry and other Land Use activities to Global Warming, 1990–2012. *Global Change Biology* 21(7), 2655–2660.

UK Foresight (2011) *The Future of Food and Farming: Final Project Report*. Government Office for Science: London, UK.

UNFCCC (2015) Compilation of information on nationally appropriate mitigation actions to be implemented by developing country parties. FCCC/SBI/2013/INF.12/Rev.3. United Nations Framework Convention on Climate Change, New York, USA.

UNSCN (2017a) Sustainable diets for healthy people and a healthy planet. Available at https://www.unscn.org/uploads/web/news/document/Climate-Nutrition-Paper-EN-.pdf (accessed 1 October 2018).

UNSCN (2017b) The UN decade of action on nutrition 2016-2025. Available at https://www.unscn.org/en/topics/un-decade-of-action-on-nutrition?idnews=1684 (accessed 3 July 2018).

Vermeulen, S.J., Campbell, B.M. and Ingram, J.S.I. (2012) Climate change and food systems. *Annual Review of Environment and Resources* 37, 195–222.

Wellesley, L., Happer, C. and Froggatt, A. (2015) Changing Climate, Changing Diets: Pathways to Lower Meat Consumption. Chatham House Report. Available at https://www.chathamhouse.org/sites/files/chathamhouse/publications/research/20151124DietClimateChangeWellesleyHapperFroggattExecSum.pdf (accessed 1 October 2018).

WHO (2004) Global recommendations on physical activity for health. Available at http://www.who.int/dietphysicalactivity/publications/9789241599979/en/ (accessed 1 October 2018).

WHO (2015) Healthy diet Fact sheet N°394. Available at http://www.who.int/mediacentre/factsheets/fs394/en/ (accessed 1 October 2018).

WHO (2016) *Strengthening Health Resilience to Climate Change*. Technical Briefing for the WHO Conference on Climate and Health 2016. World Health Organization, Geneva, Switzerland.

WCRF/AICR (2007) Food, nutrition, physical activity, and the prevention of cancer: a global perspective. Available at http://www.aicr.org/assets/docs/pdf/reports/Second_Expert_Report.pdf (accessed 1 October 2018).

WRI (2016) Shifting diets for a sustainable food future: creating a sustainable food future. Available at http://www.wri.org/sites/default/files/Shifting_Diets_for_a_Sustainable_Food_Future_0.pdf (accessed 1 October 2018).

Zhao, L.-G., Sun, J.-W., Yang, Y., Ma, X., Wang, Y.-Y. and Xiang, Y.-B. (2015) Fish consumption and all-cause mortality: a meta-analysis of cohort studies. *European Journal of Clinical Nutrition* 70(2), 155–161.

5 Biodiversity Loss: We Need to Move from Uniformity to Diversity

Emile A. Frison and Nick Jacobs

Abstract

Today's food and farming systems have succeeded in supplying large volumes of foods to global markets but are generating negative outcomes on multiple fronts: wide-spread degradation of land, water and ecosystems; high greenhouse gas emissions; biodiversity losses; persistent hunger and micronutrient deficiencies, and the rapid rise of obesity and diet-related diseases; and livelihood stresses for farmers around the world. These problems are tied to the industrial model of agriculture that is increasingly dominant around the world. The uniformity at the heart of these systems leads systematically to negative outcomes and vulnerabilities, and particularly the use of an increasingly narrow pool of animal breeds and plant varieties. The 'Green Revolution' of the post-war period left a dual legacy: huge advances in the productivity of staple crops, and the concurrent marginalization of whole swathes of foods, crop varieties – and the communities depending on them. The low-diversity industrial model is locked in place by a series of vicious cycles. Highly compartmentalized approaches to research, education and policymaking allow one-dimensional productivity-focused solutions to prevail, and obscure the links between healthy ecosystems, a healthy planet and healthy people. Meanwhile, the way food systems are currently structured allows value to accrue to a limited number of actors, reinforcing their economic and political power, and thus their ability to influence the governance of food systems. To break these cycles, a fundamentally different model of agriculture is required, based on diversifying farms and farming landscapes, replacing chemical inputs, optimizing biodiversity and stimulating interactions between different species, as part of holistic strategies to build long-term fertility (i.e. 'diversified agroecological systems'). There is growing evidence that these systems keep carbon in the ground, support biodiversity, rebuild soil fertility and sustain yields over time, providing a basis for secure farm livelihoods and diverse healthy diets.

We Need a Systemic Approach

The food systems we inherit in the 21st century have allowed major inroads to be made against hunger through huge advances in productivity. However, they also represent some of the greatest threats to our continued health and prosperity. Indeed, the outcomes of these food systems are poor on many counts, and in many countries and regions of the world. The very foundations on which these systems were built are becoming increasingly fragile.

Despite decreases in the percentage of the global population going hungry over recent decades, 795 million people still suffered from hunger in 2015 (FAO *et al.*, 2015). Expanding the lens to take in those who are malnourished, the failures are far starker. In addition to acute hunger, two billion are afflicted by the 'hidden hunger' of micronutrient deficiencies (Bioversity International,

2014), and over 1.9 billion are obese or overweight (WHO, 2015). Indeed, one of the greatest paradoxes of our time is the coexistence of the different faces of malnutrition within the same region or even the same household (Graziano da Silva, 2014). Non-communicable diseases (NCDs) associated with imbalanced diets have increased so rapidly as to have overtaken infectious diseases as the number one cause of global mortality (WHO, 2012; Murray *et al.*, 2015). In addition, while food-borne illnesses persist in a variety of settings, food scares affecting large numbers of people are emerging in increasingly globalized food markets, threatening to unravel the historical progress on food safety.

The environmental outlook is equally troubling. Today, food systems contribute between 19% and 29% of global anthropogenic greenhouse gas (GHG) emissions (Vermeulen *et al.*, 2012). Upstream of agriculture, major contributions are made by the fossil fuel-intensive production of chemical fertilizer and pesticides (Gilbert, 2012). Downstream, emissions arise from food processing and retail sectors that rely increasingly on abundant synthetic packaging and soaring 'food miles' in order to deliver the highly processed and unseasonal products to which consumers have become accustomed (Schnell, 2013). Meanwhile, 70% of all water withdrawn from aquifers, streams and lakes is used for agriculture – often at unsustainable rates (FAO, 2013). The agricultural sector is responsible for nitrate, phosphorus, pesticide, soil sediment and pathogen pollution in soil and water (Parris, 2011). Furthermore, agricultural systems have contributed significantly to land degradation as well as to the destruction of natural habitats and losses of wild biodiversity around the world (Scherr and McNeely, 2012).

Food systems are also failing food producers themselves. Many small farmers, especially women, struggle to emerge above subsistence level, often lacking access to credit, technical support and markets – or facing the uncertainties of volatile prices on global commodity markets (FAO, 2004). Globalization has brought new challenges in terms of downward price pressures and costly regulatory burdens for farmers. As a result, the world faces the irony of small-scale farming communities making up about 50% of the hungry (World Economic Forum, 2015). Even in wealthier countries, farmers continue to face high risks and uncertainties, with farming incomes showing little prospect of rising durably (European Commission, 2017). This leaves many farmers reliant on government subsidies or living in abject poverty. Meanwhile, labour conditions are systematically poor at various nodes of the food chain – particularly for hired and migrant farm labourers (ILO, 2015). While food and agriculture generate increasing value for grain traders and global retail giants, decent livelihoods remain out of reach for many of those employed in food systems.

The problems in food systems are deeply interconnected and mutually reinforcing. Some 35% of global cultivated crops depend on pollination (WHO *et al.*, 2015). The global decline in insect pollinators – driven in large part by the use of pesticides in agriculture (van Lexmond *et al.*, 2015) – now threatens the very basis of agriculture and its future crop yields. Meanwhile, the livelihoods of many food producers are being pushed to breaking point by climate change and environmental degradation. Nearly one billion people who derive their livelihoods primarily from agriculture are presently living in vulnerable environments, and these are the populations that will bear the brunt of large-scale environmental change in the near future (Fischer *et al.*, 2002). In other words, modern agriculture is failing to sustain the people and resources on which it relies and has come to represent an existential threat to itself.

Agricultural Diversity and Dietary Diversity: Victims of the Green Revolution

Modern agriculture is increasingly synonymous with industrial agriculture, that is, modes of farming that are analogous to industrial processes in their scale and task segregation and seek to derive productivity gains from specialization and intensification of production. This model, relying on universal crop applications and synthetic fertilizers and pesticides, is the dominant form of agriculture in many developed countries and is increasingly the pathway proposed for developing countries. The prevailing approach to delivering food security, based on delivering maximum volumes of uniform crop commodities for global markets, is what industrial agriculture is designed to deliver.

However, as described above, food systems are now generating negative and costly outcomes on a range of fronts: the environmental impacts are particularly severe (IPES-Food, 2016), while the troubling implications for human health are increasingly being documented (IPES-Food, 2017).

Many of these problems trace back to industrial agriculture, and the uniformity at the heart of this model. One of its main characteristics is a dramatic loss of genetic diversity in farmers' fields and domestic animal populations. The shift towards industrial agriculture has been characterized by loss of agrobiodiversity, and the use of an increasingly narrow pool of animal breeds and plant varieties. Furthermore, the erosion of entire production systems has occurred alongside the mass production of a handful of staple crops: 'underutilized' or minor crop species (e.g. indigenous leafy vegetables, small-grained cereals, legumes, wild fruits and tree crops) are disappearing in the face of competition with industrially produced varieties of rice, maize and wheat (Jacobsen *et al.*, 2013). By 1970, 20% of the wheat area and 30% of the rice area in low-income countries were planted with a small number of 'high yielding varieties', and by 1990, the share had increased to about 70% for both crops. The 'Green Revolution' of the post-war period therefore left a dual legacy: huge advances in the productivity of staple crops, and the concurrent marginalization of whole swathes of foods, crop varieties – and the communities depending on them.

For livestock, a few highly productive breeds adapted to industrial production systems have now replaced most local breeds across the world (Groeneveld *et al.*, 2010). The Food and Agriculture Organization's Global Databank for Animal Genetic Resources for Food and Agriculture contains 7616 livestock breeds; 6536 of these are purely local breeds, meaning that they are found in only one country. Of this total, 20% are classified as at risk. Between 2001 and 2007, 62 breeds became extinct – amounting to the loss of almost one breed per month (FAO, 2007).

While these approaches respond to short-term productivity objectives, they entail a general reduction in practical applications of genetic diversity, potentially limiting the genetic pool available to future generations of farmers and limiting the options in terms of adapting to changing environments (Vigouroux *et al.*, 2011). The implications of this genetic erosion could be huge, given the unpredictability of future stresses.

This drastic reduction in agricultural diversity has also contributed to a reduction in dietary diversity, with major health implications. A diverse and balanced diet can ensure exposure to a broader set of nutrients and non-nutrients that have antioxidant, anti-cancer and other beneficial properties (Fanzo *et al.*, 2013). Furthermore, the association between the diversity of a child's diet and his/her nutritional status operates independently of other socioeconomic factors (Arimond and Ruel, 2004). There is a strong link between a low monthly Diet Diversity Score and underweight among children under two (Fanzo *et al.*, 2011).

While promoting dietary diversity may be the subject of broad agreement, there are widely divergent views of how to get there. The pathway offered by industrial agriculture is through highly specialized and productive agriculture around the world, combined with well-functioning trading systems that allow a variety of different foodstuffs to be accessible to consumers in a given place. The viability of this channel is thus contingent on people's ability to access this array of foods. To date, the diversity of produce delivered by international trade has mainly benefited wealthy consumers in high-income countries, while poor people in low-income countries struggle to afford the diversity available via global markets (Sibhatu *et al.*, 2015).

The parallel neglect of traditional crops (e.g. in research programmes and development schemes) has meant that poorer populations have struggled either to access internationally traded products or to obtain a diverse diet on the basis of local traditional foods. On a global level, of the 7000 plants that have been used as food by humans, just three of them – rice, maize and wheat – provide more than 50% of the world's plant-derived food energy intake (FAO, 1995). Wheat, rice, maize and other ubiquitous crop commodities were among those with the greatest gains in both relative and absolute abundance in national per capita food supplies over the past 50 years (Khoury *et al.*, 2014). The promotion of energy-rich staple cereals has helped to drive a decline in consumption of pulses and other minor crops with high nutritional value (Hawkes, 2007).

In some cases, the general trend has been compounded by government policies with an explicit focus on monocrops – often for export. For example, since 2009, the Rwandan government has promoted the monocropping of modern, selected varieties together with input intensification. As a result, intercropping and crop diversity have declined substantially in recent years, falling from 9–11 crops per farm to 3–4, with potentially highly negative consequences for household dietary diversity (Isaacs, 2014; Snapp and Fischer, 2014). For many years, Indian agricultural policies favoured specialization in major cereal production through crop-specific subsidies, with the effect of exacerbating micronutrient deficiencies (World Bank, 2006). In general, cash crop production – sometimes for non-food purposes – helps to push out more diverse food cropping at the expense of nutritionally important foodstuffs. For example, tobacco farming is considered to have displaced vegetables and pulses in Bangladesh, as well as cassava, millet and sweet potatoes in Kenya (Lecours *et al.*, 2012).

While recent efforts to 'biofortify' staple crops have led to improved content of specific nutrients, this has not compensated for the general decrease in nutritional density of modern varieties of staple crops. Indeed, the specialization of agricultural systems has also had negative impacts on this front (AFSSA, 2003; Barański *et al.*, 2014). Breeding programmes for the major crops have focused mainly on productivity increases by altering plant height or disease resistance (Tadele and Assefa, 2012), resulting in varieties that are rich in energy but have a lower content of various macronutrients and micronutrients (Jones *et al.*, 2014).

As a result, the theoretical diversification of diets facilitated by industrial agriculture and global trade has not managed to remedy the problem of micronutrient deficiencies, which continue to undermine the health status and development of over two billion people (Hunt, 2005; Sibhatu *et al.*, 2015). Meanwhile, the prevalence of energy-rich crops and foodstuffs continues to be a major factor in the explosion of overweight and obese populations and the associated health impacts (Wallinga, 2010). Primarily through their contribution to NCDs, these trends have incurred huge financial costs to society (Alwan, 2011) and are also responsible for the biggest increases in mortality rates over recent years, predominantly in low- and middle-income countries (WHO, 2015).

We Need a Change in Paradigm: from Uniformity to Diversity

We need to move away from the currently dominant paradigm of industrial agriculture towards one that takes a systemic approach and can deliver on multiple fronts: economic, environmental, social, cultural, nutritional and health. In its 2016 report, IPES-Food proposed a paradigm shift from industrial agriculture to *diversified agroecological systems* (IPES-Food, 2016). This refers to diversifying farms and farming landscapes, replacing chemical inputs with organic inputs, optimizing biodiversity and stimulating interactions between different species, as part of holistic strategies to build long-term fertility, healthy agro-ecosystems and secure livelihoods. In other words, this approach nurtures diversity at all levels and builds on this basis.

The growing body of evidence on these systems suggests major potential to deliver mutually reinforcing benefits, and to sustain them over time (for a detailed synthesis of this evidence, see IPES-Food, 2016). Evidence is particularly strong on the ability of diversified agroecological systems to deliver strong and stable outputs on the basis of building environmental resilience and highly functioning agro-ecosystems – and to limit losses and enable recovery in the face of environmental stresses and shocks (Holt-Giménez, 2002; Badgley *et al.*, 2007; Cardinale *et al.*, 2008; IAASTD *et al.*, 2009; Picasso *et al.*, 2008; Tirado and Cotter, 2010; Rosset *et al.*, 2011; Pretty *et al.*, 2011; Mijatović *et al.*, 2013; Altieri *et al.*, 2015; Prieto *et al.*, 2015; Rodale Institute, 2015). The environmental outcomes represent a virtuous cycle with reduced GHG emissions, improved water- and resource-use efficiency, greater biodiversity and ecosystem services and restoration of degraded land (Gliessmann, 2007;

Alonso and Guzmán, 2010; Altieri *et al.*, 2012; Aguilera *et al.*, 2014). In particular, diversified systems have shown the capacity to raise productivity in the places where additional food is desperately needed (Badgley *et al.*, 2007; Pretty *et al.*, 2011).

There is also growing evidence of positive linkages between agricultural diversity and nutritional diversity at the household and local level, through the increased availability of nutrient-rich diverse foods throughout the year (Herforth, 2010; Oyarzun *et al.*, 2013; Jones *et al.*, 2014; Carletto *et al.*, 2015; Kumar *et al.*, 2015; Shively and Sununtnasik, 2015). Agricultural diversity has been linked specifically to increased consumption of a range of key nutritional elements often missing in diets based around staple cereal crops. Polycultures and mixed crop–livestock farming systems help to ensure that key nutrients are available throughout the year (Remans *et al.*, 2011; Jones *et al.*, 2014), and improved health outcomes have been observed in relation to diversified food production and its dietary benefits.

In addition, a significant health benefit of diversified agroecological systems is the reduced exposure to pesticides and other harmful chemicals used in agriculture (Reganold and Wachter, 2016). Meanwhile, health-giving qualities have been identified in foods not treated with chemical pesticides. For example, concentrations of a range of antioxidants such as polyphenols have been found to be substantially higher in organic crops/organic crop-based foods which have not been sprayed with pesticides. Many of these compounds have been linked to a reduced risk of chronic diseases (Barański *et al.*, 2014). Polyphenol intakes have also been associated with decreased mortality (Zamora-Ros *et al.*, 2013).

What is Preventing a Shift Towards Diversified Agroecological Systems?

If the evidence stacks up in favour of a major paradigm shift, what then is keeping industrial food and farming systems in place? IPES-Food identified a series of feedback loops or 'lock-ins' helping to reinforce the current model and keep the alternatives off the table.

Lock-in 1: path dependency

Industrial agriculture requires significant up-front investments, in terms of equipment, training, networks and retail relationships. To see a return on these investments, farmers are often required to scale up to deliver sufficiently high volumes of (low-value) uniform commodity crops. Once these structural shifts have been made, it is difficult for farmers to change course.

Lock-in 2: export orientation

Over decades, production subsidies, energy subsidies, trade liberalization and a range of other measures have been put in place with a view to producing large volumes of cheap commodity crops for global markets. Specific supply chains (e.g. for animal feed or processed food ingredients) have become increasingly export oriented and export dependent. Supporting these chains has often been prioritized over other interests (e.g. ensuring resources for local food production) and in spite of the risks (e.g. price volatility, declining terms of trade, environmental degradation, competition for land).

Lock-in 3: the expectation of cheap food

Industrial agriculture and shifting consumer habits have helped to facilitate the emergence of mass food retailing, characterized by the abundance of relatively cheap highly processed foods, and the year-round availability of a wide variety of foods. In many countries, consumers have become accustomed to spending less on food – and increasingly detached from the realities of how that food is produced. The food industry has therefore become increasingly reliant on the cheap and flexible supply of uniform commodities that industrial agriculture is uniquely positioned to provide.

Lock-in 4: compartmentalized thinking

Highly compartmentalized structures continue to govern the setting of priorities in politics, education, research and business, allowing the solutions offered by industrial agriculture to remain at centre stage. Agricultural ministries, committees and lobbies retain a privileged position relative to other constituencies (e.g. environment, health) in determining the policies that shape food systems (i.e. agricultural policies, trade policies). Increasingly privatized agricultural research and development programmes remain focused on the handful of commodities for which there is a large enough market to secure significant returns. Educational silos remain in place, and sectoral 'value chain' organizations share knowledge vertically (by product) rather than encouraging food systems approaches.

Lock-in 5: short-term thinking

The advantages of diversified agroecological systems are not immediately visible, given the time needed to rebuild soil health and fertility, to increase biodiversity in production systems, and to reap the benefits of enhanced resilience. Unfortunately, key players in food systems are often required to deliver short-term results. Politicians are locked into short-term electoral cycles that encourage and reward policies that deliver immediate returns. Meanwhile, publicly traded agribusiness firms are required to deliver rapid returns to shareholders.

Lock-in 6: 'feed the world' narratives

Food security continues to be framed by many prominent actors as a question of how to 'feed the world', or in other words, how to produce sufficient dietary energy at the global level – particularly in the wake of the 2007–2008 food price spikes. These narratives predispose us to approach the question in terms of global production volumes of mainly energy-rich, nutrition-poor crop commodities. This reinvents industrial agriculture as the solution, while sidelining the questions of nutritional quality, poverty, access, power and equity that have been recognized as essential pieces of the food security puzzle.

Lock-in 7: measures of success

Diversified agroecological systems are by definition geared towards producing diverse outputs, while delivering a high degree of resource efficiency, reducing GHG emissions and producing a range of environmental services and social benefits on and off the farm. Narrowly defined indicators of agricultural performance (e.g. yields of specific crops or productivity per worker) reward large-scale industrial monocultures while failing to capture the benefits of alternative systems. Current systems will be held in place insofar as they continue to be measured in terms of what industrial agriculture is designed to deliver, at the expense of the many other outcomes that really matter to society.

Lock-in 8: concentration of power

The way food systems are currently structured allows value to accrue mainly to a limited number of actors, reinforcing their economic and political dominance, and thus their ability to influence the governance of food systems. This influence is brought to bear in ways that reinforce the status quo and ward off transition.

How to Shift Towards Diversified Agroecological Systems

Sparking a transition is therefore a major challenge, requiring key incentives in agriculture, food systems and beyond to be fundamentally revisited and a series of cycles to be broken. Diversified agroecological systems provide the guiding principles and the direction of travel – but there is no simple recipe for getting there. The specific obstacles and the specific solutions will differ from context to context. However,

IPES-Food was able to identify a series of steps that, collectively, could shift the centre of gravity in food systems, allowing harmful dependencies to be cut, the agents of change to be empowered, and alliances to be forged in favour of change.

Recommendation 1: develop new indicators for sustainable food systems

It is essential to adopt a broader range of indicators, covering long-term ecosystem health; total resource flows; sustainable interactions between agriculture and the wider economy; the sustainability of outputs; nutrition and health outcomes; livelihood resilience; and the economic viability of farms with respect to debt, climate shocks, and so on.

Recommendation 2: shift public support towards diversified agroecological production systems

Governments must shift public support (e.g. agricultural subsidies) away from industrial production systems, while rewarding the array of positive outcomes in diversified agroecological systems. Governments must implement measures that allow farms to diversify and transition towards agroecology. In particular, they must support young people to enter agriculture and adopt diversified agroecological farming – before they are locked into the cycles of industrial agriculture.

Recommendation 3: support short supply chains and alternative retail infrastructures

Governments should support and promote short circuits in order to make them a viable, accessible and affordable alternative to mass retail outlets, e.g. by repurposing infrastructure in cities to favour farmers' markets. More attention should also be paid to the role of informal markets and policy measures must be put in place that empower emerging initiatives linking farmers to consumers (e.g. community-supported agriculture schemes).

Recommendation 4: use public procurement to support diverse local agroecological produce

Public procurement should be used with increasing ambition in order to ensure sales outlets for diversified agroecological farms, while providing fresh, nutritious food and diversified diets for the users of public canteens, particularly schoolchildren.

Recommendation 5: strengthen movements that unify diverse constituencies around agroecology

Governments can support farmers' groups, community-based organizations and social movements which encourage diversification and the spread of agroecological practices and advocate for sustainable food systems, and ensure the participation of diverse civil society groups from the global North and South in global governance processes and forums. To become effective, alliances must reach across various divides (e.g. between producers and consumers, between farmers and researchers), demands must be made operational, and strong and unified messages must emerge to counter the 'feed the world' narratives which currently hold sway.

Recommendation 6: mainstream diversification, agroecology and holistic food systems approaches into education and research agendas

Public research agendas must be redefined around different priorities. Investments must be redirected towards equipping farmers to shift their production. The mission of university research should be redefined around the delivery of public goods. The Food and Agriculture Organization and other international agencies should mainstream agroecology into all of their work, in order to spread existing knowledge and plug the remaining gaps in our understandings. Research conducted by the CGIAR Centres should be refocused around diversified agroecological systems and farmer participatory research.

Recommendation 7: develop joined-up food policies at multiple levels

Governance structures and policy processes must be up to the task of managing the systemic challenges we face. Integrated food policies and food strategies are required to overcome the traditional biases in sectoral policies (e.g., export orientation in agricultural policy) and to align various policies with the objective of delivering environmentally, socially, and economically sustainable food systems. Integrated food policies allow trade-offs to be weighed up, while providing a forum for long-term systemic objectives to be set (e.g., managing scarce resources in the face of competing demands; ensuring a sequenced rebalancing away from export orientation). Crucially, food systems planning must be based on broad participation of various constituencies and groups with a stake in food systems reform. At the global level, the Committee on World Food Security should advocate for coherent food policies and contribute to strengthening diversified agroecological food systems.

References

AFSSA (2003) Evaluation nutritionnelle et sanitaire des aliments issus de l'agriculture biologique. Agence française de sécurité sanitaire des aliments, Maisons-Alfort Cedex, France.

Aguilera, E., Guzmán, G. and Alonso, A. (2014) Greenhouse gas emissions from conventional and organic cropping systems in Spain. II. Fruit tree orchards. *Agronomy for Sustainable Development* 35, 725–737. DOI: 10.1007/ s13593-014-0265-y

Alonso, A.M. and Guzmán, G.J. (2010) Comparison of the efficiency and use of energy in organic and conventional farming in Spanish agricultural systems. *Journal of Sustainable Agriculture* 34, 312–338. DOI: 10.1080/10440041003613362

Altieri, M., Nicholls, C., Henao, A. and Lana, M. (2015) Agroecology and the design of climate change-resilient farming systems. *Agronomy for Sustainable Development* 35, 869–890.

Altieri, M.A., Funes-Monzote, F.R. and Petersen, P. (2012) Agroecologically efficient agricultural systems for smallholder farmers: contributions to food sovereignty. *Agronomy for Sustainable Development* 32, 1–13. DOI: 10.1007/ s13593-011-0065-6

Alwan, A. (2011) *Global Status Report on Noncommunicable Diseases 2010*. World Health Organization, Geneva, Switzerland.

Arimond, M. and Ruel, M.T. (2004) Dietary diversity is associated with child nutritional status: evidence from 11 demographic and health surveys. *Journal of Nutrition* 134, 2579–2585.

Badgley, C., Moghtader, J., Quintero, E., Zakem, E., Chappell, M.J., *et al.* (2007) Organic agriculture and the global food supply. *Renewable Agriculture and Food Systems* 22, 86–108. DOI: 10.1017/ S1742170507001640

Barański, M., Srednicka-Tober, D., Volakakis, N., Seal, C., Sanderson, R., *et al.* (2014) Higher antioxidant and lower cadmium concentrations and lower incidence of pesticide residues in organically grown crops: a systematic literature review and meta-analyses. *British Journal of Nutrition* 112, 794–811. DOI: 10.1017/S0007114514001366

Bioversity International (2014) *Bioversity International's 10-year Strategy 2014–2024: Agricultural Bio-Diversity Nourishes People and Sustains the Planet*. Bioversity International, Rome, Italy.

Cardinale, B.J., Wright, J.P., Cadotte, M.W., Carroll, I.T., Hector, A., *et al.* (2007) Impacts of plant diversity on biomass production increase through time because of species complementarity. *Proceedings of the National Academy of Sciences USA* 104, 18123–18128. DOI: 10.1073/pnas.0709069104

Carletto, G., Ruel, M., Winters, P. and Zezza, A. (2015) Farm-level pathways to improved nutritional status: Introduction to the special issue. *The Journal of Development Studies* 51, 945–957. DOI: 10.1080/00220388.2015.1018908

European Commission (2017) Prospects for EU agricultural markets and income 2014–2024. Available at https://ec.europa.eu/agriculture/sites/agriculture/files/markets-and-prices/medium-term-outlook/2017/2017-fullrep_en.pdf (accessed 16 June 2018).

Fanzo, J., Hunter, D., Borelli, T. and Mattei, F. (eds) (2013) *Diversifying Food and Diets: Using Agricultural Biodiversity to Improve Nutrition and Health*. Earthscan from Routledge, London, UK.

Fanzo, J., Remans, R., Pronyk, P.M., Negin, J., Warie-ro, J., Mutuo, P., *et al.* (2011) A 3-year cohort study to assess the impact of an integrated food-and livelihood-based model on undernutrition in rural Western Kenya. In: Thompson, B. and Amoroso, L. (eds) *Combating Micronutrient Deficiencies: Food-Based Approaches*. The Earth Institute at Columbia University, New York, USA, p. 76.

FAO (1995) *Dimensions of Need: An Atlas of Food and Agriculture*. Food and Agriculture Organization of the United Nations, Santa Barbara, California, USA.

FAO (2004) *The State of Agricultural Commodity Markets: 2004*. Food and Agriculture Organization of the United Nations, Rome, Italy.

FAO (2007) *The State of the World's Animal Genetic Resources for Food and Agriculture: In Brief*. Food and Agriculture Organization of the United Nations, Rome, Italy.

FAO (2013) *Water and Food: The post 2015 Water Thematic Consultation – Water Resources Management Stream Framing Paper*. Food and Agriculture Organization of the United Nations, Rome, Italy.

FAO, IFAD, WFP (2015) State of Food Insecurity in the World – SOFI. Available at http://www.fao.org/3/a-i4646e.pdf (accessed 16 June 2018).

Fischer, G., Shah, M. and Velthuizen, H. (2002) *Climate Change and Agricultural Vulnerability*. International Institute for Applied Systems Analysis, Vienna, Austria.

Gilbert, N. (2012) One-third of our greenhouse gas emissions come from agriculture. *Nature News*, 31 October 2012. DOI: 10.1038/nature.2012.11708

Gliessman, S.R. (2007) *Agroecology: The Ecology of Sustainable Food Systems*. CRC Press, Boca Raton, Florida, USA.

Graziano da Silva, J. (2014) *Better Nutrition – Better Lives. Addressing Today's Major Nutrition Challenges*. Ministers Reference Book: Commonwealth 2014. Henley Media Group, London, UK.

Groeneveld, L.F., Lenstra, J.A., Eding, H., Toro, M.A., Scherf, B., *et al.* (2010) Genetic diversity in farm animals. *Animal Genetics* 41, 6–31. DOI: 10.1111/j.1365-2052.2010.02038.x

Hawkes, C. (2007) Promoting healthy diets and tackling obesity and diet-related chronic diseases: what are the agricultural policy levers? *Food and Nutrition Bulletin* 28, S312–S322.

Herforth, A. (2010) *Promotion of Traditional African Vegetables in Kenya and Tanzania: A Case Study of an Intervention Representing Emerging Imperatives in Global Nutrition*. Doctoral thesis, Cornell University, New York, USA.

Holt-Giménez, E. (2002) Measuring farmers' agroecological resistance after Hurricane Mitch in Nicaragua: a case study in participatory, sustainable land management impact monitoring. *Agriculture, Ecosystems and Environment* 93, 87–105. DOI: 10.1016/S0167- 8809(02)00006-3

Hunt, J.M. (2005) The potential impact of reducing global malnutrition on poverty reduction and economic development. *Asia Pacific Journal of Clinical Nutrition* 14, 10–38.

IAASTD (2009) *Synthesis Report: A Synthesis of the Global and Sub-Global IAASTD Reports, Agriculture at a Crossroads*. Island Press, Washington, DC, USA.

ILO (2015) *Combating Forced Labour: A Handbook for Employers and Business*. Available at http://www.ilo.org/wcmsp5/groups/public/---ed_norm/---declaration/documents/publication/wcms_101171.pdf. (accessed 16 June 2018).

IPES-Food (2016) *From Uniformity to Diversity: A Paradigm Shift from Industrial Agriculture to Diversified Agroecological Systems*. International Panel of Experts on Sustainable Food Systems, Brussels, Belgium.

IPES-Food (2017) *Unravelling the Food-Health Nexus: Addressing Practices, Political Economy and Power Relations to Build Healthier Food Systems*. The Global Alliance for the Future of Food and IPES-Food, Brussels, Belgium.

Isaacs, K.B. (2014) *Rediscovering the Value of Crop Diversity in Rwanda: Participatory Variety Selection and Genotype by Cropping System Interactions in Bean and Maize Systems*. Doctoral thesis, Michigan State University, East Lansing, Michigan, USA.

Jacobsen, S.-E., Sørensen, M., Pedersen, S.M. and Weiner, J. (2013) Feeding the world: genetically modified crops versus agricultural biodiversity. *Agronomy for Sustainable Development* 33, 651–662. DOI: 10.1007/ s13593-013-0138-9

Jones, A.D., Shrinivas, A. and Bezner Kerr, R. (2014) Farm production diversity is associated with greater household dietary diversity in Malawi: Findings from nationally representative data. *Food Policy* 46, 1–12. DOI: 10.1016/j.foodpol.2014.02.001

Khoury, C.K., Bjorkman, A.D., Dempewolf, H., Ramirez-Villegas, J., Guarino, L., *et al.* (2014) Increasing homogeneity in global food supplies and the implications for food security. *Proceedings of the National Academy of Sciences* 111, 4001–4006.

Kumar, N., Harris, J., Rawat, R. (2015) If they grow it, will they eat and grow? Evidence from Zambia on agricultural diversity and child undernutrition. *The Journal of Development Studies* 51, 1060–1077. DOI: 10.1 080/00220388.2015.1018901

Lecours, N., Almeida, G.E.G., Abdallah, J.M. and Novotny, T.E. (2012) Environmental health impacts of tobacco farming: a review of the literature. *Tobacco Control* 21, 191–196. DOI: 10.1136/tobaccocontrol-2011-050318

Mijatović, D., Van Oudenhoven, F., Eyzaguirre, P. and Hodgkin, T. (2013) The role of agricultural biodiversity in strengthening resilience to climate change: towards an analytical framework. *International Journal of Agricultural Sustainability* 11, 95–107. DOI: 10.1080/14735903.2012.691221

Murray, R., Godfrey, K.M. and Lillycrop, K.A. (2015) The early life origins of cardiovascular disease. *Current Cardiovascular Risk Reports* 9, 1–8. DOI: 10.1007/ s12170-015-0442-9

Oyarzun, P.J., Borja, R.M., Sherwood, S. and Parra, V. (2013) Making sense of agrobiodiversity, diet, and intensification of smallholder family farming in the Highland Andes of Ecuador. *Ecology of Food and Nutrition* 52, 515– 541. DOI: 10.1080/03670244.2013.769099

Parris, K. (2011) Impact of agriculture on water pollution in OECD countries: recent trends and future prospects. *International Journal of Water Resources Development* 27, 33–52. DOI: 10.1080/07900627.2010 .531898

Picasso, V.D., Brummer, E.C., Liebman, M., Dixon, P.M. and Wilsey, B.J. (2008) Crop species diversity affects productivity and weed suppression in perennial polycultures under two management strategies. *Crop Science* 48, 331. DOI: 10.2135/cropsci2007.04.0225

Pretty, J., Toulmin, C. and Williams, S. (2011) Sustainable intensification in African agriculture. *International Journal of Agricultural Sustainability* 9, 5–24. DOI: 10.3763/ijas.2010.0583

Prieto, I., Violle, C., Barre, P., Durand, J.-L., Ghesquiere, M. and Litrico, I. (2015) Complementary effects of species and genetic diversity on productivity and stability of sown grasslands. *Nature Plants* 1, 15033. DOI: 10.1038/nplants.2015.33

Reganold, J.P. and Wachter, J.M. (2016) Organic agriculture in the twenty-first century. *Nature Plants* 2, 15221. DOI: 10.1038/nplants.2015.221

Remans, R., Flynn, D.F.B., DeClerck, F., Diru, W., Fanzo, J., *et al.* (2011) Assessing nutritional diversity of cropping systems in African villages. *PLoS ONE* 6, e21235. DOI: 10.1371/ journal.pone.0021235

Rodale Institute (2015) The farming systems trial. Avialable at http://rodaleinstitute.org/assets/FSTbooklet-FINAL.pdf (accessed on 16 June 2018).

Rosset, P.M., Sosa, B.M., Jaime, A.M.R. and Lozano, D.R.Á. (2011) The Campesino-to-Campesino agroecology movement of ANAP in Cuba: social process methodology in the construction of sustainable peasant agriculture and food sovereignty. *The Journal of Peasant Studies* 38, 161–191. DOI: 10.1080/03066150.2010.538584

Scherr, S.J. and McNeely, J.A. (2012) *Farming with Nature: The Science and Practice of Ecoagriculture*. Island Press, Washington, DC, USA.

Schnell, S.M. (2013) Food miles, local eating, and community supported agriculture: putting local food in its place. *Agriculture and Human Values* 30, 615–628. DOI: 10.1007/ s10460-013-9436-8

Shively, G. and Sununtnasik, C. (2015) Agricultural diversity and child stunting in Nepal. *Journal of Development Studies* 51.

Sibhatu, K.T., Krishna, V.V. and Qaim, M. (2015) Production diversity and dietary diversity in smallholder farm households. *Proceedings of the National Academy of Sciences USA* 112, 10657–10662. DOI: 10.1073/pnas.1510982112

Snapp, S.S. and Fisher, M. (2014) "Filling the maize basket" supports crop diversity and quality of household diet in Malawi. *Food Security* 7, 83–96. DOI: 10.1007/ s12571-014-0410-0

Tadele, Z. and Assefa, K. (2012) Increasing food production in Africa by boosting the productivity of under-studied crops. *Agronomy* 2, 240–283. DOI: 10.3390/ agronomy2040240

Tirado, R. and Cotter, J. (2010) *Ecological Farming: Drought-Resistant Agriculture* (No. GRL-TN 02/2010). Greenpeace Research Laboratories, University of Exeter, UK.

Van Lexmond, M.B., Bonmatin, J.-M., Goulson, D. and Noome, D.A. (2015) Worldwide integrated assessment on systemic pesticides: Global collapse of the entomofauna: exploring the role of systemic insecticides. *Environmental Science and Pollution Research* 22, 1–4. DOI: 10.1007/s11356-014-3220-1

Vermeulen, S.J., Campbell, B.M. and Ingram, J.S.I. (2012) Climate change and food systems. *Annual Review of Environment and Resources* 37, 195–222. DOI: 10.1146/annurev-environ-020411-130608

Vigouroux, Y., Barnaud, A., Scarcelli, N. and Thuillet, A.-C. (2011) Biodiversity, evolution and adaptation of cultivated crops. *Comptes Rendus Biologies* 334, 450–457. DOI: 10.1016/j. crvi.2011.03.003

Wallinga, D. (2010) Agricultural policy and childhood obesity: a food systems and public health commentary. *Health Affairs* 29, 405–410. DOI: 10.1377/ hlthaff.2010.0102

World Bank (2006) *Repositioning Nutrition as Central to Development: A Strategy for Large Scale Action*. World Bank Publications, Washington, DC, USA.

World Economic Forum (2015) Why are most of the world's hungry people farmers? Available at https://www.weforum.org/agenda/2015/05/why-are-most-of-the-worlds-hungry-people-farmers/ (accessed 16 June 2018).

WHO (2012) NCD mortality and morbidity. Available at http:// www.who.int/gho/ncd/mortality_morbidity/en/ (accessed 8 December 2015).

WHO (2015) Obesity and overweight. Available at http:// www.who.int/mediacentre/factsheets/fs311/en/ (accessed 30 November 2015).

World Health Organization, Convention on Biological Diversity (Organization), United Nations Environment Programme (2015) *Connecting Global Priorities: Biodiversity and Human Health: A State of Knowledge Review*. World Health Organization, Geneva, Switzerland.

Zamora-Ros, R., Rabassa, M., Cherubini, A., Urpí- Sardà, M., Bandinelli, S., *et al.* (2013) High concentrations of a urinary biomarker of polyphenol intake are associated with decreased mortality in older adults. *The Journal of Nutrition* 143, 1445–1450. DOI: 10.3945/jn.113.177121

6 Agroecology and Nutrition: Transformative Possibilities and Challenges

Rachel Bezner Kerr, Maryam Rahmanian, Ibukun Owoputi and Caterina Batello

Abstract

Agroecology is a holistic approach to agriculture, which takes into account the ecological, social, political and economic dimensions of producing food in order to build sustainable and resilient food systems that ensure food security and nutrition. It is thus an approach that resonates closely with sustainable diets. Positive nutritional outcomes should be one important outcome of such an approach; however, there has been limited research to date on the relationship between agroecology and nutrition. Building on a series of dialogues on agroecology hosted by the Food and Agriculture Organization, as well as relevant scientific literature, this chapter presents several dimensions of agroecology that seem to be relevant for nutrition. On the technical side of agroecology, some promising studies point to the role that biodiverse farming systems and agroforestry have in ensuring positive nutritional outcomes. Other studies contend that agroecology, when linked to questions of social inequality such as gender or class, can lead to improvements in nutrition. Areas of interest and further investigation are outlined in this chapter: biodiverse production systems, social empowerment, local knowledge, culture and diets, livelihoods and rights. An agroecological approach to nutrition will likely critically engage with several known underlying determinants of malnutrition, which include the political, economic and social environment. This approach critiques the concept of nutritionism, a reductionist approach to nutrition science that is focused on single nutrients rather than food systems, and has often dominated efforts to link agriculture to nutritional outcomes.

What is Agroecology?

From a scientific and technical perspective, agroecology applies ecological concepts and principles to farming systems, focusing on the interactions between plants, animals, humans and the environment, to ensure food security and nutrition for all, now and in the future (HLPE, 2016). Emphasizing sustainable production, cultural acceptability and economic fairness, agroecology is an approach that resonates closely with sustainable diets.

At the centre of agroecology lies the sustainable production system. Agroecological approaches stress the importance of diversification, enhancing energy, water and nutrient flows, supporting soil health, and maximizing beneficial interactions between components of an agroecosystem in order to minimize toxic external inputs and ensure intergenerational benefits (Gliessman, 2015). Yet agroecology goes beyond environmentally sustainable production to contribute to all dimensions of food security, linking sustainable production to economic and social goals. It emphasizes, for example, the need to strengthen the co-creation of knowledge, to strengthen farmers' organizations and collective action, and to transform markets to deliver multiple benefits.

It is important to recognize that the term agroecology is understood in multiple ways and is a concept in contention (Méndez *et al.*, 2013). Therefore, agroecology has been described as taking many forms: it refers to a scientific discipline, a specific set of agricultural practices, and a political or social movement (Wezel *et al.*, 2009). A more technical approach of agroecology focuses on agronomic and ecological dimensions, while more practical and political agroecological approaches incorporate social, economic, ethical and political dimensions of the food system (Méndez *et al.*, 2013). Many social movements and peasant organizations, for example, advocate that agroecology is not possible without food sovereignty (Pimbert, 2015). Agroecology in the context of food sovereignty means taking a critical food systems approach that looks not only at the practices that support forms of agriculture that have negative environmental and social consequences, but also at the actors who benefit from these practices (through fossil fuels, agrochemicals, production, processing, retailing, etc.). The aim is to re-configure power relations throughout the food system by identifying the winners and losers of current food system dynamics (IPES-Food, 2016).

While there is growing attention to issues related to markets, consumers and local food systems within agroecology, specific attention to nutrition as such is not very well developed, despite being highlighted as an important dimension.[1] A Food and Agriculture Organization (FAO) seminar on agroecology and nutrition provided a rare opportunity to discuss this issue at the global level. It highlighted possible linkages between agroecology and nutrition, including biodiverse production systems, nutrient composition of various crops and varieties, markets for agroecological produce, community approaches to nutrition, and the role of Farmer Field Schools in holistic nutrition education.[2] Despite the relative lack of attention on nutrition, there are both linkages and transformative possibilities between agroecology and nutrition, as this chapter seeks to uncover.

Nutritional Concepts, Measures and Implications for Agroecology

Although the link between nutrition and agroecology has not been well researched, the link between nutrition and agriculture in general has been explored extensively over the last two decades (Berti *et al.*, 2004; Haddad, 2013). While some research studies detailing the link between agriculture and nutrition use food security as a proxy for nutrition, other studies have more narrow measures of nutrition, such as child growth outcomes, for which there are limited studies showing strong linkages (Masset *et al.*, 2012).

Food security is necessary for nutrition security but not sufficient.[3] Food security and nutrition security can occur together or separately. While they are highly correlated, malnutrition can exist with adequate food security, for example, due to the prevalence of disease or unsafe water. Nutrition security tends to encompass more biological factors, in that nutritional status is a function of both food intake and health status (Weingartner, 2009). In many regions, all types of malnutrition – overweight and obesity, micronutrient deficiencies, and underweight – can exist together, known as the triple burden of malnutrition (Labadarios, 2005).

The underlying direct causes that lead to malnutrition include food insecurity, inadequate care, insufficient health services and an unhealthy environment, while broader structural causes include poverty and inequality (UNICEF, 1991). The social, political, economic, and environmental structure has been identified as a crucial piece of optimal growth and development for infant and child nutrition (Black *et al.*, 2013). Nutrition-sensitive interventions, or interventions that tackle these deeper, underlying determinants of nutrition, have been highlighted as an area of promise for improving nutritional outcomes. These include agricultural and food interventions, which have the potential to address all facets of food security including acceptability, and may benefit local farmers (Ruel *et al.*, 2013).

Although the importance of social and economic policies to address poverty, agriculture, and trade has been highlighted to combat malnutrition (Bryce *et al.*, 2008), nutritional science has been critiqued for using a narrow, reductionist framework to examine nutritional outcomes. The term '*nutritionism*' was coined by Scrinis (2008) to apply to this medicalized, nutrient-level focus rather than a broader consideration of food systems. A nutritionism approach to food systems supports a functional approach to food,

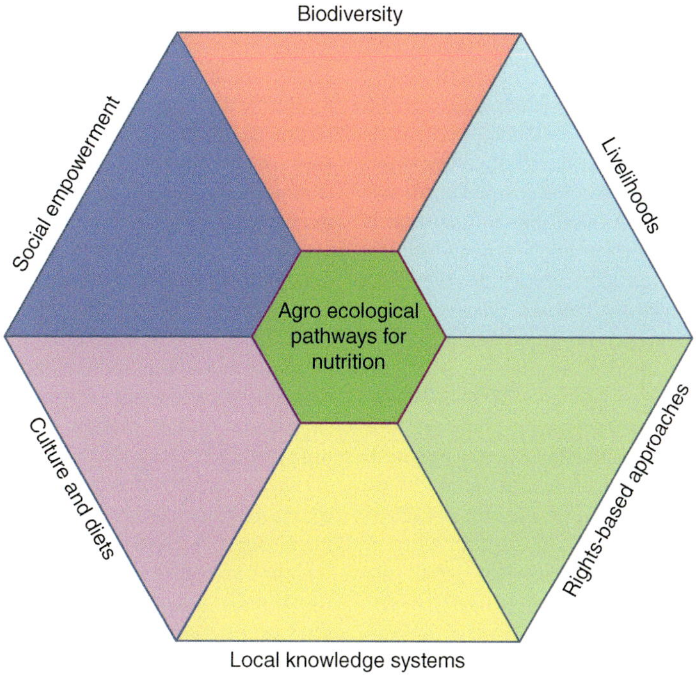

Fig. 6.1. Potential links between agroecology and nutrition.

rather than a holistic consideration of the political, economic, sociocultural and ecological contexts in which diets are shaped (Scrinis, 2008). Together, all of these broader structural factors are important in an agroecological approach to nutrition, as agroecology has the potential to address many of these underlying determinants of nutrition and food system contexts.

To date, the question of whether agroecological methods address both nutrition and food security has not been adequately examined. There is limited research highlighting the connection between agroecology and food and nutrition security; however, there is much promise within existing research linking agroecology and nutrition (Fig. 6.1).

Biodiversity

Agroecology is built on the cornerstone of biodiversity. Agroecological systems are rich in a range of species including pollinators, wild animals and plants, and soil microorganisms. Through practices such as intercropping, rotations, livestock integration and minimizing external inputs, these systems enhance biotic diversity (e.g. natural herbivore predators, beneficial soil organisms, edible weeds) and improve environmental conditions through higher nutrient availability, a range of habitats, and increases in soil organic matter and soil structure (Gliessman, 2015). In addition, agroecological farming systems deploy intraspecies diversity (e.g. a range of varieties of the same crop) for a number of reasons, including local adaptation, cultural significance, taste and health.

There is growing scientific evidence of the important contribution of biodiversity to nutrition (Frison *et al.*, 2011).[4] While much of this research focused initially on nutrient composition of species and varieties that directly make up the human diet, more recently there have been calls for a more holistic approach, arguing that human nutrition is dependent on numerous ecosystem services (de Clerck, 2013). This view takes into account the range of significant contributions of biodiversity to nutrition, such as soil microorganisms, pollinators, beneficial insects and fuel. For example, researchers estimate that the majority of several micronutrients – vitamin A, vitamin C, and most carotenes and tocopherols – come from crops that partially depend on animal pollinators (Eilers *et al.*, 2011). Others have pointed to the link between soil microbial diversity and gut microbial composition (Miller, 2015). This expanded frame for considering biodiversity's contributions to nutrition is an appropriate starting point for understanding the contribution of agroecology to nutrition through biodiversity.

Scientific evidence for the contribution of biodiverse agroecological systems to nutrition exist across a range of agroecosystems. Numerous studies have concluded that home gardens are associated with better household nutrition (Mitchell and Hanstad, 2004). In Nepal, home gardens in rural areas can provide up to 60% of the total vegetables and fruit requirements of the family (Gautam *et al.*, 2005). Forests and agroforests provide important ecosystem services (Sunderland *et al.*, 2013; Vira *et al.*, 2015), and also contribute significantly to healthy diets globally (Rowland *et al.*, 2017). Studies have documented a wide range of aquatic species in rice fields. Some of these aquatic species are used as predators in integrated pest management strategies in rice production (FAO, 2001). A study from Lao PDR showed that a broad diversity of aquatic plants and animals (about 200 species) are frequently used by villagers, and that fish and other aquatic animals make up the main animal protein sources in peoples' diets (Meusch *et al.*, 2003). Agroecological landscapes can provide a range of foods from uncultivated sources. A key principle in agroecology is the diversification of the agroecosystem, favouring in-field diversity as well as landscape heterogeneity. The landscape scale plays an important role as it involves a wider range of biodiversity that perform important functions, such as habitats for wildlife and natural enemies of agricultural pests, leaf litter to enhance organic matter and residues for mulching for fields, and so on (Altieri *et al.*, 1987). The source, type and relative importance of wild foods varies greatly from one socioecological setting to another. In some cases, wild foods make up a significant portion of the diet, especially for non-staple food items rich in micronutrients, and are particularly important during the 'lean season' (Powell *et al.*, 2015).

Social empowerment

As a holistic system of production, it is often implicitly assumed that agroecology will lead to improved household food security and nutrition. However, considerable research has demonstrated that without attention to intra-household power dynamics, such as the division of labour, decision making and distribution of food, changes to agricultural production may have no positive impact, and in fact could have negative impacts (Berti *et al.*, 2004). Gender equity and women's empowerment are increasingly considered a key link between agricultural production and nutritional outcomes (see, for example, Smith and Haddad, 2015; Malapit and Quisumbing, 2015). Agricultural activities, for example, that increase women's labour may take them away from other activities, such as breastfeeding, which have significant impacts on children's nutritional status. Furthermore, increases in food production at the household level can be diverted elsewhere or distributed unevenly within households. Other forms of social inequity, such as discrimination based on ethnicity, health status or class, can put particular households at a disadvantage. These households may then be unable to access adequate resources, knowledge or control over those resources in order to implement agroecological practices. Some scholars have pointed to the critical link between agroecology and food sovereignty, namely addressing inequality and affording greater power and control over the food system to vulnerable groups, which includes access to land, seeds, and organic inputs (Martínez-Torres and Rosset, 2014).

There have been few studies explicitly linking agroecological production to gender equity or other forms of social empowerment. One study in Malawi involving farmer participatory research in which smallholder farming households experimented with legume intercropping found significant differences in legume intercropping choices between different types of households (such as female-headed versus married households) (Bezner Kerr *et al.*, 2007). Power dynamics, including domestic violence and unequal division of labour operating within households made it difficult for women to apply particular agroecological methods, such as incorporation of crop

Fig. 6.2. Men cooking during a community 'recipe day' in Bwabwa, Malawi in 2012. Recipe days bring communities together to share different healthy food recipes and child care practices, while at the same time encouraging men and women to share workloads more equitably (Photo: Rachel Bezner Kerr).

residue soon after harvest (Bezner Kerr *et al.*, 2008). The researchers integrated participatory nutrition education with explicit attention to gender inequalities into the intervention activities, and found evidence for significant differences in child growth outcomes between households who used legume intercropping and also participated in the educational activities compared to those households who did not (Bezner Kerr *et al.*, 2010).

In a follow-up study in Malawi, the researchers worked with 400 households explicitly selected for high levels of food insecurity and poor health, including AIDS-affected households. Participating households received agroecology training and could choose which agroecological method they wanted to test (e.g. crop diversification, legume intercrops and livestock integration). Research showed significant improvements in household food security, household dietary diversity and reported health status (Nyantakyi-Frimpong *et al.*, 2016, 2017).

Oliver (2016) examined a herb cooperative of women farmers in Uruguay that used feminist and agroecological principles. Diversification and participatory methods were named by Uruguayan women farmers as key efforts to foster their leadership, interest and success in the cooperative. While agroecology and women's empowerment were viewed by the women as complementary, the study did not include any examination of the food security or nutritional impacts of the agroecological cooperative.

Local knowledge systems

Agroecology underlines the importance of context-specific and adapted knowledge to find solutions for complex ecological and human systems. Within complex agroecological systems, there are no ready-made silver bullets: solutions strongly rely on learning and innovation processes among local actors, with farmers and their innovations systems and networks at the centre (Hainzelin, 2015). Farmers' knowledge of managing local natural resources form the foundations of agroecology (Pimbert, 2015). By valuing and making active use of this knowledge, agroecological approaches encourage the survival of these knowledge systems.

While much agroecology research has focused on the importance of peasant and indigenous knowledge regarding knowledge of local ecological contexts (e.g. Altieri and Toledo, 2011), there has been less attention to nutrition-related knowledge. Nevertheless, there are indications that rural peoples have extensive understanding of nutritional issues from the field to the plate, including knowledge of the health and nutrition qualities of different crop varieties (Powell *et al.*, 2015). It has been shown that among a range of criteria that farmers have applied to varieties they have selected and cultivated, some may be kept for their dietary or nutritional value (Jarvis *et al.*, 2011). For example, sorghum landraces identified by Ethiopian farmers with names such as 'milk in my mouth' and 'squirts out like honey' were found to have high levels of lysine and protein (Gebrekidan and Kebede, 1979).

Knowledge of food preparation, combinations, processing and preservation are an important part of the biocultural knowledge of many communities (Johns and Sthapit, 2004). Fermented foods provide a good example of traditional knowledge that enhances nutrition. Fermentation processes have been developed in order to preserve food for times of scarcity, to impart desirable flavour to foods, and to reduce toxicity and have enabled traditional societies to survive harsh weather such as drought (Marshall and Meija, 2011). Lactic fermentation of vegetables (e.g. sauerkraut) or of meat (e.g. the Inuit-fermented fish, walrus and other sea animals) adds nutritional and microbiological diversity to the diet that can have significant health impacts (Selhub *et al.*, 2014).

Culinary knowledge and food skills are essential to healthy diets and good nutrition, but there is evidence of a loss of culinary skills often due to increased reliance on processed and prepared foods. In such cases, the local agrobiodiversity and culinary knowledge that has been lost, devalued or forgotten, can be brought back with active efforts at revival (Bezner Kerr, 2014). Farmer participatory research methodologies have been used effectively in Malawi as a way to 'amplify' agroecological learning. Combined with participatory nutrition education and attention to gender, these agroecological practices have led to improvements in food security and nutrition (Bezner Kerr *et al.*, 2010; Kangmennaang *et al.*, 2017).

Culture and diets

Producers' organizations have linked agroecological approaches with promotion of communities' pride in their cultures, values and knowledge systems (International Planning Committee for Food Sovereignty, 2015). This connection highlights how food production, based on local knowledge, culture and values can lead to reviving nutritious traditional diets. More industrialized modern food systems are associated with lower rates of undernutrition and higher rates of overweight, obesity and non-communicable diseases than traditional food systems. The rise in chronic non-communicable diseases associated with transitions to industrial diets have been avoided precisely in those parts of the world that have managed to keep a strong traditional food system, one in which the health, cultural and ecological roles of diets are appreciated, such as Japanese and Mediterranean diets (Johns and Sthapit, 2004). In Mali, a cooperative of women agroecological farmers, COFERSA (Convergence des Femmes Rurales pour la Souverainite Alimentaire), are creating new markets for their products by raising awareness about the nutritional benefits of local foods, such as fonio, millet and sorghum, and encouraging consumers to avoid imported products with low nutrition value, such as white bread. Pride in local biodiversity, based on traditional knowledge and culture and manifested in local cuisines, is a driving force of their work (BEDE and COFERSA, 2015).

The important link between culture and agroecological approaches that value local biodiversity is particularly noticeable among indigenous communities, where diets are heavily influenced by Western industrial dietary patterns of high fat, salt, sugar and processed food, resulting in high rates of diet-related chronic disease such as diabetes. An in-depth review of indigenous peoples' food systems (Kuhnlein et al., 2013) found that pride in local culture was one of the intervention strategies that improved health and nutrition among indigenous peoples.

A project with the Ainu indigenous peoples in Japan, for instance, aimed to improve the social and cultural health of the community by promoting Ainu food culture. In the United States, the Inter-Tribal Bison Cooperative assists tribes in returning bison to their lands, a culturally significant animal that was the cornerstone of indigenous autonomy, and which was decimated by colonizers. The bison provides cultural enhancement, spiritual revitalization, ecological restoration and economic development alongside potential nutrition and health benefits by reintroducing lean sources of meat.

Livelihoods

Another means by which agroecology can improve food and nutrition security is through improving livelihoods. Agroecology emphasizes building relationships with local markets and developing alternative economic models that provide viable livelihoods for producers, such as farmers' markets and participatory guarantee systems (Loconto *et al.*, 2016). Although there are few studies, there is some evidence that building more direct and local market linkages can have positive impacts on food security. In Mexico, El Salvador and the Dominican Republic, agroecological methods of smallholder coffee production, including increasing biodiversity, have been shown to improve livelihoods, food security and ecosystem services (Méndez, 2008; Méndez *et al.*, 2013; Gross *et al.*, 2014). Farmers in Thailand using organic farming methods in rice production formed a cooperative adhering to values of promotion of household food security alongside spiritual and community values. Their approach showed improvements in household food security, household dietary diversity as well as improvements in physical health (Kaufman and Petpha, 2016). In Uganda, innovations in marketing to urban consumers using 'participatory guarantee systems' by organic producers led to improvements in household food security for both the farmers and urban consumers (Nakalanda and Kugonza, 2016). Malawian farmers using agroecological methods improved wealth and food security over a 3-year period compared to those using conventional methods (Kangmennaang *et al.*, 2017). However, such livelihood and related food and nutrition security impacts from agroecology are threatened by broader economic and environmental factors, such as volatility of global markets and climate change, which threaten the long-term viability of smallholder production (Bacon *et al.*, 2008; Méndez *et al.*, 2013; Gross *et al.*, 2014).

A rights-based approach to nutrition

Adopting a human rights approach could strengthen and guide agroecological approaches that seek to improve nutrition and ensure sustainable diets. Producers' organizations have pointed to the importance of integrating a human rights approach with agroecology (International Planning Committee for Food Sovereignty, 2015). The former UN Special Rapporteur on the Right to Food (de Schutter, 2010) has outlined a conceptual connection between agroecology and the right to food. A rights-based approach would imply participation of food insecure groups in the design and implementation of the policies that most affect them (de Schutter, 2010). Guideline 10 of the Voluntary Guidelines to Support the Progressive Realization of the Right to Adequate Food in the Context of National Food Security addresses several dimensions of nutrition, including dietary diversity, availability and sustainability, nutrition education, inclusive participation and non-discrimination, in particular with respect to women and girls, and culture in dietary and eating patterns (FAO, 2005). As shown in this chapter, agroecology could contribute to all of these dimensions. Several General Recommendations of the Convention on the Elimination of All Forms of Discrimination Against Women, and the Convention on the Rights of the Child, give some basis for elaborating the importance and implications of adopting a human rights approach to nutrition (Bellows *et al.*, 2015). Further work in this area could benefit from greater understanding of how agroecology, linked with a rights-based approach, could contribute to improved nutritional outcomes.

Conclusion

As a holistic and ecological approach to food production, agroecology shows potential for addressing nutrition through multiple pathways.

Increasing diet quality, diversity and sustainability, enhancing knowledge and pride in indigenous foods that foster nutritious diets, reducing exposure to toxic substances and fostering synergistic interactions between different components of the agroecosystem are all ways that producers may enhance food and nutrition security. In addition, some streams of agroecology encourage addressing political, social, economic and environmental dimensions of agriculture that impact nutrition, such as who benefits from and controls the different dimensions of the food system, and how to ensure decent livelihoods from agricultural production. There is limited research to date on the relationship between agroecology and nutritional outcomes. Promising studies point to the role that biodiverse farming systems, supporting more diverse landscapes including forests and aquatic systems, can play in leading to positive nutritional outcomes. Other studies contend that agroecology, when linked to questions of social inequality, such as gender or class, can lead to improvements in nutrition. Linking agroecology to nutrition may foster more holistic approaches to diets that consider political, social and economic contexts rather than a primary focus on individual nutrients. Studies that have incorporated some dimensions of power dynamics, such as gender inequality or limited knowledge of nutritional issues, provide evidence for the transformative possibilities of linking agroecology to nutrition, but more research is needed in this area. Rights-based approaches to nutrition could enhance such approaches.

Notes

[1] The FAO for example mentions it under the element of 'culture and food traditions': http://www.fao.org/agroecology/knowledge/10-elements/culture-food-traditions/en/.

[2] FAO Technical Seminar, Achieving nutrition for all: What role for agroecology?, 26 July 2016, FAO Headquarters, Rome.

[3] Food security by definition exists 'when all people, at all times, have physical and economic access to sufficient, safe, and nutritious food to meet their dietary needs and food preferences for an active and health life' (World Food Summit, 1996).

[4] The scientific evidence is starting to have an impact on policy debates, with recent decisions from the Convention on Biological Diversity, the FAO Commission on Genetic Resources for Food and Agriculture, and the Second International Conference on Nutrition, promoting policy support for biodiversity for nutrition.

References

Altieri, M.A. and Toledo, M.V. (2011) The agroecological revolution in Latin America: rescuing nature, ensuring food sovereignty and empowering peasants. *Journal of Peasant Studies* 38(3), 587–612.

Altieri, M.A., Anderson, M.K. and Merrick, L.C. (1987) Peasant agriculture and the conservation of crop and wild plant resources. *Conservation Biology* 1, 49–58.

Bacon, C.M., Méndez, V.E., Gliessman, S.R., Goodman, D. and Fox, J.A. (2008) *Confronting the coffee crisis: fair trade, sustainable livelihoods and ecosystems in Mexico and Central America*. MIT Press, Cambridge, Massachusetts, USA.

BEDE and COFERSA (2015) Consommer la biodiversité locale pour mieux se nourrir. IMPRIM Services, Bamako, Mali, 32 pp.

Bellows, A.C., Valente, F.S.L. and Lemke, S. (2015) *Gender, Nutrition, and the Human Right to Adequate Food: Toward an Inclusive Framework*. Routledge, Taylor and Francis Group, London, UK.

Berti, P., Krasevec, J. and Fitzgerald, S. (2004) A review of the effectiveness of agriculture interventions in improving nutrition outcomes. *Public Health Nutrition* 7(5), 599–609.

Bezner Kerr, R. (2014) Lost and found crops: agrobiodiversity, indigenous knowledge, and a feminist political ecology of sorghum and finger millet in northern Malawi. *Annals of the Association of American Geographers* 104(3), 577–593. DOI: 10.1080/00045608.2014.892346

Bezner Kerr, R., Berti, P. and Chirwa, M. (2007) Breastfeeding and mixed feeding practices in Malawi: timing, reasons, decision makers, and child health consequences. *Food and Nutrition Bulletin* 28(1), 90–99.

Bezner Kerr, R., Dakishoni, L. and Shumba, L. (2008) 'We grandmothers know plenty': breastfeeding, complementary feeding and the multifaceted role of grandmothers in Malawi. *Social Science and Medicine* 66(5), 1095–1105.

Bezner Kerr, R., Berti, P.R. and Shumba, L. (2010) Effects of participatory agriculture and nutrition project on child growth in northern Malawi. *Public Health Nutrition* 14(8), 1466–1472. DOI: 10.1017/S1368980010002545

Black, R.E., Victora, C.G., Walker, S.P. and the Maternal and Child Nutrition Study Group (2013) Maternal and child undernutrition and overweight in low-income and middle-income countries. *The Lancet* 382(9890), 427–451.

Bryce, J., Coitinho, D., Darnton-Hill, I., Pelletier, D., Pinstrup-Andersen, P. and Maternal and Child Undernutrition Study Group (2008) Maternal and child undernutrition: effective action at national level. *The Lancet* 371(9611), 510–526.

de Clerck, F. (2013) Harnessing biodiversity: from diets to landscapes. In: Fanzo, J., Hunter, D., Borelli, T. and Mattei, F. (eds) *Diversifying Food and Diets: Using Agricultural Biodiversity to Improve Nutrition and Health.* Routledge, London, UK.

de Schutter, O. (2010) Agroecology and the right to food. Report submitted by the Special Rapporteur on the right to food [A/HRC/16/49]. Available at http://www.srfood.org/en/report-agroecology-and-the-right-to-food (accessed 28 March 2018).

Eilers, E.J., Kremen, C., Smith Greenleaf, S., Garber, A.K. and Klein, A.M. (2011) Contribution of pollinator-mediated crops to nutrients in the human food supply. *PLoS One* 6(6), e21363. DOI: 10.1371/journal.pone.0021363

FAO (2001) *Integrated Agriculture-Aquaculture: A Primer*. Food and Agriculture Organization of the United Nations, Rome, Italy.

FAO (2005) *Voluntary Guidelines to Support the Progressive Realization of the Right to Adequate Food in the Context of National Food Security*. Food and Agriculture Organization of the United Nations, Rome, Italy.

Frison, E.A., Cherfas, J. and Hodgkin, T. (2011) Agricultural biodiversity is essential for a sustainable improvement in food and nutrition security. *Sustainability* 3(1), 238–253. DOI: 10.3390/su3010238

Gautam, R., Suwal, R. and Basnet, S.B. (2005) *Enhancing contribution of home gardens to on-farm management of plant genetic resources and to the improvement of the livelihoods of Nepalese farmers: findings of baseline survey of four project sites (Jhapa, Ilam, Rupandehi and Gulmi)*. Local Initiatives for Biodiversity, Research and Development, Pokhara, Nepal.

Gebrekidan, B. and Kebede, Y. (1979) The traditional culture and yield potential of the Ethiopian high lysine sorghums. *Ethiopian Journal of Agricultural Science* 1, 29–40.

Gliessman, S.R. (2015) *Agroecology: The Ecology of Sustainable Food Systems*, 3e. CRC Press, Boca Raton, Florida, United States.

Gross, L.H., Erickson, J.D. and Méndez, V.E. (2014) Supporting Rural Livelihoods and Ecosystem Services Conservation in the Pico Duarte Coffee Region of the Dominican Republic. *Agroecology and Sustainable Food Systems* 38(9), 1078–1107, DOI: 10.1080/21683565.2014.932883

Haddad, L. (2013) From nutrition plus to nutrition driven: how to realize the elusive potential of agriculture for nutrition? *Food and Nutrition Bulletin* 34(1), 39–44. DOI: 10.1177/156482651303400105

Hainzelin, E. (2015) Enhancing the function and provisioning of ecosystem services in agriculture: agroecological principles. In: *Agroecology for Food Security and Nutrition: Proceedings of the FAO International Symposium*, 18–19 September 2014, Rome, Italy.

HLPE (2016) Sustainable Agricultural Development for Food Security and Nutrition: What Roles for Livestock? A report by the High-Level Panel of Experts on Food Security and Nutrition of the Committee on World Food Security. Food and Agriculture Organization of the United Nations, Rome, Italy.

IPES-Food (2016) From uniformity to diversity: a paradigm shift from industrial agriculture to diversified agroecological systems. Available at http://www.ipes-food.org/images/Reports/UniformityToDiversity_FullReport.pdf (accessed 28 March 2018).

International Planning Committee for Food Sovereignty (2015) Declaration of Nyeleni 2015 (Declaration of the International Forum for Agroecology). Available at http://www.foodsovereignty.org/wp-content/uploads/2015/10/NYELENI-2015-ENGLISH-FINAL-WEB.pdf (accessed 28 June 2018).

Johns, T. and Sthapit, B.R. (2004) Biocultural diversity in the sustainability of developing-country food systems. *Food and Nutrition Bulletin* 25(2), 143–155.

Kangmennaang, J., Bezner Kerr, R., Lupafya, E., Dakishoni, L., Katundu, M. and Luginaah, I. (2017) Impact of a participatory agroecological development project on household wealth and food security in Malawi. *Food Security* 9, 561–576. DOI: 10.1007/s12571-017-0669-z

Kaufman, A. and Petpha, N. (2016) Moral Rice Network, Dharma Garden Temple, Yasothon province, Northeast Thailand. In: Loconto, A., Poisot, A.S. and Santacoloma, P. (eds) *Innovative Markets for Sustainable Agriculture: How Innovations in Market Institutions Encourage Sustainable Agriculture in Developing Countries*. FAO/INRA. Rome, Italy.

Kuhnlein, H.V., Erasmus, B., Spigelski, D. and Burlingame, B. (2013) *Indigenous Peoples' Food Systems and Well-being Interventions and Policies for Healthy Communities*. Food and Agriculture Organization of the United Nations, Rome, Italy.

Labadarios, D. (2005) Malnutrition in the developing world: the triple burden. *South African Journal of Clinical Nutrition* 18 (2), 119–121.

Loconto, A., Poisot, A.S. and Santacoloma, P. (2016) *Innovative Markets for Sustainable Agriculture: How Innovations in Market Institutions Encourage Sustainable Agriculture in Developing Countries*. FAO/INRA. Rome, Italy.

Malapit, H.J.L. and Quisumbing, A.R. (2015) What dimensions of women's empowerment in agriculture matter for nutrition in Ghana? *Food Policy* 52, 54–63. DOI: 10.1016/j.foodpol.2015.02.003

Marshall, E. and Mejia, D. (2011) *Traditional Fermented Foods and Beverages for Improved Livelihoods*. Food and Agriculture Organization of the United Nations, Rome, Italy.

Martínez-Torres, M.E. and Rosset, P.M. (2014) Diálogo de saberes in La Vía Campesina: food sovereignty and agroecology. *Journal of Peasant Studies* 41(6), 979–997. DOI: 10.1080/03066150.2013.872632

Masset, E., Haddad, L., Cornelius, A. and Isaza-Castro, J. (2012) Effectiveness of agricultural interventions that aim to improve nutritional status of children: systematic review. *British Medical Journal* 344(7843), 1–16.

Méndez, V.E. (2008) Farmer livelihoods and biodiversity conservation in a coffee landscape of El Salvador. In: Bacon, C.M., Mendez, V.E., Gliessman, S.R., Goodman, D. and Fox, J.A. (eds) *Confronting the Coffee Crisis: Fair Trade, Sustainable Livelihoods and Ecosystems in Mexico and Central America*. MIT Press, Cambridge, Massachusetts, USA, pp. 207–236.

Méndez, V.E., Bacon, C.M. and Cohen, R. (2013) Agroecology as a transdisciplinary, participatory, and action-oriented approach. *Agroecology and Sustainable Food Systems* 37(1), 3–18.

Meusch, E., Yhoung-Aree, J., Friend, R. and Funge-Smith, S.J. (2003) The role and nutritional value of aquatic resources in the livelihoods of rural people – a participatory assessment in Attapeu Province, Lao PDR. FAO Regional Office Asia and the Pacific, Bangkok, Thailand, Publication No. 2003/11, 34 pp.

Mitchell, R. and Hanstad, T. (2004) *Small Homegarden Plots and Sustainable Livelihoods for the Poor*. Food and Agriculture organization of the United Nations, Rome, Italy.

Miller, D. (2015) Rediscovering our lost "farmacy": what protective health factors are lost when moving from an agroecological to an industrial model of agriculture? In: *Agroecology for Food Security and Nutrition: Proceedings of the FAO International Symposium*, 18–19 September 2014, Rome, Italy.

Nakalanda, J.M. and Kugonza, I.B. (2016) Facilitating social networks by linking smallholder organic farmers in Uganda to markets for sustainable products: the fresh veggies participatory guarantee system. In: Loconto, A., Poisot, A.S. and Santacoloma, P. (eds) *Innovative Markets for Sustainable Agriculture: How Innovations in Market Institutions Encourage Sustainable Agriculture in Developing Countries*. FAO/INRA. Rome, Italy.

Nyantakyi-Frimpong, H., Kangmennaang, J., Bezner Kerr, R., Luginaah, I., Dakishoni, L., *et al.* (2017) Agroecology and healthy food systems in semi-humid tropical Africa: Participatory research with vulnerable farming households in Malawi. *Acta Tropica* 175, 42–49. DOI: 10.1016/j.actatropica.2016.10.022

Nyantakyi-Frimpong, H., Mambulu, F.N., Bezner Kerr, R., Luginaah, I. and Lupafya E. (2016) Agroecology and sustainable food systems: Participatory research to improve food security among HIV-affected households in northern Malawi. *Social Science and Medicine* 164, 89–99. DOI: 10.1016/j.socscimed.2016.07.020

Oliver, B. (2016) The Earth gives us so much, agroecology and rural women's leadership in Uruguay. *Culture, Agriculture, Food and Environment* 38(1), 38–47. DOI: 10.1111/cuag.12064

Pimbert, M. (2015) Agroecology as an Alternative Vision to Conventional Development and Climate-smart agriculture. *Development* 58(2–3), 286–298.

Powell, B., Thilsted, S.H., Ickowitz, A., Termote, C., Sunderland, T. and Herforth, A. (2015) Improving diets with wild and cultivated biodiversity from across the landscape. *Food Security* 7(3), 535–554.

Rowland, D., Ickowitz, A., Powell, B., Nasi, R. and Sunderland, T. (2017) Forest foods and healthy diets: Quantifying the contributions. *Environmental Conservation* 44(2), 102–114. DOI: 10.1017/S0376892916000151

Ruel, M.T., Alderman, H. and Maternal and Child Nutrition Study Group (2013) Nutrition-sensitive interventions and programmes: how can they help to accelerate progress in improving maternal and child nutrition? *The Lancet* 382(9891), 536–551.
Scrinis, G. (2008) On the ideology of nutritionism. *Gastronomica* 8(1), 39–48. DOI: 10.1525/gfc.2008.8.1.39
Selhub, E.M., Logan, A.C. and Bested, A.C. (2014) Fermented foods, microbiota, and mental health: ancient practice meets nutritional psychiatry. *Journal of Physiological Anthropology* 33(1), 2. DOI: 10.1186/1880-6805-33-2
Smith, L.C. and Haddad, L. (2015) Reducing child undernutrition: past drivers and priorities for the post-MDG era. *World Development* 68, 180–204. DOI: 10.1016/j.worlddev.2014.11.014
Sunderland, T., Powell, B., Ickowitz, A., Foli, S., Pinedo-Vasquez, M., *et al.* (2013) Food security and nutrition: The role of forests. Discussion Paper. CIFOR, Bogor, Indonesia.
UNICEF (1991) *Strategy for Improved Nutrition of Children and Women in Developing Countries*. UNICEF policy review. UNICEF, New York, USA.
Vira, B., Wildburger, C. and Mansourian, S. (2015) *Forests, Trees and Landscapes for Food Security and Nutrition*. A Global Assessment Report. IUFRO World Series Volume 33. International Union of Forest Research Organizations, Vienna, Austria, 172 pp.
Weingartner, L. (2009) The concept of food security and nutrition. In: Klennert, K. (ed.) *Achieving Food and Nutrition Security: Actions to Meet the Global Challenge*. InWent, Feldafing, Germany, pp. 21–52.
Wezel, A., Bellon, S., Doré, T., Francis, C., Vallod, D. and David, C. (2009) Agroecology as a science, a movement and a practice. A review. *Agronomy for Sustainable Development* 29, 503–515.
World Food Summit (1996) Rome Declaration on World Food Security. Available at http://www.fao.org/docrep/003/w3613e/w3613e00.HTM (accessed 28 June 2018).

7 Indigenous Food Systems: Contributions to Sustainable Food Systems and Sustainable Diets

Harriet Kuhnlein, Paul Eme and Yon Fernandez de Larrinoa

> Nana and Baba the creators told us: We have given you everything, you will not be poor if you are close to us, there will always be food. This is why the Guna are always respecting the Forest and the Oceans, and everything created by Baba and Nana (Guna Yala Chief).
>
> (López, 2017)

Abstract

Indigenous food systems are remarkable reservoirs of unique cultural knowledge grounded in historical legacy and spirituality that acknowledge the inextricable link of people with their sustainably managed resources. These sustainable food systems can provide essential understanding about sustainable diets and their importance to many of the Sustainable Development Goals. Unique practices of land and plant and animal management are now threatened by extreme weather and overall climate variability that compound the risks of a long list of environmental assaults upon indigenous lands. Despite vast knowledge of the world's territories and guardianship of 80% of global species diversity, indigenous peoples experience extreme disparities with greater population obesity, undernutrition and micronutrient malnutrition, as well as other health gaps that are grounded in poverty and marginalization. This contributes to the inability of many indigenous peoples to realize sustainable diets known with traditional knowledge. Indigenous food system knowledge is incorporated in both cultivated and wild foods, synergies with the natural environment and biodiversity, adaptation to local conditions and knowledge how these conditions are changing, light carbon footprints, and minimal use of external inputs as fuel and environmentally sensitive technologies. Indigenous food systems across the world demand recognition and protection for their valuable knowledge not only for the benefit of populations of the knowledge holders, but as part of the collective global heritage. Governments, universities, research centers, and United Nations agencies must make Indigenous food systems a priority in their work to document the scientific and cultural benefits of these resources, and to promote more sustainable food systems and, with them, to develop more sustainable global diets.

Introduction

In recent years, there has been a shift in technical discussions from agriculture production towards food systems. The concept of food systems continues to evolve, encompassing different interpretations. There is an ongoing divide between practitioners, some of whom interpret food systems from the value-chain approach while others advocate for the inclusion of factors such as the environment, byproducts, energy and cash (FAO, 2017).

The discussion on sustainable food systems and expanding the food base has been fueled by the challenge of how to feed humankind by 2050 with an estimated population of 9 billion, 70% living in urban areas, and food demand increasing by 40–60%. In an attempt to address the projected demand and supply for food, experts, scientists and policymakers are exploring what makes a food system sustainable.

There are several definitions of food systems, for example, the agroecology of food systems, traditional food systems and farmer-based food systems. However, indigenous food systems introduce a series of conceptual considerations accumulated from empiric evidence that render them unique. Figure 7.1 is an example of a unique indigenous food system in Java.

For the purpose of this chapter, we refer to indigenous food systems as the set of indigenous peoples' management, knowledge and traditional practices that generate food from their territories. Sustainable diets for indigenous peoples are derived and maintained from indigenous sustainable food systems. As defined by the FAO:

> Sustainable diets are those with low environmental impacts which contribute to food and nutrition security and to healthy life

Fig. 7.1. Indigenous food system in Java. Source: Y. Fernandez de Larrinoa.

> for present and future generations. Sustainable diets are protective and respectful of biodiversity and ecosystems, culturally acceptable, accessible, economically fair and affordable; nutritionally adequate, safe and healthy; while optimizing natural and human resources.
> (Burlingame, 2012).

Indigenous peoples in the world

Ten years have passed since the adoption of the 2007 United Nations Declaration on the Rights of Indigenous Peoples (UNDRIP), marking a turning point in acknowledging the existence and rights of millions of indigenous peoples (UNDRIP, 2007). Despite this recognition, there is no final figure for the number of indigenous people in the world. This is partly due to the fact that different countries are at different stages in terms of recognizing indigenous peoples.

The United Nations Department of Economic and Social Affairs and the United Nations Permanent Forum on Indigenous Issues (UNPFII) estimate indigenous peoples to number more than 370 million (United Nations, 2007). They live in seven regions throughout 90 countries and constitute more than 5000 groups. They speak 4000 out of the 7000 existing languages and make up 15% of the poor, despite being only 5% of the world's population (United Nations, 2007). Their richness in culture, spirituality and traditional knowledge contrasts sharply with their poorness in financial terms. It is generally believed that these estimates are conservative. More precise statistics are difficult to obtain since many countries have not yet included intercultural components in their statistical censuses and surveys.

Whenever intercultural components are included into censuses, new light emerges. For example, India's population census accounted Adivasis and Tribal Peoples at 104.3 million in 2011. While El Salvador in 2014, after years of no recognition, modified its constitution to recognize indigenous peoples for the first time (FAO, 2016).

The Sustainable Development Goals approach of 'no-one will be left behind' (General Assembly resolution 70/1) (United Nations, 2015) provides a new impetus to include interculturality in the statistics and work of governments, which will improve the overall data on indigenous peoples.

Indigenous food systems

The last few years has witnessed increased attention to indigenous food systems and their holistic approach towards food. An indigenous food system can be described as 'all food within a particular culture available from local natural resources and culturally accepted. It also includes the sociocultural meanings, acquisition/processing techniques, use, composition and nutritional consequences for the people using the food' (Kuhnlein and Receveur, 1996).

> These systems share several important characteristics; for example, they do not exhaust the natural resources base, their main focus is not commercial and they have low access to markets. For instance, Indigenous food systems tend to be people-centered with many resources managed sustainably. They also combine the consumption of produce with the purchase and sale of food, avoiding a fully commercial orientation. Foods with these features have appeared only recently in large distribution chains, through production systems such as organic farming, perma culture and biodynamic agriculture, which reflect to some extent the philosophical approaches of traditional societies.
> (FAO, 2017: p. 110)

Indigenous food systems do not differentiate between the environment and the people, perceiving that living beings and territory are interconnected, and embedded with spirituality. This holistic view does not place humankind, nor the production of food itself, at the center of the food system. Instead, maintaining the equilibrium between the environment and the beings inhabiting it is the central focus. The key concept is equilibrium between the different parts that make up the system. This is significantly different from other food systems interpretations, which place food production at the center. The milpa system of intercropping is shown in Fig 7.2 (Fragniere, 2007).

> Indigenous food systems have characteristics that make them particularly attractive, including the use of both cultivated crops and gathered wild plants, synergies with the natural

Fig. 7.2. Milpa system of intercropping 1 (Fragniere, 2007).

> environment and biodiversity, close adaptation to local conditions, a high level of diversification, a light carbon footprint, fewer 'negative externalities' and reduced use of external inputs. They are closely tied to culture and social and religious activities.
>
> (FAO, 2017)

Being highly adapted to their environment, indigenous food systems have the capacity to generate food in soils with low fertility, are resistant to stress factors (reduced rainfall, increased temperatures) and have a low demand for inputs.

Some indigenous crops are climate resilient, and their cultivation systems (such as waru-waru, milpa, terra preta and shifting or swidden cultivation) contribute to the management of the environment while generating food.

Since most of the foods consumed in indigenous food systems are not cultivated but harvested/hunted or fished, consumption patterns depend on seasonality and on availability, leading to increased sustainability (see the example shown in Fig. 7.3). In addition, over-exploitation results in depletion of the source and eventual disappearance of the food itself. In most indigenous groups, this is considered not only bad practice, but it carries negative spiritual connotations for the community, and is avoided as much as possible. By focusing on the quality and utilization of foods rather than on production quantities, indigenous food systems can influence the reshaping of the current food systems thinking (FAO, 2017, p. 110).

How are Indigenous Food Systems Important in the Context of Sustainable Diets?

The role of the sustainable diets concept in the context of the Sustainable Development Goals (SDGs) is cross-cutting to many of the seventeen goals (UNSDG, 2015). Furthermore, the centrality of sustainable diets was anticipated in the call for action from the Door of Return to achieve several relevant Millennium Development Goals in Africa (AFROFOODS, 2013).

Indigenous peoples' knowledge emanates from their collective experience in managing 22% of the world's ecosystem and land mass and preserving the majority of the planet's biodiversity. Indigenous peoples understand how their local foods are resilient and adapted to their local environments, even when climate challenged. They know the animals and plants that are natural resources in the world's forests, pastures, riverine lands and waters, lakes, and seas, which contain the genetic material of the world's biodiversity. The knowledge of these resources is grounded in their culture, spirituality and historical legacy. Those who can relate and express such knowledge can help us develop, realize and enjoy the benefits of indigenous food systems, which are essential for sustainable diets.

There are three ways in which indigenous food systems can contribute to the present food challenges:

- expand the available food base;
- present empirical evidence of effective food generation capacity while maintaining the resource base in contrasting climatic areas; and
- include indigenous peoples' knowledge in the sustainable food systems debate linked to climate change.

Fig. 7.3. Forest legumes in indigenous areas of Indonesia. Source: Y. Fernandez de Larrinoa.

Expanding the present food base

Throughout human history, around 7000 of the approximately 250,000 existing plant species in the world have been either cultivated or gathered for consumption. Of these, 150 are commercial, of which 120 are cultivated and 103 presently provide 90% of human foods on the planet. Wheat, maize, rice and potatoes provide about 56% of global human caloric consumption. The remaining thousands of edible plants are either neglected or underutilized (UNESCO and Tudor Rose, 2015). There is extensive literature relating how agricultural production has contributed over time to prime yield at the expense of diversity. Over time, this has resulted in a drastic reduction of the species food base. The 2007–2008 global food price crisis best illustrated the existing dependency of humankind on the global trade and production of a handful of plant staples that provide the majority of the calorie intake in the world (FAO, 2011).

Recording the world's unique food species use, nutritional properties, and other scientific facets is daunting. This has been identified as a priority by the United Nations Food and Agriculture Organization (FAO) because of the progressive dietary simplification resulting from agricultural industrialization in the world's food supplies at the expense of diversity and micronutrient malnutrition (Demment *et al.*, 2003; Khouri *et al.*, 2014).

In contrast, indigenous peoples' diets are diversified and benefit from the utilization of different non-commercial species. 'While modernizing food systems rely heavily on few edible plant species and varieties, indigenous food systems make use of several hundreds of edible and nutritious plants' (FAO, 2017).

Thanks to the recent work of culinary chefs and organizations such as the FAO and Slow Food, more and more indigenous food items have been incorporated into the menus in several countries. Foods such as quinoa (*Chenopodium quinoa*), moringa (*Moringa oleifera*), Amaranthus and bread fruit (*Artocarpus altilis*) have been recently joined by a new wave of other 'superfoods'. These items from indigenous communities have become part of the food menu of middle-income customers valuing local and diverse products, as shown in recent field notes from FAO staff (Fig. 7.4).

Superfoods are generally defined as foods with a low caloric but high micronutrients content. Commercial food systems, driven by yield maximization, have overlooked superfoods, which are produced at small scale, locally and are well adapted to the environment. In most cases, they either grow naturally in the wild or are cultivated in intercropping or shifting cultivation systems. Stevia (*Stevia rebaudiana*), chia (*Salvia hispanica*), kañiwua (*Chenopodium pallidicaule*), kiwicha (*Amaranthus caudatus*), olluco, maca (*Lipidium meyenii*), goji berries (*Lycium barbarum*), guaraná (*Paullinia cupana*), sato palm (*Cycas revolute*), sachainchi (*Plukenetia volubilis*), azai (*Euterpe oleracea*), yarsagumbu (*Ophiocordyceps sinensis*), tara (*Alpinia nigra*) and mahua flowers (*Madhuca longifolia*) are some of the examples of indigenous peoples' foods that have broadened the world's available food base (see Figs 7.5 and 7.6).

The bias towards commercial food products and yield maximization has conditioned not only the research agendas of the agriculture centers and universities, but also the seeds and agro-input markets. This focus on quantity and yield at the expense of diversity affects the available genetic pool of species cultivated and therefore the sustainability of the food systems in the context of mounting climate change pressure. This entire trend has a direct impact on the dietary diversity and the nutritional status of consumers.

Generating food while preserving biodiversity and the natural resource base

Indigenous food systems have provided communities with food for millennia. These systems have the ability to generate food and by-products (shelter, clothes, medicines, housing materials) while maintaining the environment, the resource base and biodiversity. In 2016, IUCN and National Geographic completed a map of biodiversity and forests in Central America, clearly documenting the overlap with indigenous peoples' territories. The map identifies 948 recognized terrestrial and marine protected areas. In fact, 39 percent of those areas - some 96,432 square kilometers - are also home to indigenous peoples. Forty-four percent of Central American forests are located inside areas inhabited and used by indigenous peoples; much of this land still contains intact ecosystems (IUCN and National Geographic, 2016).

Similar results are found in other regions of the world. This is why it is estimated that indigenous peoples are the guardians of 80% of the remaining global biodiversity. Indigenous territorial management and food systems are closely interrelated, as documented in the GIAHS catalogue of global heritage systems, such as the Ifugao rice terraces and the Andean agriculture systems (FAO *et al.*, 2015).

Shifting or swidden cultivation is a good example of how the management of a territory is linked with the food system. Three characteristics define shifting cultivation: the removal of natural vegetation by cutting and burning; an alternation between short cultivation and long fallow; and the shifting or moving of the fields. Practiced in forested areas by millions of indigenous peoples in Latin America, Africa and Asia, this territorial management

Fig. 7.4. Indigenous food recall from women in Paraguay. Source: FAO field notes, 2016; Y. Fernandez de Larrinoa.

technique combines rotation and shifting in the use of the forest with intercropping, harvesting, fishing and hunting as food generation techniques.

Shifting cultivation is one of the most misunderstood and controversial forms of land management (FAO et al., 2015). Stigmatized for the last 70 years for combining fire with the slashing of forest, shifting cultivation is reclaiming new attention by practitioners. Scientists have started to review their perceptions about shifting cultivation, given the fact that areas where it has been practiced are forested years later, whereas other areas under different management techniques are sometimes degraded or deforested.

Although there is need for more research on shifting cultivation and its ability to generate food and preserve the environment, the system itself has changed in recent years. It is believed that demographic growth, market incentives for cash generation, migration of

Fig. 7.5. Indigenous superfood: Guaranaá (*Paullinia cupana*). (Fortis, 2006).

Fig. 7.6. Indigenous superfood: Sachainchi (*Plukenetia volubilis*). Source: M. Herman.

indigenous youth, pressure from extractive industries, and intensive agriculture and livestock have somewhat altered the fallow cycles of shifting cultivation.

Territorial management and the indigenous food system form an interdependent symbiosis. Unfortunately, this linkage between territory and food, along with the difficulties practitioners have had in fully grasping nomadic livelihoods, has resulted in insufficient research. Besides the importance of the territory–food linkage, spiritual and cultural practices, community response mechanisms, and traditional knowledge, there are other factors important to preserve ecosystems. These are:

- adaptability to the climate and environment;
- energy use within the system; and
- nodular relationships and 'reticular space'.

Adaptability to the climate and environment

Pastoralists in Central Asia, Inuit and Sami in Arctic regions, hunter gatherers in Central Africa, and agroforestry farmers in the Amazon, all practice a remarkable array of coping strategies to adapt to the extreme weather conditions in which they live. These coping strategies are implemented through changes in the territorial and livelihoods management, use of buffer areas in their territory for times of crisis, or use of 'emergency foods' in certain circumstances. Despite this wealth of accumulated and empirical traditional knowledge, indigenous peoples across the world report that their traditional coping mechanisms are now under threat due to new episodes of climatic variability and extreme weather (FAO, 2008).

Energy use within the system

Indigenous food systems have a low energy use in terms of fossil fuels, coal and other sources of energy. Whereas global agriculture's reliance on nitrogen energy continues to grow – 20% growth from 2002 to 2009 (Marsden and Morley, 2014) – indigenous systems rely on the capacity of the environment to generate food and on the sun as the primary source of energy from which secondary forms of energy are generated (e.g. firewood, compost, firewood coal, manure). The low use of fossil fuel and other external energy sources is directly linked to the sustainability of indigenous food systems and their success in preserving the environment. At the same time, these systems generate food with a good conversion rate from energy inputs to food output. With externalities accounted for, the conversion rate provided by forests, rivers, oceans and pastures is more efficient than the present global agriculture synthetic substitutes that govern intensive production systems. It is necessary, however, to undertake more research to improve the understanding of energy use in indigenous food systems.

Nodular relations and reticular space

Cartesian systems of borders and relations have not been able to depict the richness of roles, behaviours and management practices that individuals and communities have in place to operate their indigenous food systems. The 'border view' approaches the relations between the environment and humans with the constraint of imaginary geographic boundaries. This does not correspond with how indigenous peoples manage their food systems and territories. Even less in cases where the communities are pastoralists, mobile hunters, or nomadic fishers.

The research done by Dr Dounias and colleagues in Central Africa and Asia (Dounias and Bahuchet, 2000; Dounias, 2017) shows how hunters and gatherers have a complex reticular territorial management system composed of several nodular relations that are activated according to a code that is deeply rooted in the community's traditional knowledge. For instance, hunters manage their territory by combining a network of reticular spaces with different functions (e.g. provision of tubers, harvesting of fruits, hunting areas, sacred and spiritual areas, camping areas, emergency feeding). At the same time, these communities cannot be understood without considering the relations and exchanges, whether through barter or cash, they maintain with neighboring communities.

The reticular spaces and nodular relations are maintained through complex and stratified knowledge-exchange mechanisms between elders, children and adults in reproductive age, with different tasks and knowledge associated to each age group. For instance, the information about which poisonous tubers can be consumed in times of food scarcity and how they should be treated to make them edible is passed from elders to children.

Indigenous peoples' knowledge and the sustainable food systems debate

It is essential to include the wisdom of indigenous peoples on sustainable diets as we address several pressing issues of planetary health.

Indigenous peoples in rural homelands retain the most knowledge and wisdom about food biodiversity in diverse ecosystems, preserved throughout generations by strong cultural identities. The biodiversity of species in indigenous food systems has been recorded in several cultures to contain as many as 390 unique species from their local territory that are recognized, harvested and used (see, for example, Kuhnlein *et al.*, 2009). However, many of these food species, while used and enjoyed in local preparations and dishes, do not yet have scientific identifications or nutrient data (Okeke *et al.*, 2009).

The FAO (through the INFOODS data base), in collaboration with several research institutions and laboratories, is organizing the nutrition composition data of thousands of food items, of which several come from indigenous food systems (INFOODS, 2018).

However, it is important to increase the research on indigenous food systems and enhance the documentation process of food species and traditions. Without documentation, it is not possible to record, share and save this knowledge for future generations, except through cultural inter-generational transfer, which is steadily decreasing.

How Indigenous food systems are lost

Sociocultural factors

Indigenous peoples everywhere are exposed to and adapting to, rapidly changing sociocultural and economic circumstances which invariably affect their decreasing use and transmission of traditional food system knowledge. For examples, see reference to the diverse regions and cultures of the Nuxalk in Canada (Kuhnlein, 1989, 1992) and the Maylayalis in India (Huang et al., 2016). Culture change is manifest in similar ways, including:

- acceptance of the need to pursue income (for clothing, electronics, education, health service, medicines, etc.);
- exposure to and purchase of industrially processed foods stimulated by income generation, with concomitant decline in use of local species;
- rapid advances in media and use of cell phones, advertising and social media, driving ultra-processed food purchasing;
- change in taste preferences by younger generations;
- change in attitudes to food availability and work required to harvest and prepare local foods;
- reduced knowledge transfer about local food species by elders to younger generations.

As is the case for all populations, indigenous peoples also experience urban migration, removing them from the traditional territories where use of local biodiversity is practiced. For these and other reasons, there is gradual and often rapid loss in ecological literacy and the ever increasing untapped potential of local food biodiversity (de Schutter, 2011).

Ecosystem and climate change

Change in use of traditional food systems is also driven by loss of integrity of the rural ecosystems in which indigenous peoples developed their vast cultural knowledge. The long list of environmental assaults to indigenous peoples' traditional lands includes such well-known topics as atomic testing (near small communities in Pacific nation states and New Mexico), oil and gas extraction (Amazon, Arctic), mining (Amazon, all continents), agriculture and processing of illicit drugs (South America), overexploitation of natural populations of fish and birds (Pacific, North Atlantic, New Zealand), drilling and pipelines across traditional lands (Alaska, Canada, US), massive deforestation for timber and agriculture (Amazon, Indonesia), commercial herding over traditional grasslands (Africa, Scandinavia), hydroelectric dams (Japan, US, China, India), land contamination from animal and livestock waste (all continents), industry originated pesticides, organochlorine and heavy metal contamination in lands and waters (Canada, USA, Seychelles, Africa).

All of these concerns and difficulties rest against the background of impending climate change, which is especially noted to affect territories inhabited by indigenous peoples. For example, in small Pacific Island states (rising sea levels) and in the Arctic (unstable land and sea ice), circumstances impede food fishing, hunting and harvest. The resilience of ecosystems that support indigenous peoples and their food systems is all too often stressed beyond legally healthy limits (see, for example, Turner *et al.*, 2013).

What Happens when Indigenous Food Systems are Lost

It is paradoxical that, while indigenous peoples still hold a wealth of knowledge on the breadth of food biodiversity on the planet, they experience the greatest disparities in diet and diet-related health circumstances in the countries where they live, largely driven by the poverty and disenfranchisement that indigenous peoples are pushed into.

However, it is abundantly clear that when the diets of indigenous peoples transition from sustainable local resources to include ever-increasing amounts of poor-quality commercial foods purchased with limited income, the nutritional status and health of populations declines (Kuhnlein *et al.*, 2009; Kuhnlein and Burlingame, 2013). This is in part because of the changes that drive loss of traditional food system knowledge and use, urban migration, and climate change affecting ecosystem function.

On the other hand, considerable research reveals that local foods provide many benefits, especially those related to dietary quality and health (Kuhnlein *et al.*, 2004; Golden *et al.*, 2011; Powell *et al.*, 2013).

In high-income countries such as the US, Canada, Australia and New Zealand, indigenous peoples experience poor diet quality with excess energy and greater obesity than the general populations. For example, in the US, the percentage of obese and overweight American Indian and Alaska Natives in 2015 was 81%, in contrast to 69% for all US adults (Centers for Disease Control and Prevention, 2015). In low- and middle-income countries (e.g. Brazil and India), it was shown to be indigenous peoples who have greater undernutrition than is experienced in the total population; stunting in Brazilian indigenous children less than 5 years of age was 25.7%, whereas the national population figure was 7% (Anderson *et al.*, 2016; Egeland and Harrison, 2013).

Disparities extend beyond obesity and undernutrition. In Australia, about 26% of the Aboriginal and Torres Strait Islander people residing in isolated communities represent 40% of the health gaps in Australia (Vos *et al.*, 2009). Out of the seven top risk factors identified in this health gap, poor nutrition was one of them, which could be attributed to the fact that over 95% of the calorie intake of the indigenous communities was obtained from purchased, poor-quality foods, with the rest from traditional foods gathered through hunting and gathering (Henryks *et al.*, 2017).

Food consumption in remote Australian Aboriginal communities is characterized by highly processed foods containing high levels of sugar, salt and refined grains, and low intakes of fruits and vegetables (Brimblecombe *et al.*, 2013). The low consumption of fruits and vegetables resulted from lack of access due to their high cost (Harrison *et al.*, 2007; Council of Australian Governments, 2009).

Throughout developing regions where indigenous peoples reside, traditional markets are rapidly disappearing and being replaced by commercialized food from infiltration of mega- and supermarket food enterprises (Reardon *et al.*, 2003). This major source of food in the developing world is being replaced by large convenience store chains that are subsidiaries of multinational companies (Walmart, Shoprite, etc.) (Popkin *et al.*, 2012). Indigenous people who migrate to urban areas are exposed, along with all those living in urban poverty, to commercial foods that are highly processed, energy dense and nutrient poor (Asfaw, 2011).

Some countries provide food subsidies to their poor in both urban and rural areas, and while carefully designed subsidy programmes are impactful and can be monitored, some programmes are not without controversy. For example, during the 1970s, the United States food commodity programme of the United States Department of Agriculture made an unpopular cornmeal food subsidy available to traditional maize-farming American Indian tribes in the Southwest (Calloway *et al.*, 1974).

Political instability and unequal power relationships among indigenous population can impede and create an unbalance in food subsidy initiatives (Holden and Lunduka, 2013; Mason *et al.*, 2013). Some food policies focus exclusively on energy/calorie availability without attention to persistent problems of micronutrient malnutrition and non-communicable diseases (Gómez *et al.*, 2013; Pingali and Sunder, 2017). The use of the Public Distribution System (PDS) in India that distributes rice and sugar to the poor has been criticized for many years for undermining the use of local grains such as millets and sorghum, which have higher nutrient content (FAO, 2016).

The Way Forward

Indigenous food systems provide ingenious answers to several of the questions scientists are asking today about what makes a food system sustainable and what makes a diet sustainable – issues such as:

- the ability to generate food while maintaining the resource base, the environment and its biodiversity;
- the ability to use energy within the system in an efficient way; and
- the capacity to generate byproducts, medicines, and shelter through multipurpose strategies.

Building activities in indigenous peoples' communities that foster sustainable diets begins with community members and their indigenous values, priorities, and knowledge. Commitment to food system protection for community health and well-being was evident in the twelve case studies highlighted in the CINE-FAO publications (Kuhnlein *et al.*, 2009; Kuhnlein and Burlinghame, 2013). In these areas of diverse ecosystems and cultures, the activities developed were also diverse, but with consistent threads to focus on children's and women's health, including elders as respected reservoirs of food system knowledge and seeking support from a diversity of friends and stakeholders. These case-study projects were in rural areas where traditional knowledge was still expressed for the biodiversity of species present in the local ecosystems. It is important to reflect on the capacity of the local ecosystem to make provision for local species and how commercial foods can be part of the sustainable diet dyad of traditional and commercial food. In this regard, indigenous peoples can be a model for exploring how local food and sustainable commercial foods can form sustainable diets for larger populations (Hunter *et al.*, 2016). Problematic issues to be solved within indigenous food systems include generation of cash to purchase high-value goods outside the system (vehicles, cell phones, computers, etc.). In most instances, accumulation processes revolve around the effective management of the natural resource base and its related food system. Whenever an indigenous food system is geared towards accumulating and generating cash, it runs into issues similar to other non-indigenous food systems (FAO and IWGIA, in press).

Another pressing issue is leveraging indigenous food systems to retain the youth in the community. The migration of youth to urban centers in search of education and job opportunities is threatening the intergenerational knowledge transmission that is fundamental for the survival of the indigenous food systems, and indeed, the entire cultural fabric.

Overall, it is accepted that indigenous food systems are at risk of disappearing due to: the destruction of habitats and displacement of indigenous peoples from their territories; the loss of languages and cultures by indigenous communities; the loss of traditional seeds; the shift in food habits; and the decrease of intergenerational exchange coupled with youth migration. When the indigenous food system is abandoned, the health of the community deteriorates, and the traditions and culture associated with food are progressively lost.

The impetus of the spirit of the SDGs of 'leaving no-one behind' presents an opportunity that cannot materialize unless governments, universities, research centers and UN organizations all make indigenous food systems a priority in their work. A first step would be to make available resources to document traditional food resources through non-governmental organizations and governments at all levels (Hunter *et al.*, 2016). Building on this information to provide livelihoods within their communities is clearly within the mandate of FAO and the Convention on Biological Diversity (CBD) to meet the SDGs (FAO, 2013; UNEP-CBD, 2016).

There is overall acceptance that there is a need for additional research, data and information. Traditional food system information, including scientific identifications, nutrient composition data and ecosystem requirements to maintain the species for defined population levels, needs to be included in data repositories. With critical information on food biodiversity data disappearing from the living knowledge of indigenous peoples, and the world's food supplies depending on fewer and fewer crops, there is increased demand for this knowledge. In the public sector, advisories for increased dietary diversity should be encouraged by governments to increase demand for diverse and sustainable foods in available food markets. Metrics for measuring the inclusion of more sustainable agricultural species, animals and plants in world food supplies for food security and dietary sustainability should be derived, confirmed and monitored, and food producers nudged by governments to provide more nutritious and affordable commercial food. (Berry *et al.*, 2015; Fanzo, 2017; Lartey, 2016). Indigenous peoples are willing and able partners in the conversation and planning of increasing documentation, agriculture and marketing of their diverse foods.

Governments should be responsive to the need for oversight of the commercial food sector to make provision for diverse supplies of healthy food and to reduce ultra-processed food and food waste, thus ensuring stocks of biodiverse food resources originating from traditional knowledge. With policies like these, there should be a return of livelihood income to the indigenous partners. These goals can be realized with leadership, commitment, and hard work in networking, communication, and partnership.

The FAO, together with Bioversity International, the Center for International Forestry Research and The Indigenous Partnership are collaborating to understand sustainable food systems by profiling different indigenous food systems across the world (forest hunter–gatherers, fishers, shifting cultivators and pastoralists) to recognize the threats and opportunities indigenous food systems present. Policy recommendations based on this understanding can then influence the preservation of these millenary systems.

The outcomes of initiatives to protect indigenous food systems will determine not only their survival, but also their ability to contribute to the quest for sustainable diets that are essential in the context of the SDGs.

References

AFROFOODS (2013) Call for action from the door of return for food renaissance in Africa, Appendix 4. In: Kuhnlein, H.V., Erasmus, B., Spigelski, D. and Burlingame, B. (eds) *Indigenous Peoples' Food Systems and Wellbeing: Interventions and Policies for Healthy Communities*. Food and Agriculture Organization of the United Nations and Centre for Indigenous Peoples' Nutrition and Environment, Rome, Italy.

Anderson, I., Robson, B., Connolly, M., Al-Yaman, F., Bjertness, E., *et al.* (2016) Indigenous and tribal peoples' health. *The Lancet* 388, 131–157.

Asfaw, A. (2011) Does consumption of processed foods explain disparities in the body weight of individuals? The case of Guatemala. *Health Economics* 20, 184–195.

Berry, E.M., Dernini, S., Burlingame B., Meybeck, A. and Conforti, P. (2015) Food Security and sustainability: can one exist without the other? *Public Health Nutrition* September, 2293–2302. DOI: 10.1017/S136898001500021X

Brimblecombe, J., Ferguson, M., Liberato, S. and O'Dea, K. (2013) Characteristics of the community-level diet of Aboriginal people in remote northern Australia. *Medical Journal of Australia* 198, 380–384. DOI:10.5694/mja12.11407

Burlingame, B. (2012) Preface. In: Burlingame, B., and Dernini, S. (Eds) Sustainable Diets and Biodiversity: Directions and Solutions for Policy, Research and Action. *Proceedings of the International Scientific Symposium Biodiversity and Sustainable Diets United Against Hunger*, 3–5 November 2010, FAO Headquarters, Rome. http://www.fao.org/docrep/016/i3004e/i3004e.pdf (accessed 28 August 2018).

Calloway, D.H., Giauque, R.D. and Costa, F.M. (1974) The superior mineral content of some american indian foods in comparison to federally donated counterpart commodities. *Ecology of Food and Nutrition* 3, 203–211.

Centers for Disease Control and Prevention (2015) CDC growth charts. Available at http://www.cdc.gov/growthcharts/cdc_charts.htm (accessed 21 September 2016).

Council of Australian Governments (2009) National strategy for food security in remote indigenous communities. Available at https://www.coag.gov.au/node/92 (accessed 3 March 2018).

Demment, M.W., Young, M.M. and Sensenig, R.L. (2003) Providing micronutrients through food-based solutions: a key to human and national development. *Journal of Nutrition* 133, 3879–3885. DOI: 10.1007/s12571-017-0701

De Schutter, O. (2011) The right of everyone to enjoy the benefits of scientific progress and the right to food: from conflict to complementarity. *Human Rights Quarterly* 33, 304–350.

Dounias, E. and Bahuchet, S. (2000) Habitat semi-permanent en forêt d'Afrique centrale. In: Brun, B., Dufour, A.-H., Picon, B. and Ribéreau-Gayon, M.D. (eds) *Cabanes, Cabanons et Campements. Formes Sociales et Rapports à la Nature en Habitat Temporaire*. Editions de Bergier, Châteauneuf de Grasse, France, pp. 161–181.

Dounias, E. (2017) Collaborating with the wild. Management of wild food resources and taming of the tropical rainforest by modern-day foraging societies [PowerPoint Presentation]. Presented during the FOA-ESN Seminar 'Wild but edible and nutritious!' Exploring new (and old) ways to contribute to the UN Decade of Action on Nutrition and the SDGs. FAO, Rome, 25 May 2017.

Egeland, G.M. and Harrison, G.G. (2013) Health disparities: promoting indigenous peoples' health through traditional food systems and self-determination. In: Kuhnlein, H.V., Erasmus, B., Spigelski, D. and Burlingame, B. (eds) *Indigenous Peoples' Food Systems and Wellbeing: Interventions and Policies for Healthy Communities*. Food and Agriculture Organization of the United Nations and Centre for Indigenous Peoples' Nutrition and Environment, Rome, Italy, pp. 9–22.

Fanzo, J.C. (2017) Decisive decisions on production compared with market strategies to improve diets in rural Africa. *The Journal of Nutrition* 147(1), 1–2.

FAO (2008) Análisis del impacto de los eventos fríos del 2008 en la agricultura y ganadería altoandina en el Perú. Available at http://www.fao.org/fileadmin/user_upload/emergencies/docs/1_Peru_ESTUDIO_FINAL_FRIAJE_OCT_13_2008.pdf (accessed 27 February 2018).

FAO (2011) En Enseñanzas de la crisis alimentaria Mundial 2006–2008. In: El estato de la inseguridad alimentaria en el mundo ¿Cómo afecta la volatilidad de los precios internacionales a las economías nacionales y la seguridad alimentaria? Organización de las Naciones Unidas para la alimentación y la agricultura, Roma, Italia, pp. 23–34. Available at http://www.fao.org/docrep/014/i2330s/i2330s.pdf (accessed 20 February 2018).

FAO (2013) Fourteenth regular session of the Commission on Genetic Resources for Food and Agriculture, 15–19 April 2013, FAO Rome – Item 2.5 of the Provisional Agenda: Review of Key Issues on Biodiversity and Nutrition. CGRFA-14/13/8. Rome, Italy, 7 pp. Available at http://www.fao.org/docrep/meeting/027/mf917e.pdf (accessed 1 March 2018).

FAO, IWGIA and AIPP (2015) Shifting Cultivation, livelihood and food security: New and Old for Indigenous Peoples in Asia. Available at http://www.fao.org/3/a-i4580e.pdf (accessed 20 February 2018).

FAO (2016) El Salvador's Constitution of 1983 with Amendments through 2014. Available at http://extwprlegs1.fao.org/docs/pdf/els127410E.pdf (accessed 20 February 2018).

FAO (2017) The future of food and agriculture – trends and challenges. Available at www.fao.org/3/a-i6583e.pdf (accessed 20 February 2018).

FAO and IWGIA (2018) Issues of Brum bush in Megalaya to generate cash for the Khasi Indigenous communities. Analysis of the matrifocal society of the Khasi people (in press).

Fortis A. (2006) Own work. [CC BY-SA 3.0 (https://creativecommons.org/licenses/by-sa/3.0)], via Wikimedia Commons. Available at https://upload.wikimedia.org/wikipedia/commons/2/29/Guaran%C3%A1_02.jpg (accessed 1 March 2018).

Fragniere I. (2007) Own work. [GFDL (http://www.gnu.org/copyleft/fdl.html) or CC BY-SA 3.0 (https://creativecommons.org/licenses/by-sa/3.0)], via Wikimedia Commons. Available at https://upload.wikimedia.org/wikipedia/commons/3/3a/Association_culturale_ma%C3%AFs-Haricot-Courge_dans_la_r%C3%A9gion_du_Mixtepec_au_Mexique.JPG (accessed 1 March 2018).

Golden, C.D., Fernald, L.C.H., Brashares, J.S., Rasolofoniaina, B.J.R. and Kremen, C. (2011) Benefits of wildlife consumption to child nutrition in a biodiversity hotspot. *Proceedings of the National Academy of Sciences USA* 108, 19653–19656.

Gómez, M.I., Barrett, C.B., Raney, T., Pinstrup-Andersen, P., Meerman, J., *et al.* (2013) Post-green revolution food systems and the triple burden of malnutrition. *Food Policy* 42, 129–138.

Harrison, M.S., Coyne, T., Lee, A.J., Leonard, D., Lowson, S., *et al.* (2007) The increasing cost of the basic foods required to promote health in Queensland. *Medical Journal of Australia* 186, 9–14.

Henryks, J., Brimblecombe, J. and Bidstrup, G. (2017) Supporting healthier food choices in remote indigenous communities: developing a food choice app. *Journal of Food Products Marketing* 23(6), 609–620. DOI: 10.1080/10454446.2015.1048028.

Holden, S.T. and Lunduka, R. (2013) Who benefits from Malawi's targeted farm input subsidy program? *Forum for Development Studies* 40(1), 1–25.

Huang T., Farmer P.A., Goddard E., Willows N. and Subhans F. (2017) An ethnographic exploration of perceptions of changes in dietary variety in the Kolli hills, India. *Food Security* 9, 759–771.

Hunter, D., Özkan, I., Oliveira Beltrame, D.M., Samarasinghe, W.L.G., Wasike, V.W., et al. (2016) Enabled or disabled: is the environment right for using biodiversity to improve nutrition? *Frontiers in Nutrition* 3(14), 1–6. DOI: 10.3389/fnut.2016.00014

INFOODS (2018) About Infoods. Available at http://www.fao.org/infoods/infoods/en/ (accessed 27 February 2018).

IUCN and National Geographic (2016) Map shows Indigenous peoples as guardians of Central American ecosystems. Available at https://www.iucn.org/content/map-shows-Indigenous-peoples-guardians-central-american-ecosystems (accessed 27 February 2018).

Khoury, C.K., Bjorkman, A.D., Dempewolf, H., Ramirez-Villegas, J., Guarino, L., *et al.* (2014) Increasing homogeneity in global food supplies and the implications for food security. *Proceedings of National Academy of Sciences USA* 111 (11), 4001–4006. DOI: 10.1073/pnas.13134901117

Kuhnlein, H.V. (1989) Factors influencing use of traditional foods among the Nuxalk people. *Journal of Canada Dietetic Association* 50(2), 102–106.

Kuhnlein, H.V. (1992) Change in the use of traditional foods by the Nuxalk Native people of British Columbia. *Ecology of Food Nutrition* 27 (3–4), 259–282.

Kuhnlein, H.V. and Burlingame, B. (2013) Why do Indigenous Peoples' food and nutrition interventions for health promotion and policy need special consideration? In: Kuhnlein, H.V., Erasmus, B., Spigelski, D. and Burlingame, B. (eds) *Indigenous Peoples' Food Systems and Wellbeing: Interventions and Policies for Healthy Communities*. Food and Agriculture Organization of the United Nations and Centre for Indigenous Peoples' Nutrition and Environment, Rome, Italy, pp.3–8.

Kuhnlein, H.V. and Receveur, O. (1996) Dietary change and traditional food systems of indigenous peoples. *Annual Review Nutrition* 16, 417–442.

Kuhnlein, H.V., Erasmus, B. and Spigelski, D. (2009) *Indigenous Peoples' Food Systems: The Many Dimensions of Culture, Diversity and Environment for Nutrition and Health*. United Nations Food and Agriculture Organization, Rome, Italy.

Kuhnlein, H.V., Receveur, O., Soueida, R. and Egeland, G.M. (2004) Arctic Indigenous Peoples experience the nutrition transition with changing dietary patterns and obesity. *Journal of Nutrition* 134, 1447–1453.

Lartey, A. (2016) Linking agriculture with nutrition within SDG2: making a case for a dietary diversity indicator. *Field Exchange* 50, 51.

López, B. (2017) Cosmovisión san de la comarca indígena Guna Yala [video]. Available at https://www.youtube.com/watch?v=JIVF7Nvf-mo (accessed 27 February 2018).

Marsden, T.K. and Morley, A.S. (2014) Sustainable Food Systems: Building a New Paradigm. Earthscan Food and Agriculture/Routledge, London, UK.

Mason, N.M., Jayne, T.S. and Walle, N.V. (2013) Fertilizer subsidies and voting pattern: political economy dimensions of input subsidy programs. Paper presented at the Agricultural and Applied Economics Association's 2013 AAEC & CAES Joint Annual Meeting, Washington, DC, USA, 4–6 August 2013.

Okeke, E.C., Ene-Ebong, H.N., Uzuegbunam, A., Ozioko, A., Umeh, S. and Chukwuone, N. (2009) The Igbo traditional food system documented in four states in Southern Nigeria. In: Kuhnlein, H.V., Erasmus, B. and Spigelski, D. (eds) *Indigenous Peoples' Food Systems: The Many Dimensions of Culture, Diversity and Environment for Nutrition and Health*. United Nations Food and Agriculture Organization, Rome, Italy, pp. 251–281.

Pingali, P. and Sunder, N. (2017) Transitioning toward nutrition sensitive food systems in developing countries: a review. *Annual Review of Resource Economics* 1–28. DOI: 10.1146/annurev-resource-100516-053552

Popkin, B.M., Adair, L.S. and Ng, S.W. (2012) Now and then: the global nutrition transition: the pandemic of obesity in developing countries. *Nutrition Review* 70 (1), 3–21. DOI: 10.1111/j.1753-4887.2011.00456.x

Powell, B., Maundu, P., Kuhnlein, H.V. and Johns, T. (2013) Wild foods from farm and forest in the East Usambara Mountains, Tanzania. *Ecology of Food and Nutrition* 52(6), 451–478.

Reardon, T., Timmer, C.P., Barrett, C.B. and Berdegue, J.A. (2003) The rise of supermarkets in Africa, Asia, and Latin America. *American Journal of Agricultural Economics* 85, 1140–1146.

Turner, N.J., Plotkin, M. and Kuhnlein, H.V. (2013) Global environmental challenges to the integrity of indigenous peoples' food systems. In: Kuhnlein, H.V., Erasmus, B., Spigelski, D. and Burlingame, B. (eds) *Indigenous Peoples' Food Systems and Wellbeing: Interventions and Policies for Healthy Communities*. Food and Agriculture Organization of the United Nations and Centre for Indigenous Peoples' Nutrition and Environment, Rome, Italy, pp. 23–28.

UNDRIP (2007) United Nations Declaration on the Rights of Indigenous Peoples. Available at http://www.un.org/esa/socdev/unpfii/documents/DRIPS_en.pdf (accessed 27 February 2018).

UNEP-CBD Secretariat (2016) UN Biodiversity Conference results in significant commitments for action on biodiversity, Agreements reached on actions to integrate biodiversity in forestry, fisheries, agriculture, and tourism sectors and to achieve the 2030 Agenda on Sustainable Development. Web Announcement, 18 December 2016, pp 1–6. Available at https://www.cbd.int/doc/press/2016/pr-2016-12-18-un-bidov-conf-en.pdf (accessed 1 March 2018).

UNESCO and Tudor Rose (2015) Agree to differ, UNESCO publishing and Tudor Rose. Available at http://unesdoc.unesco.org/images/0023/002326/232657e.pdf (accessed 27 February 2018).

United Nations (2007) State of world's indigenous peoples. Available at http://www.un.org/esa/socdev/unpfii/documents/SOWIP/en/SOWIP_web.pdf (accessed 27 February 2018).

United Nations (2015) Transforming our world: the 2030 Agenda for Sustainable Development. Available at undocs.org/A/RES/70/1 (accessed 27 February 2018).

UNSDG (2015) Sustainable development goals, 17 goals to transform our world. Available at http://www.un.org/ga/search/view_doc.asp?symbol=A/RES/70/1&Lang=E (accessed 27 February 2018).

Vos, T., Barker, B., Begg, S., Stanley, L. and Lopez, A.D. (2009) Burden of disease and injury in Aboriginal and Torres Strait Islander peoples: the Indigenous health gap. *International Journal of Epidemiology* 38, 470–477. DOI:10.1093/ije/dyn240

Warner, K. (1991) Shifting cultivators: local technical knowledge and natural resource management in the humid tropics. Available at http://www.fao.org/docrep/u4390e/u4390e00.htm (accessed 27 February 2018).

8 Can Cities from the Global South be the Drivers of Sustainable Food Systems?

Jorge M. Fonseca, Jane Battersby and Luis Antonio T. Hualda

Abstract

Urbanization has been associated with significant transformations in our society, with paramount influence in agriculture and the world food industry, and subsequently in consumers' diets. Arguably, the current food consumption trend is non-sustainable given the non-regenerative, and rather disruptive, ways of using natural resources for meeting the growing food demand and the growing inequality for food affordability across regions. Cities have been an easy target to promote non-sustainable consumption, due to a lifestyle that encourages it and where 'convenience' is the prominent sought-after feature in food. Moreover, the food systems feeding urban populations need to be not only environmentally sustainable, but also socially and economically sustainable, and these pillars of sustainability are inextricably linked. It is within this context that this chapter asks: *how can cities be drivers of food system sustainability?* It specifically focuses on cities of the South due to their rapid urbanization and particular persistent challenges of poverty and food insecurity. Indeed, in cities of the global South, population in slums, where poverty is prevalent, constitute nearly four out ten of the total urban dwellers in developing countries, and as high as seven out of ten in African countries, revealing cities can no longer afford to treat slums as an excluded part or 'exception' to the rest of the city. We reviewed the global context and identify current opportunities that cities can exercise to drive what can be the sustainable food systems of the future. It is highlighted that social and environmental inclusion in city-linked food systems can be effectively articulated through: (i) participatory governance; (ii) solidarity schemes; (iii) inclusive value chain collaborations; and (iv) food system planning. Importantly, interventions in cities of the South require improved coherence given the inter-cross of jurisdictions of pertinent institutions, evidencing the need for a territorial approach where the different levels of government engage in dialogue.

Introduction

The recent decades have brought structural changes in society that has focused attention on the role of cities, and their governments, as potential way to address many of the sustainability issues faced by humanity today. Urbanization is now regarded as one of the most complex phenomena of our times, impacting food in multiple ways, including the way food is produced, conceptualized and valued for human life (McNicoll, 1984). The pace of the urban growth will result in two-thirds of the global population living in urban areas by 2050 (UN DESA, 2014).

The potential role of cities as drivers of sustainability through food system interventions is increasingly acknowledged by global development analysts. The Paris Agreement, and the 2030 Agenda are prompting action in many cities. Moreover, The New Urban Agenda (NUA) signed in 2016 by UN member countries, serves as guidance for intervention on urban-related matters for the following 20 years. The NUA emphasizes the need for fostering sustainable food systems, development of inclusive value chains, improved natural resource management and stronger rural–urban links that can support an integrated development. However, implementation of the NUA is the true challenge.

Local governments commonly prioritize other basic needs such as public transportation, decent housing and tap water, and urban planners commonly overlook food systems in their projections (Greenstein *et al.*, 2015). However, signs of local governments assigning budgetary resources to food systems are surfacing, for example, the Milan Urban Food Policy Pact. Launched in 2015, the pact has now been signed by over 160 cities from both developed and developing countries looking to boost sustainable approaches to provide healthy food to all. City Networks (e.g. ICLEI, UCLG, C-40) are also starting to advocate for more attention to food systems among their members, as well as creating alliances within cities and technical institutions to provide advisory services to their city members.

It is within this context that this chapter asks: how can cities be drivers of food system sustainability? It specifically focuses on cities of the South due to their rapid urbanization and particular ongoing challenges of poverty and food insecurity.

The Era of Urbanization Across the Regions in the Global South

We discuss three regions in the South to give a background for the discussion on how the cities can implement improved food systems. In these three regions, with the exception of the East Asia subregion and three countries of Central Asia, the individual income per capita has increased at a very low rate compared to the increased rate of food prices during the period 2000–2014 (FAOSTAT, 2016).

This situation poses a threat for poor urban residents who mainly rely on cash to avoid food insecurity. In many countries in the global South, the urban poor spend more than half of their income on food (Ruel *et al.*, 2017). Based on data collected in 146 countries in 2014–2015 by Food and Agriculture Organization (FAO) using the Food Insecurity Experience Scale in the Gallup World Poll, nearly 50% of urban populations in the world are moderately or severely food insecure in the least developed countries compared to approximately 43% in rural areas (Fig. 8.1). In informal urban settlements across the world, however, the prevalence of food insecurity increases to between 70% and 95% of the population (Battersby, 2013; Tacoli *et al.*, 2013), a large percentage of whom are severely food insecure.

Urbanization trends and intersections of the urban and the food systems vary considerably across and within regions of the global South. The United Nations, Department of Economic and Social Affairs, provides an overview of historical, current and projected proportions of regional populations resident in urban areas (UN DESA, 2014). Africa and Asia are the fastest urbanizing regions of the world, with Latin America and the Caribbean's urban transition having occurred much earlier. These regional figures mask significant intraregional variation. Increasingly, urbanization in Africa and Latin America is centred on towns and intermediate centres, while demographic growth in large cities continues.

While cities are centres of wealth and power, poor people in cities often face living conditions that are less than ideal, in the slums. In Latin America and the Caribbean, half of the poor live in urban areas, with around 23% of the poor population living in overcrowded informal settlements or tenements that lack adequate provision for basic utilities and public services (Magalhães, 2016). In sub-Saharan Africa, two-thirds of urban residents live in slums (UN Habitat, 2016). In some countries (Angola, Ethiopia, Madagascar, Niger, Sudan) this proportion is as high as 80%.

These high levels of urbanization and urban poverty play important roles in shaping the food systems of these regions. First, in both Africa and Asia, remittance from urban households and earning from non-farm activities play a major role in financing innovation and intensification of farming. In Asia, traders often turn to the local processing of agricultural produce, which diversifies the economic base of large villages and aids their transformation into small urban centres (Satterthwaite *et al.*, 2010). Large villages and small centres have emerged as hubs for services, post-harvest operations and trade, whereas rural and urban areas are better linked by infrastructure

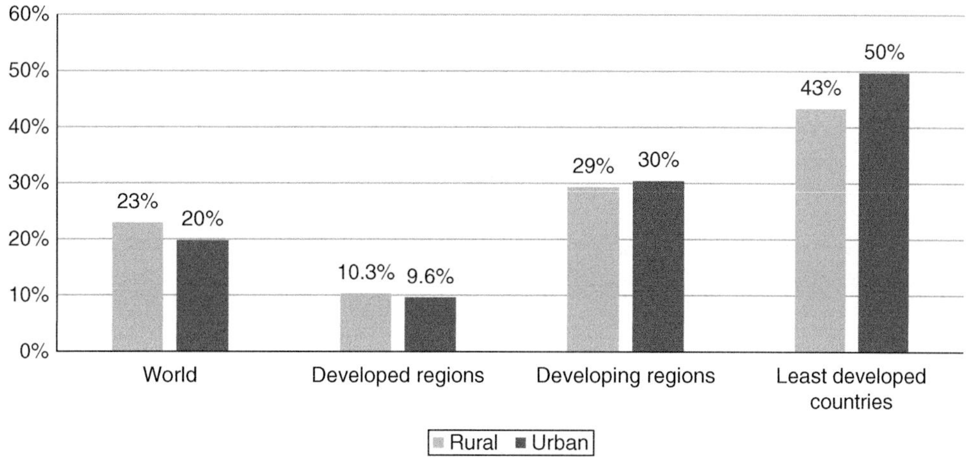

Fig. 8.1. Prevalence of moderate or severe food insecurity among rural and urban populations based on the Food Insecurity Experience Scale, 2014–2015. Source: Tefft *et al.*, 2017, based on disaggregated data from FAO's Project 'Voices of the Hungry' (http://www.fao.org/in-action/voices-of-the-hungry/). Accessed June 1, 2017.

and information networks (Tacoli and Vorley, 2016). Second, urban residents are the dominant purchasers of food within the regions. In Africa, urban markets absorb two-thirds of the market for non-staple products. In Eastern and Southern Africa, spending on processed foods among urban households was 56% of their expenditures, while in Asia it accounts for as high as two-thirds of the total food expenditure (Reardon, 2016). Third, in light of these market opportunities, new value chain opportunities have arisen. Central to this transformation has been the emergence of the supermarket, first in Latin America, then in Asia and now in Africa. While this has brought some benefits to food security and nutrition, there are concerns about the impact of these changing value chains on food system sustainability, equity and consumer health.

An Insight of Sustainability in the Context of Food Systems

Sustainable Development Goal (SDG) 12 calls for sustainable consumption and production, but is poorly aligned to SDG 2, which calls for an end to hunger, the achievement of food security and improved nutrition, and the promotion of sustainable agriculture. This chapter contends that sustainable production will not lead to sustainable consumption without urban food system governance, and yet SDG 11, the urban SDG, fails to mention food.

The overwhelming focus of food system sustainability policy is on production to the exclusion of the wider food system (Affognon *et al.*, 2015), and the absence of food in urban policy has led to profoundly unsustainable urban food systems. This manifests in the form of food waste and the urban coexistence of hunger, micronutrient deficiency and obesity (Marvin and Medd, 2006; Alvez *et al.*, 2011). However, loss and waste can be reduced, access to healthier and safer foods improved, and increased economic viability of food businesses can be enhanced through more food-conscious design of retail spaces, sanitation and pro-poor transport infrastructure with the objective of a sustainable food system at its core (Development Initiatives, 2017).

Three Pillars of Sustainability, and a Call to Focus on Slums

The sustainable food agenda has largely focused on pre-harvest production, engaging sustainability primarily from an environmental standpoint. However, by viewing the food system from an urban perspective, it is clear that for a food system to be sustainable, the policy and programme focus needs to extend beyond production. It is equally clear that a food system cannot be considered sustainable if it is not environmentally, economically and socially sustainable, and that these three pillars of sustainability are inextricably linked. Food consumption patterns are the driver of the environmental, economic and social sustainability of the food system, but food consumption patterns are in turn shaped by urban governance and policy. The following section is far from comprehensive, but is designed to be illustrative of the intersecting nature of the three pillars of sustainability.

Environmental sustainability

National agricultural policies across Africa have been directed towards the production of staple crops (maize) to ensure availability and affordability for urban populations (McMichael and Schneider, 2011). This has led to environmentally unsustainable agricultural practices across the region. Additionally, urban populations have high demands for protein, which requires increasing production of meat, fish and dairy products. Production of feeds for livestock may include the monocropping activities that may not preserve biodiversity and sustainable use of natural resources (Alkemade *et al.*, 2013). The prevailing logic of food systems feeding urban populations is profoundly environmentally unsustainable, and is organized along principles that are economically and socially unsustainable too. Furthermore, the growing trend of geographically longer supply chains, while decreasing seasonality of food supply, may be undermining food system resilience to shocks by undermining local supply chains. It is essential to ensure a diversity of supply chains in order to increase food system resilience (Battersby, 2012).

Economically viable food systems

Economic sustainability entails inclusive economies, equity along supply chains and the generation of employment and livelihood opportunities. What is required in the global South, therefore, is a food system that is inclusive of informal sector activities that dominate the food systems of the global South. Increased urbanization will enhance preference for convenience, unfolding diverse opportunities of entrepreneurship from small and informal to large and formal. However, it will likely result in an increased share of modern retail markets, which may be difficult for smallholder producers and processors to enter. The role of governance in shaping economically inclusive food systems is fundamental. This implies direct food system interventions, such as inclusive institutional procurement programmes, but also consideration of spatial planning, trading by laws and regulation that support the sustainability of small-scale food system actors, all of which fall with the existing mandates of local government.

Social sustainability

There are increasing concerns that the transitioning food systems of the global South are not socially sustainable, with concerns about the health consequences of changing systems and the loss of traditional foods and food cultures. There is increasing evidence of the triple burden of malnutrition (hunger, micro-nutrient deficiency and obesity) coexisting in cities of the global South (Global Panel, 2017). Supermarkets and the fast-food outlets frequently associated with them may be providing cheaper food, but they are also contributing to dietary shifts (Demmler *et al.*, 2017). Diets are often shaped by poor urban housing conditions, which necessitate consumption of processed, non-perishable goods over healthier, unprocessed foods (Global Panel, 2017). Those working from a food sovereignty perspective highlight the ways in which transforming food systems are resulting in the loss of agency for consumers and producers, and a decline in access to traditional foods and food practices. While there have been attempts to develop alternative food systems, such as schemes linking small farmers and consumers in a box scheme in Nairobi, Freidberg and Goldstein (2010) found that uptake of such schemes is limited. It is perhaps more instructive to consider supporting existing, less formal supply chains that are being eroded by the transition towards the formalization of the food system.

The urban slums: an essential focal point

An urban food system cannot be considered to be sustainable if it is not meeting the environmental, economic and social needs of the poor. In cities of the global South, poverty is concentrated in slum areas. Slum dwellers constitute 36.5% of the urban population in developing countries, and as high as 70% in Africa, 62% in sub-Saharan Africa and 43% in Southern Asia (Reardon, 2016; UN Habitat, 2016). Cities can no longer afford to treat slums as excluded part or 'exception' to the rest of the city.

Food represents the greatest share of expenditure for slum residents (Satterthwaite *et al.*, 2010), and yet food insecurity remains high. Food production, processing and retail serving the slum areas are associated with poor environmental conditions and infrastructure deficiencies (Development Initiatives, 2017). This shapes what kind of businesses are viable and what kind of food is available and accessible to poor residents. These same factors shape household food utilization. Most urban slums are essentially clusters of small living spaces with very little or inexistent kitchen space, expensive fuel costs, lack of or inadequate refrigeration and storage, and no garden production. In the case of Nairobi, a great share of slum households rely on ready-made or street food (IIED, 2016) as a means to overcome infrastructural challenges that affect the safety of household food storage, preparation and consumption (Kimani-Murage, 2014).

It is therefore essential to consider what forms of urban planning, programming and management might be required to enable the development of sustainable food system components within slums, as well as what interventions may be required at the national and global scale to develop sustainable value chains from production to consumption.

A Challenging yet Promising Outlook

Although food systems in cities of the global South appear to be following the trajectory of cities in the North (e.g. proliferation of corporate grocery stores), there are potential spaces for interventions that can enhance the sustainability of current and emergent food systems.

The control of the supply of processed food to and within cities is increasingly dominated by firms supplying retailers, the food service/HORECA (hotels, restaurants and catering) and the public institutional sectors because these groups have established advanced logistics systems in the course of the last few decades (Fraser, 2013; Reardon, 2016). Acknowledging the influential role of this axis of intermediaries across the food system is important because while some food supplied is considered healthy ('raw' frozen foods), these food suppliers distribute much of the high calorie, dense food seen today in cities, and have an important influence in diet nutrient composition.

Extremely poor urban dwellers are often excluded from the benefits of efficient 'last mile' perishable food supplying systems (e.g. logistics systems within the cities), that improve access to affordable end products. This exclusion happens for different reasons, including poor or absent food infrastructure, bad coverage of public transportation and lack of security. The impact of 'walking distance' to the different food assets in cities (e.g. wet markets) on healthy diets has been well documented (Coveneya and O'Dwyer, 2009). The concepts of food deserts (low availability of affordable fresh food in high population densities), food swamps (plethora of excess of non-nutritious food) and food tundras (lack of affordable fresh food but also excess of non-nutritious food) were first described in cities of the global North, but these concepts are now a growing issue for the South.

However, there are emerging social entrepreneurship interventions acknowledged as innovative ways of reaching underserved communities through cross-sectional alternative distribution schemes. An inclusive process in place is the bicycle-based egg sellers of Dar es Salaam (Wegeriff, 2014). These initiatives require support and appropriate infrastructure development. For innovation in food businesses, appropriate capacity building and improved coordination for logistics is needed (Fonseca and Vergara, 2015). Successful interventions reveal the significance of government support to enable an adequate (institutional) environment for improved food systems (e.g. research, extension). Unfortunately, in most developing countries where advisory services exist, pre-harvest production is the only emphasis (Affognon *et al.*, 2015). Fostering knowledge for post-production aspects of nutritious food supply chain, including cross cutting issues such as gender and environmental sustainability, is not yet emphasized.

Foreseen – *Needed* – Areas of Interventions in the South for Global Sustainability

The intrinsic situation of countries in the global South, including the indicated socioeconomic inequality and the current unsustainable consumption trends, suggest global sustainability requires solid steps in the South toward: (i) stable plurality of actors across the food supply chain and related sectors; (ii) a holistic approach that truly constructs a resilient social and environment world; and (iii) a viable economic environment. Selected approaches (Fig. 8.2) that could be key to keep the diversity (bio, stakeholders, diet) needed and the actions to keep social and environment involved in a just and regenerative way are presented in this section. The first four examples are commonly systemic by nature, integrating supply with consumption, with the aim to internalize the social and environmental costs. The last example described has entry points that are not largely systemic and often miss the connection between production–consumers or the full sustainability dimension. However, these practices have constituted solid first steps to create social and political momentum that has prompted more systemic approaches.

Solidarity schemes

Social and solidarity economy has been ignored by many on the basis that it makes little sense if the economic returns are lower when re-investing in the sustainability of the system. For city-linked food systems, this makes a lot of sense as people need income to improve wellbeing both in rural areas and in the city. As an example, the government of Peru brought Peruvian cuisine to the status of National Heritage after times of political instability in 2007. This prompted chefs to establish the Peruvian Society of Gastronomy, also known as APEGA, a year after to showcase authentic Peruvian dishes. This evolved into connecting actors across the food system and started a movement of solidarity between chefs and small producers, and disseminating 'know

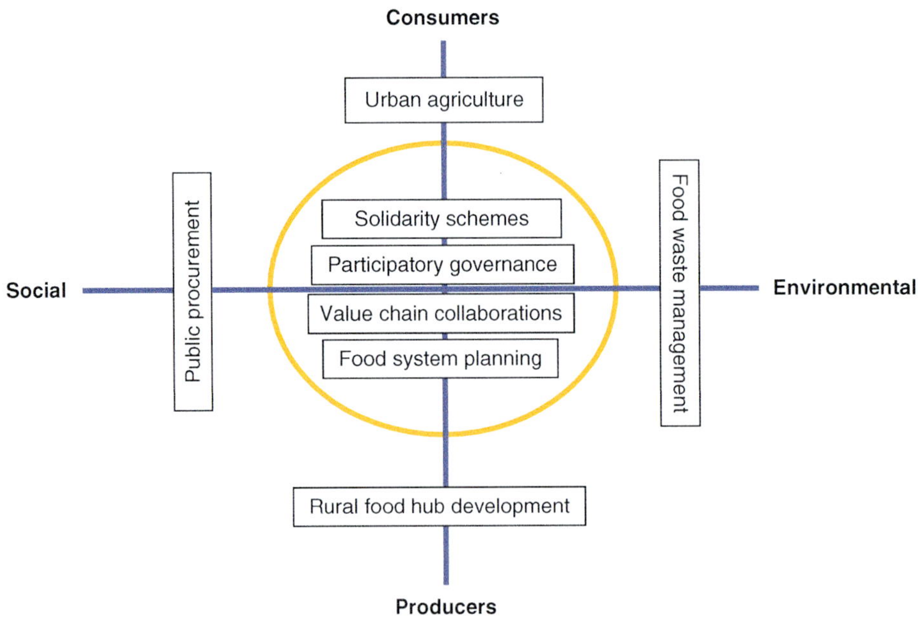

Fig. 8.2. Possibilities for local government in the global South to boost efforts towards sustainable food systems. Areas of intervention outside the core circle are options initially perceived as not as holistic actions compared to those within the circle, but are often triggers of systemic solutions in the long term (e.g. urban agriculture promotion has resulted in cities re-connecting with food and later stimulating a participatory process on food system planning).

how' to school cafeterias on food preparation of indigenous food.

Inclusive value chain collaborations

Value chain collaboration (VCC) focuses on the integration of smallholders and building on market niches, often connecting consumers to the suppliers. While some VCCs have shown decline of biological and dietary diversity with the social insertion and higher productivity, a good selection of collaborators can alleviate that and possibly stimulate a sustainable food system approach (Ros-tonen *et al.*, 2015).

In many parts of the global South, modern supermarkets are seen as the alternative to traditional food systems, and what are considered 'alternative food systems' in the global North have not been widely successful so far, as the example of the box scheme in Nairobi illustrated. What may be appropriate is greater transparency and efficiency along traditional supply chains. In Zimbabwe, the eMKambo platform run by Knowledge Transfer Africa serves to provide market information and finance to agriculture value chain actors to enable economically sustainable agricultural and market sectors (www.emkambo.co.zw). This kind of intervention, along with fostering an environment for smallholders to associate (e.g. cooperatives) may support pathways to sustainable food systems.

Food system planning

Food system planning was only introduced early in this millennium as a 'new subject' for urban planners. Food planning is seen as a holistic analysis and implementation-based strategy for addressing both policy and investment needs with the goal of improving sustainability of the food system linked to cities and surrounding territories. The poor food utilization plus the reduction of access to food as result of climate change aggravate the situation of the most vulnerable in urban areas (Tacoli *et al.*, 2013). The evidence from food system responses to extreme weather events, which are becoming increasingly prevalent as a result of climate change, indicates significant food system vulnerability and a need to plan from field to table (Fonseca, 2009). Food system planning has been included within the NUA and speaks of the need to plan across a range of scales.

Participatory processes

Multi-stakeholder processes are increasingly considered to be an important element of policy design, action planning and implementation, by directly addressing systemic learning in society (Buchanan *et al.*, 2013). Smit (2016) concluded that cities and central governments need to better understand existing urban governance processes and the competing interests of actors to be able to collaboratively design interventions to improve urban food security and overall food systems. The role of different actors involved in governing urban food systems is key, often prompting something like the food policy councils found in North America. The most successful stories to reduce urban health problems involved participatory processes where the poor were engaged to change their own conditions. The challenge seen was to bring these examples at a speed and scale that can cope with the growing rate of slums dwellers.

Selected 'cascade trigger' entry points

While the previous approaches are systemic in nature, specific sectoral urban actions may also serve to trigger holistic approaches in the long term. *Urban agriculture* has emerged as focus of research in sub-Saharan Africa as a potential path towards food security and poverty alleviation strategy. However, it is often hindered by a failure to take a systemic approach (Battersby, 2013). In parts of the Philippines, school and community gardens are implemented as part of a food security and nutrition strategy in an urban area (http://www.searca.org.). It has shown an effect on people's re-connection with food that often results in larger food system programmes through social movements. The connection with food has proven effective with children (FAO, 2010). *Public procurement*, another important entry point, shows potential not only for garden production, but for smallholder economic development and nutrition education, and forming ample food system programmes. Cities struggle with waste, and this is an opportunity for promoting a circular economy, through recycling and composting. Stimulating waste prevention and management creates awareness raising on environmental and people needs, and has also inspired larger more holistic movements. Furthermore, *food safety* can be enhanced by coordination along the supply chain, appropriate infrastructure to reduce transport time and contamination, and good handling and processing practices (Fonseca, 2006). This does not mean formalization, as multiple studies have demonstrated food sold by informal traders is often as safe as that from formal retailers in African cities (von Holy and Makhoane, 2006). *Investment on rural hubs*, while initially being only a target for better logistics and market, should serve to raise awareness among consumers on the safety and capability for tracing product.

Final Remarks

Cities, food systems and sustainability, are explicitly described in SDGs 2 (food security), 11 (resilient cities) and 12 (sustainable production and consumption). For the growing urbanization of the global South, achieving these three goals requires an integrated set of actions, with a concerted focus on the slum areas that house the most vulnerable urban populations. A world that promotes sustainable diets must prioritize actions for the wellbeing of inhabitants in slum areas.

Sustainable food systems of the future in the global South may also rely on how small cities and towns are fostered in the next decades. The so-called 'missing links' in the rural–urban interface may be the key to providing the balanced and harmonized exchange of product and services. Consumers in secondary cities may have a different 'relationship' towards food. Moreover, there is growing evidence that the poor in rural areas find a more efficient way to improved economic situation through the rural non-farm economy and secondary towns, rather than migrating to large cities (Christiaensen *et al.*, 2013). Investment for enhanced value addition interventions in small towns, in addition to the aim of balanced economic development, can open opportunities to highlight 'credence values' or perceived values (e.g. freshness, local) in urban dwellers and promote regional foods, and very importantly the underutilized indigenous species, which are often resilient to climate change and add immensely to the diversification of diets. An initiative in Asia, called *Future Smart Food*, is a good example.

The solidarity of downstream food system actors (consumers, retailers, HORECA) with producers is an important driver that can also retain culture, and one that can impact enormously a food system. Above all, participatory processes seem to be the ultimate way to enable sustainability, first because the voices of the most vulnerable can be heard, and second because the changes toward environmental sustainability appear to have continuity only when beneficiaries are also the ones on command in the planning phase.

Unfortunately, our research found very few successful food system interventions by local authorities in the global South, and particularly only a handful of cases with examples that comprehensively address all pillars of sustainability (e.g. Belo Horizonte) similar to those in cities from the North (e.g. Vancouver, Bristol, Toronto). We found the global South is still lacking broad interventions that are directed ultimately towards a coherent commitment at global, national, subnational and the local levels. This is not a simple task, not even within city geographical boundaries, as often policies impacting food systems feeding cities are established with different institutions inter-crossing in their jurisdiction, which highlights the need for a territorial approach where the different levels of governments engage in dialogue.

References

Affognon, H., Mutungi, C., Sanginga, P. and Boergemeister, C. (2015) Unpacking postharvest losses in sub-Saharan Africa: a meta-analysis. *World Development* 66, 49–68.

Alkemade, R., Reid, R.S., van den Berg, M., de Leeuw, J. and Jeuken, M. (2013) Assessing the impacts of livestock production on biodiversity in rangeland ecosystems. *Proceedings of the National Academy of Sciences USA* 110, 20900–20905.

Alvez, J.G., Falcao, R.W., Pinto, R.A. and Correia, J.B. (2011) Obesity patterns among women in a slum area in Brazil. *Journal of Health, Population and Nutrition* 29, 286–289.

Battersby, J. (2012) Urban food security and climate change: a system of flows. In: Moser, C., Frayne, B. and Ziervogel, G. (eds) *Climate Change, Assets and Food Security in Southern African Cities*. Earthscan, Abingdon, UK, pp. 35–56.

Battersby, J. (2013) Hungry cities: a critical review of urban food security research in sub-Saharan African cities. *Geography Compass* 7, 452–463.

Buchanan, K., Brouwer, H., Klerks, L., Schaap, M., Brouwers, J. and Le Borgne, E. (2013) Editorial. Facilitating multi-stakeholder processes: balancing internal dynamics and institutional politics. *Knowledge Management For Development Journal* 9, 3–10.

Christiaensen, L., De Weerdt, J.C. and Yasuyuki, T. (2013) Urbanization and poverty reduction: the role of rural diversification and secondary towns. *Agricultural Economics* 44, 447–459.

Coveneya, J. and O'Dwyer, L. A. (2009) Effects of mobility and location on food access. *Health and Place* 15 (1), 45–55.

Demmler, K.M., Klasen, S., Nzuma, J.M. and Qaim, M. (2017) Supermarket purchase contributes to nutrition-related non-communicable diseases in urban Kenya. *PLoS One* 12(9), 0185148.

Development Initiatives (2017) *Global Nutrition Report 2017: Nourishing the SDGs*. Development Initiatives, Bristol, UK.

FAO (2010) A New Deal for School Gardens. Available at http://www.fao.org/docrep/013/i1689e/i1689e00.pdf (accessed 16 June 2018).

FAOSTAT (2016) Food and agriculture data. Available at http://faostat3.fao.org/ (accessed 16 June 2018).

Fonseca, J.M. (2006) Postharvest handling and processing: sources of microorganisms and impact of sanitizing procedures. In: Matthews, K.R. (ed.) *Microbiology of Fresh Produce*. ASM Press, Washington, DC, USA, pp. 85–120.

Fonseca, J.M. (2009) Postharvest handling under extreme weather conditions. In: Florkowski, W., Prussia, S., Brueckner, B. and Shewfelt, R. (eds) *Postharvest Handling: A Systems Approach* 2e. Academic Press, Oxford, UK, pp 539–560.

Fonseca, J.M. and Vergara, N. (2015) *Logistics in the Horticulture Supply Chain in Latin America and the Caribbean*. Food and Agriculture Organization of the United Nations, Rome, Italy, 85 pp.

Fraser, B. (2013) Latin American countries crack down on junk food. World Report. *The Lancet* 382, 385–386.

Freidberg, S. and Goldstein, L. (2011) Alternative food in the global south: Reflections on a direct marketing initiative in Kenya. *Journal of Rural Studies* 27(1), 24–34. DOI: 10.1016/j.jrurstud.2010.07.003

Global Panel (2017) *Urban Diets and Nutrition: Trends, Challenges and Opportunities for Policy Action*. Policy Brief No. 9. Global Panel on Agriculture and Food Systems for Nutrition, London, UK.

Greenstein, R., Jacobson, A., Coulson, M. and Morales, A. (2015) Innovations in the pedagogy of food system planning. *Journal of Planning Education and Research* 35, 489–500.

Kimani-Murage, E., Schofield, L., Wekesah, F., Mohamed, S., Mberu, B., *et al.* (2014) Vulnerability to food insecurity in urban slums: Experiences from Nairobi, Kenya. *Journal of Urban Health* 91, 1098–1113.

Magalhães, F. (2016) Slum upgrading and housing in Latin America. Available at https://publications.iadb.org/bitstream/handle/11319/7879/Slum-Upgrading-and-Housing-in-Latin-America.pdf?sequence=1 (accessed 18 June 2018).

Marvin, S. and Medd, W. (2006) Metabolisms of obesity: flows of fat through bodies, cities and sewers. *Environment and Planning A* 38, 313–324.

McMichael, P. and Schneider, M. (2011) Food security politics and the Millennium Development Goals. *Third World Quarterly* 32, 119–139.

McNicoll, G. (1984) Consequences of rapid population growth: An overview and assessment. *Population and Development Review* 10, 177–240.

Reardon, T.A. (2016) *Growing Food for Growing Cities*. The Chicago Council on Global Affairs, Chicago, Illinois, USA, 140 pp.

Ros-tonen, M.A.F., Van Leynseele, Y.-P., Laven, A. and Sunderland, T. (2015) Landscapes of social inclusion: inclusive value-chain collaboration through the lenses of food sovereignty and landscape governance. *The European Journal of Development Research* 27, 523–540.

Ruel, M., Garrett, J. and Yosef, S. (2017) Growing cities, new challenges. In: *Global Food Policy Report*. IFPRI, Washington, DC, USA, pp. 24–33.

Satterthwaite, D., McGranahan, G. and Tacoli, C. (2010) Urbanization and its implications for food and farming. *Philosophical Transactions of the Royal Society B* 365, 2809–2820.

Smit, W. (2016) Urban governance and urban food systems in Africa: examining the linkages. *Cities* 58, 80–86.

Tacoli, C. and Vorley, B. (2016) Food consumption, urbanisation and rural transformations in Southeast Asia. Available at http://pubs.iied.org/pdfs/17335IIED.pdf (accessed 18 June 2018).

Tacoli, C., Bukhari, B. and Fisher, S. (2013) Urban poverty, food security and climate change. Human Settlements Working Papers No. 37. IIED, London, UK, 29 pp.

Tefft, J., Jonasova, M., Adjao, R. and Morgan, A. (2017) Food systems for an urbanizing world. Available at http://documents.worldbank.org/curated/en/454961511210702794/pdf/Food-Systems-for-an-Urbanizing-World.pdf (accessed 18 June 2018).

UN DESA (2014) *World Urbanization Prospects: The 2014 Revision, Highlights* (ST/ESA/SER.A/352). United Nations, Department of Economic and Social Affairs, Population Division, New York, USA.

UN Habitat (2016) *Slum Almanac 2015/2016. Tracking Improvement in the Lives of Slum Dwellers*. United Nations, Habitat, Nairobi, Kenya, 88 pp.

von Holy, A. and Makhoane, F.M. (2006) Improving street food vending in South Africa: achievements and lessons learned. *International Journal of Food Microbiology* 111, 89–92.

Wegerif, M.C. (2014) Exploring sustainable urban food provisioning: the case of eggs in Dar es Salaam. *Sustainability* 6(6), 3747–3779.

9 Consumer-level Food Waste Prevention and Reduction Towards Sustainable Diets

Silvia Gaiani, Rosa Rolle and Camelia Bucatariu

Abstract

Agenda 2030 for Sustainable Development is a global commitment that includes a set of 17 sustainable development goals (SDGs) and 169 targets. Food systems are at the heart of this agenda. SDG 12 seeks to 'ensure sustainable consumption and production patterns'. The third target under this goal, target 12.3, calls for reducing by half per capita global food waste at the retail and consumer levels, and reducing food losses along production and supply chains (including post-harvest losses) by 2030. SDG target 12.3 has the potential to embed prevention and reduction of food loss and waste in public and private sector strategies and to contribute to more sustainable diets and consumption patterns around the world. Food systems today are confronted with, among other issues, increasing non-communicable diseases linked to diets as well as socioeconomic and environmental concerns related to food waste. The macro- and micro-food environment within which consumers find themselves is multidimensional and they – alongside national governments and food supply chain stakeholders – can play a role in preventing and reducing food waste and contributing to sustainable diets. This chapter identifies six major challenges related to food waste prevention and reduction and sustainable food systems. Challenges range from recognition that the global food system is impacted by the attitudes and behaviours of local, national, regional and global food supply chain actors, to the definitions of food waste, measurement methodologies, data collection, and the need for agro-industry productivity and behavioural change thinking. A matrix policy analysis – based on a combination of initiatives at macro, meso and micro-level – is then recommended as a possible approach to successful food waste prevention and reduction.

Agenda 2030: a Global Framework for Action Focused on Food Systems

Agenda 2030 for Sustainable Development (UN, 2015) is a global commitment that consists of a set of 17 sustainable development goals (SDGs) and 169 targets. Launched in 2015, it is considered to be a broad intergovernmental agreement that serves as the Post 2015 Development Agenda.

The main aim of the SDGs is to connect poverty alleviation, human wellbeing and environmental protection in an integrated way while ensuring sustainable human development, decoupling socioeconomic development from environmental impacts, and addressing the evolving food demand.

Food systems are at the core of Agenda 2030: they are both a means and an end towards achieving the SDGs. They address the entire food supply chain from production to trade, distribution, consumption and waste management.

Food systems are currently challenged by macroeconomic trends, such as the increasing global population, growing inequalities, depletion of natural resources, geopolitical dynamics (the world is currently experiencing the highest levels of displacement on record, with 65.3 million people displaced worldwide), undernourishment (815 million people are hungry), micronutrient malnutrition and over-nutrition (nearly 2 billion people are deficient in micronutrients and 2 billion people are overweight or obese) and large quantities of lost or wasted food (an estimated one-third of global food production).

Meeting the multi-faceted challenges faced by food systems necessitates a systems-level approach, as well as transformative changes in how and where food is produced and consumed. The establishment of food systems that are inclusive, sustainable, nutritious and healthy is vital.

The SDGs include key action steps towards achieving sustainable food systems. SDG 12 aims to '...ensure sustainable consumption and production patterns' and its target 12.3 states 'By 2030, halve per capita global food waste at the retail and consumer levels and reduce food losses along production and supply chains, including post-harvest losses'.

FAO defines food loss as the decrease in quantity or quality of food that takes place in the food supply chain. Quantitative food loss refers to the decrease in the mass of the food, while qualitative food loss relates to the decrease of quality attributes of food – appearance, flavour, texture, nutritive value, and so on. Food loss includes pre-harvest loss (loss between maturity and harvest), loss at harvest and post-harvest loss. Food waste refers to discarding or alternative (non-food) use of food that is safe and nutritious for human consumption. It occurs along the entire supply chain, from harvest to consumption, primarily in retail and at the consumer level.

Consumers can play an important role in the fight against food waste toward attaining sustainable food systems, as reducing food waste, consuming in a responsible way and adopting sustainable diets can generate multiple benefits for the economy, food security and the environment.

Consumer-level Food Waste: Determinants, Impacts and Potential Solutions

Consumers are the primary generators of food waste across the food supply chain in higher income economies (Stenmarck *et al.*, 2016), with the retail sector following closely behind. Pudel and Westenhofer (1988) identified four areas that form the backbone of food waste generation:

1. Devaluation – food is seen as something obvious, not something valuable.
2. Lack of knowledge about food identity – consumers no longer know about the cultural background and ingredients of the food they consume.
3. Lack of knowledge about the origin of food – globalization and loss of local food culture.
4. Loss of social and emotional linkage to food – eating together is no longer an everyday family tradition and traditional family recipes are disappearing.

These tendencies have led, in some cases, to an emotional neutralization/detachment and have given consumers the perception that food is simply a product, as any other product on the market, instead of a vital constituent for life.

Examples of strategic approaches to addressing food waste prevention and reduction are already available worldwide. The European Union is promoting initiatives and research on the topic. For instance, the EU project FUSIONS – after gathering nearly 300 factors on EU food waste – concluded that there are three main consumer-related food waste determinants: (i) social factors, such as household type, family structure and related lifestyles; (ii) individual behaviours, perceptions of and expectations towards foods; and (iii) consumers' lack of awareness, knowledge and skills about how to preserve, store and cook food (FUSIONS, 2013). Moreover, the European Commission's 'Preparatory study on food waste' from 2010 underlines the lack of awareness, attitudes or preferences as consumer-level causes of food waste alongside the societal trends of urbanization and changes in the composition of diets (European Commission, 2010). Finally, Quested *et al.*, (2013) stress that wasting food is not a conscious decision. There is a gap between the activity causing it and the consequence of wasting food. While food and eating are characterized by a complexity of habits and rituals, wasting food is mostly invisible and, thus, much less impacted by social norms or social signalling. People find it extremely difficult to realize how much food they throw away and consequently how they could take action against food waste.

Despite context-based cultural and economic differences, consumer attitudes and behaviours are becoming similar both in developed and in developing countries. In the BRICS[1] countries, for instance, affluence has resulted in consumers buying convenience foods, dedicating less time to cooking and increased out of home consumption (Godfray *et al.*, 2010). In addition, research conducted in the Philippines (Esguerra *et al.*, 2017) has shown that the main reasons for wasting fruits and vegetables at household level are forgetting to cook the produce purchased, not planning meals properly and overbuying (often poor quality food). Interviews conducted with households in Mamelodi Township in South Africa, by researchers of the University of South Africa (Ramukhwatho, 2014), have identified the main causes for food waste as preparing too much food, the close expiry date and promotional marketing.

One main issue to underline when assessing the (global) quantities of consumer food waste is that, due to the lack of a homogeneous methodology, it is only possible to present estimations; direct comparisons cannot yet be made. Generally, data originate mainly from specific food supply chains or national studies carried out by governmental organizations or research institutions that apply a range of different methodologies. In some cases, estimates are derived from: (i) diaries completed by household members who monitor and document the weight of their wasted food over a period of several days; (ii) self-reports estimations in questionnaires and interviews; and/or (iii) calculations of the caloric content of food that is physically sorted from the households' garbage bins. Furthermore, some studies have measured household food waste as a percentage of the total weight of consumed food or as a percentage of each of the consumed food items. Some of the reports make use of very small sample sizes, whereas others are performed at a more aggregate level than households (regional or national).

Although food waste may differ in terms of causes and quantities from region to region and from nation to nation, being dependent on geography, infrastructure, cultural traditions, eating habits, and so on, the impacts of consumer-level food waste on the environment, climate and economy have worldwide and interconnected consequences.

In terms of environmental impact, FAO (2013) estimated that the highest carbon footprint of food waste occurs at the consumption phase (37% of total). This is because every kilogram of food that is wasted at the end of the supply chain has a higher carbon intensity than at earlier stages, owing to harvesting, transportation, processing and distribution that accumulate additional greenhouse gases along the supply chain. In particular, meat and cold cuts, milk and dairy products, vegetables and rice contribute the most to the environmental burden from being wasted. Additionally, the disposal of food and drink waste in landfills adds to the release of greenhouse gases such as methane (Graham-Rowe *et al.* 2014) – which is 25 times more potent than carbon dioxide emissions (EPA, 2015) – along with others that contribute to climate change (Goebel *et al.*, 2015).

In terms of economic impact, food waste has a direct and negative impact on consumer incomes. A study conducted by LEI (2013) shows that reducing food waste by EU households could lead to annual household savings of €92 per capita. In addition to that, a number of reports underline the existence of complex relationships between per capita income and household behaviour. According to a study reported by Setti *et al.* (2016) based on an Italian sample of respondents, such a relationship is explained by an inverse U-shaped curve: mid-to-low income consumers tend to buy larger quantities of lower quality products and as a consequence waste more food.

In order to reduce food waste generation and its related environmental–economic impacts and promote sustainable food systems, FAO recommends the implementation a food waste prevention and reduction strategy based on the prioritization of actions as presented in Fig. 9.1. The food-use-not-loss-or-waste hierarchy, as introduced in the 'CFS Policy recommendations on Food Losses and Waste in the context of Sustainable Food Systems' (CFS, 2014) focuses on availability and accessibility of safe and nutritious food for direct human consumption, followed by food loss and waste prevention at source, recovery and redistribution of safe and nutritious food for direct human consumption, animal feed, compost and/or incineration with energy recovery (based on context variables) and ultimately disposal in landfills.

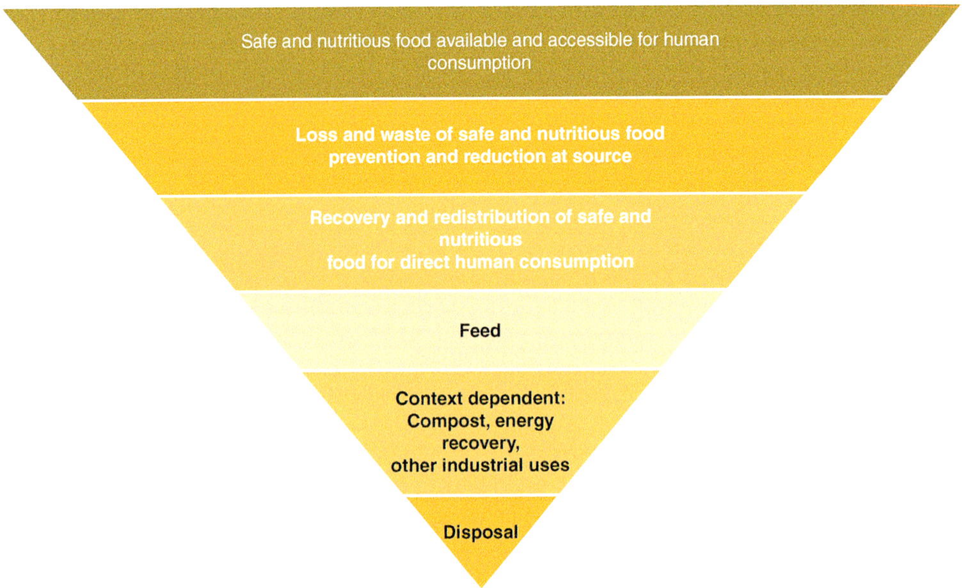

Fig. 9.1. Food-use-not-loss-or-waste hierarchy. Adapted by Bucatariu Camelia from CFS (2014).

In combination with the hierarchy principles, we also suggest that a matrix combination of macro/public sector-led approaches (based on policies, legislation, market-based instruments, provision of infrastructure), meso approaches (centred around cooperation among food chain stakeholders) and micro/consumer-led–bottom-up approaches (where changes in individual attitudes, shopping and consumption habits are pivotal) are most likely to bring about effective, long-lasting results in terms of (consumer-level) food waste reduction and prevention.

With regard to *macro approaches*, international organizations and governments are working in several areas to bring about changes. Among the main actions that need to be scaled up are: (i) the setting up of data monitoring systems and reduction targets (e.g. The Scottish Government has pledged to reduce the nation's food waste by one-third by 2025 [Zero Waste Scotland, 2016]); (ii) the enhancement of clarity concerning date marking on packaged foodstuffs (e.g. a project under the Nordic Green Growth program run by Østfoldforskning has developed common Nordic guidelines for date labelling); (iii) the provision of adequate tax measures, incentives and subsidy schemes (e.g. the South Korean government has introduced a pay-as-you-go system according to which residents and businesses are charged for the exact amount of food they throw away [Asia Today, 2013]); (iv) the promotion of awareness campaigns, education and food literacy (e.g. there are currently hundreds of food waste awareness raising campaigns running all over the world, for example South Africa to China); and (v) the support of consumption shifts towards less resource-intensive and environmentally impactful foods (e.g. the Indonesian Consumers Foundation (YLKI) has been promoting a sustainable consumption lifestyle which includes sustainable diets).

Medium- to long-term support for consumers is preferred over short-term actions.

With regard to *meso level approaches*, food producers, processors, retailers and the food service sector impact consumer choices through, for example, food packaging and labelling and the delivery of information on expiry dates, as well as on storage, freezing and packaging options.

Retailers could incentivize an optimum purchase of safe and nutritious food by adequately lowering prices in time for most effective sale and by promoting the selling of imperfect fruits and vegetables that would otherwise be thrown away. Retailers may also make use of available food items in-store by freezing or cooking them or by delivering them to secondary retail. The use of technologies designed to increase shelf-life, ensure safety and nutritional properties and warn of expiration dates is also beneficial to consumers. The food service sector could also reduce its waste – partially due to the unpredictability of consumer demand – by supporting the recovery and redistribution of safe and nutritious food for direct human consumption.

With regard to *micro-level approaches*, the understanding of date labelling on packaged food products (mainly the distinction between 'best before' and 'use by', the first indicating the date until when the food retains its expected quality and food can be still consumed past this date, the latter referring to the precise day until which food is safe to be consumed), coupled with capacity development and the provision of information on safe food handling at the household level, can represent effective options to prevent and reduce consumer-level food waste. Awareness raising and education on food planning, purchasing, handling and management can also have positive impacts on consumer attitudes and behaviours.

The synergy among these actions is of paramount importance. Accordingly, the use of a matrix that combines macro-, meso- and micro-level approaches at the local, national and international levels – in parallel with context-based strategies for food recovery and redistribution – could provide a framework toward achieving SDG 12.3.

Consumer-level Food Waste Prevention and Reduction and Sustainable Diets

A number of scientists, including researchers from LEI, the Agricultural Economics Institute of Wageningen University and Research Centre (Rutten *et al.*, 2013), suggest that the adoption of sustainable diets, alongside other valuable options such as the food packages of reduced size or reducing portion size in the food service sector, could represent an efficient step toward reducing consumer-level food waste.

Sustainable diets, as defined by Burlingame & Dernini (2012), are those diets with '...low environmental impacts which contribute to food and nutrition security and to a healthy life for present and future generations. Sustainable diets are protective and respectful of biodiversity and ecosystems, culturally acceptable, accessible, economically fair and affordable; nutritionally adequate, safe and healthy; while optimizing natural and human resources.' Sustainable diets are required to meet multiple environmental, social and developmental criteria – from health to climate change, from land use to nutrient flows, from water availability to affordability, from food security to people's security. Food waste, undernutrition, overweight and their associated non-communicable diseases linked to nutrition and diets often coexist not only in the same country but also within the same community, therefore improving food access and literacy and preventing food waste are essential dimensions of a sustainable and healthy diet.

Achieving food security and nutrition requires that food consumption and production align toward a sustainable global food system. Operational linkages with the SDGs should be ensured from local to national and from regional to global levels. A global sustainable food system, based on interconnectedness and interactions between the four food security and nutrition dimensions (availability, access, utilization and stability) can safeguard the active and adequate participation of consumers.

Challenges Ahead

It follows from the previous sections that food waste prevention and reduction in the framework of sustainable diets is confronted by a number of challenges. These range from challenges that are specific to the waste problem – including definitions, measurement methodologies, data collection, policies and individual behavioural change – to macro-level challenges that are linked to food governance, food value chains, the commodification of food and agro-industry efficiency.

Here we identify and outline six key challenges and provide a solution-oriented guidance.

Challenge related to food and food waste terminology

A key consideration is that the value of food to humans is represented by its nutrients that support life. Accordingly, it becomes relevant to highlight that when addressing food waste prevention and reduction we are also effectively preventing and reducing the waste of micronutrients and macronutrients that support human life and development.

Addressing the definition of food leads to the need to work on defining food waste as well. There is no global harmonized definition of food waste. Definitions vary in accordance with the actor of the food supply chain monitoring the waste, and the composition of food waste itself, how it is generated and managed. Culture also plays a role: what is considered waste in some countries may not be considered waste in others (e.g. offal) and this can change over time.

FAO (2017b) refers to food waste as the discarding or alternative (non-food) use of food that is safe and nutritious for human consumption. To date, the European Commission has not yet provided a definition of food waste – the position being that there is respectively a definition of 'food' in the Regulation No 178/2002 of the European Parliament and of the Council (the General Food Law) (EU, 2002) and one of 'waste' in the Waste Framework Directive (European Commission, 2016). Waste and Resources Action Programme (WRAP), in the UK, differentiates among wasted food (edible parts), and the associated inedible parts; Smil (2004) considers food waste the difference between the quantity of food that each person consumes and what s/he really needs (energetic value); and Tavill (2015) defines wasted food as the antithesis of the triple bottom line, thus referring to three pillars of sustainability – people, planet and profit.

Challenge related to food waste methodologies and data collection

Although waste is an issue of global significance, basic information is lacking on the types and quantities of food wasted. Currently available statistics provide some information but are still based on very limited data. The uncertainty relates to the debated definition of food waste, the lack of homogeneous data collection methodologies, the cost and the level of difficulty of conducting primary fieldwork. Collection of food waste data requires in-country expertise, commitment over many years, effective measurement and monitoring and funding.

One possible way to obtain quantitative information on consumer-level food waste is to launch questionnaires and interviews to seek to ascertain shopping and consumption habits. For example, a panel launched by the Ministry of Agriculture and Fisheries, Food and Environment (MAPAMA) in Spain in 2014 – originally mainly intended to collect data on consumers' weekly purchasing and consumption behaviours – has gradually expanded and has allowed the collection of relevant information on food waste for 17,506,877 Spanish households in 2016.

Challenge related to vertical and horizontal coherence in policies and food value chains

Vertical and horizontal coherence is required not only at the normative level for the food system, but also at an operational level for the entire food value chain. The current intra- and inter-disconnect across policies and stages of the supply chain generate incoherence that ultimately impact also consumer behaviour, diets, food waste prevention and reduction.

With reference to vertical and horizontal coherence in food waste prevention and reduction policy approaches, consumer-level food waste prevention and reduction require a number of different policy measures – ranging from health to infrastructure, from logistics to climate change – that go beyond the final food items delivered to the consumer. That is why the successful examples of food waste prevention and reduction, independent of their geographical implementation, always involve the public sector working along the private sector and civil society.

Interventions, such as legislation, market-based instruments, awareness campaigns, (voluntary) agreements and education can make a significant contribution when embedded in a broader and integrated food system policy with an eye on the national specific circumstances. The suggestion as highlighted earlier, is a matrix combination of macro, meso and micro approaches that reflects also the 2014

Committee on World Food Security 'food-use-not-waste' hierarchy.

Changes related to cultural habits and the promotion of a culture of active waste avoidance should be also incentivized. Consumers' capacity can be strengthened through education that engages youth along with their families and their educational institutions. An example can be drawn from the 2017 FAO education material for school children on the issue of food waste reduction (FAO, 2017a).

With reference to vertical and horizontal coherence in global food value chains, food systems around the world are impacted by multiple pressures – from burgeoning obesity to environmental degradation and farmers' livelihoods – clearly showing competing interests in different sectors: agriculture, food safety and public health, trade, environmental protection, climate and energy, economic and social cohesion, rural development and international development, employment and education. As a consequence, food value chains may sometimes suffer from fragmentation of the food system. Food value chains were defined by FAO (2014) as:

> ...the full range of farms and firms and their successive coordinated value-adding activities that produce particular raw agricultural materials and transform them into particular food products that are sold to final consumers and disposed of after use, in a manner that is profitable throughout, has broad-based benefits for society, and does not permanently deplete natural resources.

Horizontal linkages in food value chains connect operators in the same market or industry, vertical linkages connect suppliers, buyers and consumers across the entire chain in a coordinated and sustainable way and are strictly bound to food governance. When there is no coordination between horizontal and vertical linkages, value is lost. Value is also absent if the supply chain fails to integrate nutrition needs, sustainability, coordination among stakeholders and consumers' concerns into its operations. Consumer-level food waste is an additional example of value lost at the final stage of the supply chain. Across countries, there are important gains to be realized from multilateral action to provide global public goods such as a sustainable global food system.

Challenge related to the disconnect between food governance and the global food system

The notion of food governance refers to:

> the ability to design public sustainable policies (and mobilize social resources in support of them) and outline institutional, technological and financing options that should be exercised at the global, regional, national and local levels and implemented by the different actors/stakeholders involved in the process.
>
> (Rogers, 2002)

The current global food governance, based on the agro-industrial paradigm associated with the production of standardized food commodities and a focus on the production process, lacks both an effective integration of health and environmental issues in its arrangements and a holistic approach that takes into consideration food waste prevention and reduction in combination with the promotion of sustainable healthy diets. The disconnect among food governance, the way the food supply chain works, and consumer needs/attitudes, is a key issue that warrants attention. Actors in food governance should focus on implementing policies and practices that promote sustainable food consumption, improve food policy coherence, support consumer choice and access to safe and nutritious food all year long and promote value-shifting messages about health, pleasure, convenience, social interaction and taste, in order to support food waste prevention and reduction strategies.

The FAO's Global Initiative on Food Loss and Waste Reduction launched in 2011 supports an interconnected food system governance for sustainability and resilience (FAO, 2011).

Challenge related to the commodification of food

Food, over time, has evolved from a local resource held in common within a community, into a private, transnational commodity. This process of commodification, which may encompass more food miles, reduced food options to those unable to cope with transport hurdles, marketing-induced food attractiveness and all-the-year presence of seasonally produced foods, has led to the development of certain traits of food in order to fit mechanized processes put in practice for marketing purposes and not for human nutrition objectives. Safe and nutritious food should be re-conceived as a common good that supports human life in the transition toward a more sustainable global food system that is providing for all – from food producers to consumers.

Challenge related to the importance of agro-industry efficiency and access to available technologies

The agro-food industry is facing a number of challenges to promote and increase sustainable innovative growth. This requires a re-evaluation of current practices, cooperation among enterprises along the vertical supply chain as well as guidance from governments on management activities and environmental awareness.

Optimal packaging strategies, specifically, can play a role in consumer attitudes toward food, their reasons for wasting it and the psychological aspects that encourage waste prevention behaviour. Packaging can contribute to reducing the environmental and economic burdens connected to consumer-level food waste: maintenance of food quality, extension of shelf-life, enhanced food safety, delivery of product information, recognition of brand identity, convenience of pre-prepared food or portion sizes, and so on. At the same time, consumers need to have an adequate capacity for food packaging information interpretation and handling to preserve the nutrient properties as well as food safety.

In addition to that, apps, smart fridges (that allow the remote observation of food about to decay), in-store and online shopping lists, intelligent indicators of freshness or ripeness, just to name a few, all represent useful technologies. These technologies require a suitable regulatory framework before market introduction, one that considers the *ex-ante* impact analysis on all that can be affected and ensures adequate safeguard mechanisms, ownership and accountability.

Conclusion

Prevention and reduction of food waste is one of the concrete ways to improve the sustainability of the global food system. As such, reducing consumer-level food waste goes much further than just optimizing the functioning of the food system: it can be part of broader systemic change towards sustainable diets and global food security and nutrition.

A number of solutions can be implemented to fully realize the potential of food waste reduction. Based on context, this could require market-led investments in infrastructure, technological skills and knowledge, storage, transport and distribution or retailers, food services and the actions of consumer. In this chapter we have recommended a matrix approach consisting of macro, meso and micro actions to prevent and reduce food waste – going from the identification of causes and the selection of potential solutions adapted to cultural and product specificities, to the involvement of all concerned actors – that should evaluate the cost–benefit and the return on investment in the short, medium and long term.

The six identified challenges, and the suggestions for addressing them, range from the definitions of food and food waste, to vertical and horizontal systemic coherence, and agro-industry efficiency. Cross-cutting issues are represented by measurement and reporting for SDG 12.3 for which the FAO has been mandated as custodian UN agency, together with UN Environment, and the global objectives of the UN Decade of Action on Nutrition.

SDG 12.3 is the first universal goal in history set specifically to prevent and reduce post-harvest loss, food loss and food waste. This opportunity must not go without an adequate global to regional and national to local response.

Note

[1] Five major emerging national economies: Brazil, Russia, India, China and South Africa.

References

Asia Today (2013) South Korea's food waste solution: you waste, you pay. Available at: http://www.asiatoday.com/pressrelease/south-koreas-food-waste-solution-you-waste-you-pay (accessed 8 January 2018).

Burlingame, B. and Dernini, S. (Eds) (2012) Sustainable Diets and Biodiversity: Directions and solutions for policy, research and action. *Proceedings of the International Scientific Symposium Biodiversity and Sustainable Diets United against Hunger*, 3–5 November 2010, FAO Headquarters, Rome. http://www.fao.org/docrep/016/i3004e/i3004e.pdf (accessed 28 June 2018).

CFS (2014) Food losses and waste in the context of sustainable food systems. Available at http://www.fao.org/3/a-av037e.pdf (accessed 26 June 2018).

EPA (2015) Overview of greenhouse gases. Available at https://www.epa.gov/ghgemissions/overview-greenhouse-gases (accessed 12 December 2017).

Esguerra, E., del Carmen, D.R. and Rolle, R. (2017) Purchasing patterns and consumer level waste of fruits and vegetables in urban and peri-urban centers in the Philippines. *Food and Nutrition Sciences* 8(10), 961–977. DOI: 10.4236/fns.2017.810069

EU (2002) EC Regulation No 178/2002. Available at http://eur-lex.europa.eu/LexUriServ/LexUriServ.do?uri=OJ:L:2002:031:0001:0024:en:PDF (accessed 8 January 2018).

European Commission (2010) Preparatory study on food waste. Available at http://ec.europa.eu/environment/eussd/pdf/bio_foodwaste_report.pdf (accessed 12 December 2017).

European Commission (2016) EC Waste Framework Directive 2008/98/. Available at http://ec.europa.eu/environment/waste/framework/ (accessed 8 January 2018).

FAO (2011) Global food loss and food waste. Available at http://www.fao.org/docrep/014/mb060e/mb060e00.pdf (accessed 26 June 2018).

FAO (2013) Food wastage footprint: impact on natural resources. Available at http://www.fao.org/docrep/018/i3347e/i3347e.pdf (accessed 26 June 2018).

FAO (2014) Developing sustainable food value chains – Guiding principles. Available at http://www.fao.org/3/a-i3953e.pdf (accessed 26 June 2018).

FAO (2017a) The state of food security and nutrition in the world 2017. Available at http://www.fao.org/state-of-food-security-nutrition/en/ (accessed 26 June 2018).

FAO (2017b) Do good: save food! Available at http://www.fao.org/save-food/projects/educationalmaterial-fwr/en/ (accessed 26 June 2018).

FUSIONS (2013) Estimates of European food waste levels. Available at https://www.eu-fusions.org/phocadownload/Publications/Estimates%20of%20European%20food%20waste%20levels.pdf (accessed 12 December 2017).

Godfray, H.C.J., Godfray, J., Crute, I., Haddad, L. and Lawrence D. (2010) The future of the global food system. *Philosophical Transactions of the Royal Society B* 365, 1554.

Goebel, C., Langen, N., Blumenthal, A., Teitscheid, P. and Ritter G. (2015) Cutting food waste through cooperation along the food supply chain. *Sustainability* 7, 1429–1445.

Graham-Rowe, E., Jessop, D.C. and Sparks, P. (2014) Identifying motivations and barriers to minimize household food waste. Available at http://sro.sussex.ac.uk/47861/1/1-s2.0-S0921344913002711-main.pdf (accessed 26 June 2018).

Quested, T.E., Marsh, E., Stunell, D. and Parry, A.D. (2013) Spaghetti soup: the complex world of food waste behaviours. *Resources, Conservation and Recycling* 79, 43–51.

Rutten, M., Nowicki, P., Bogaardt, M.-J. and Aramyan, L. (2013) Reducing food waste by households and in retail in the EU; a prioritisation using economic, land use and food security impacts, LEI report 2013-035. Available at http://edepot.wur.nl/290135 (accessed 9 January 2018).

Pudel, V. and Westenhöfer, J. (1998) *Ernährungspsychologie: Eine Einführung*, 2e. *Verlag für Psychologie*, Göttingen, Germany.

Ramukhwatho, F.R. (2014) Household food wastage in a developing country: a case study of Mamelodi township in South Africa. Available at https://researchspace.csir.co.za/dspace/handle/10204/7757 (accessed 12 December 2017).

Setti, M., Falasconi, L., Segreè, A., Cusano, I. and Vittuari, M. (2016) Italian consumers' income and food waste behavior. *British Food Journal* 118(7), 1731–1746. DOI: 10.1108/BFJ-11-2015-0427

Smil, V. (2004) Improving efficiency and reducing waste in our food system. *Journal of Integrative Environmental Sciences* 1, 17–26.

Stenmarck, A., Jensen, C., Quested, T. and Moates, G. (2016) Estimates of European food waste levels. Report of the project FUSIONS (contract number: 311972) granted by the European Commission (FP7). Available at https://www.eu-fusions.org/phocadownload/Publications/Estimates%20of%20European%20food%20waste%20levels.pdf (accessed 29 June 2018).

Tavil, G. (2015) Wasted food: the antithesis of the triple bottom line. Available at https://www.epa.gov/sites/production/files/2015-09/documents/tavill-wadsworth_011514.pdf (accessed 12 December 2017).

UN (2015) Agenda 2030 for sustainable development. Available at http://www.un.org/sustainabledevelopment/development-agenda/ (accessed 12 December 2017).

WRAP (2009) Household food and drink waste in the UK. Report prepared by WRAP. Available at http://www.wrap.org.uk/sites/files/wrap/Household_food_and_drink_waste_in_the_UK_-_report.pdf (accessed 26 June 2018).

Zero Waste Scotland (2016) Zero Waste Scotland welcomes bold new targets on food waste and circular economy. Available at http://www.zerowastescotland.org.uk/content/zero-waste-scotland-welcomes-bold-new-targets-food-waste-and-circular-economy (accessed 8 January 2018).

10 Attaining a Healthy and Sustainable Diet

Jessica Fanzo and Haley Swartz

Abstract

The world continues to struggle with the multiple burdens of malnutrition that affect billions of individuals and the countries in which they live. One major contributor to nutrition outcomes is the consumption of diverse, safe and high-quality diets. However, diets are not static – they are changing, and rapidly so, with income growth, migration and urbanization. Unhealthy diets (those high in salt, unhealthy fats, sugar, processed red meats, and highly processed packaged foods and sugar-sweetened beverages) are considered to be one of the major risk factors for the global burden of disease, of which more people are dying of diet-related non-communicable diseases everywhere including low- and middle-income countries. Food systems and food environments serve to provide the foods that make up the diets that people eat; however, both barriers and opportunities exist across those systems and environments to accessing healthy diets. Physical proximity, affordability, marketing and acceptability all play roles in the decision-making process of consumers when purchasing and consuming food. The foods that are consumed not only impact health, but also the environment. While food choices affect the environment, the environment also impacts food choices making the consumption of sustainable diets – those diets with low environmental impacts which contribute to food security and nutrition and to healthy life for present and future generations – all the more challenging. But there are solutions by way of individual, community and institutional levels that can move us towards healthy, sustainable diets for ourselves and for the planet.

Introduction

Presently, the world is grappling with significant burdens of undernutrition, overweight and obesity. Some countries are even struggling with multiple malnutrition burdens simultaneously (IFPRI, 2016). Dietary transitions represent both an opportunity and a threat to the health and wellbeing of populations worldwide. Diets low in quality but high in energy contribute to the escalating problems of obesity and diet-related non-communicable diseases (NCDs). Recent trends show an alarming increase in these problems in countries of all income levels, highlighting the inadequacy of the global food supply, its reliance on the environment, and individual dietary and lifestyle patterns.

In aggregate, the current global agricultural system produces enough food, but access to sufficient food that is culturally acceptable, affordable, safe and nutritious remains a substantial challenge for millions around the world. Production projections in the next several decades further emphasize the need to improve diet quality and environmental sustainability. This is true especially in the context of climate change and increasing population growth with a rising appetite for environmentally costly animal-source foods.

Eating is strongly influenced by behaviours determined by culture, media and information. Consumption patterns and their consequences are a function of both income and lifestyle. Changes to the food environment, including where food is purchased and eaten, attitudes towards specific foods, taste and effect of marketing, have ripple effects throughout the food system.

This chapter outlines the key components of diets that are both healthy and sustainable. First, the chapter explains the key elements of a healthy diet and the current trends in diets around the world. Second, it describes how food environments can influence choice of dietary choices. The chapter concludes by examining the influence of diets on both the environment and future trends.

Healthy Diets

A balanced and healthy diet will vary based on specific individual needs (i.e. age, gender, lifestyle, degree of physical activity) and cultural contexts (i.e. dietary customs, locally available foods) (WHO, 2015). However, the Food and Agriculture Organization (FAO) identified the primary characteristics of food security and a healthy diet, both of which include quantity, diversity, quality and safety (FAO, 1996) (Table 10.1). A healthy diet is perceived as one that effectively prevents the onset of diet-related NCDs, such as cardiovascular disease (CVD) and diabetes, and malnutrition in all its forms (i.e. underweight, overweight/obesity and micronutrient deficiencies).

Food quality and quantity

Food quality and quantity are essential to health, nutrition and food security (Chinnakali *et al.*, 2014). A sufficient *quantity of food* refers to the ability to consume enough calories to support life, allow physical activity and maintain a healthy body weight (FAO, 2013). Attaining enough food is generally a challenge only in low- and middle-income communities (LMICs) throughout the world, where food insecurity, food deserts and undernutrition is widespread. While undernutrition occurs following below-average caloric intake, moderate food insecurity is characterized by both compromised food quality and reduced food quantity (Vuong *et al.*, 2015). In high-income areas where food is abundant, the challenge for many is refraining from consuming too many calories and higher quality foods to reduce the likelihood of overweight, obesity and diet-related diseases (Lallukka *et al.*, 2007; Mozaffarian, 2016). However, a trend increasingly apparent in low-income communities situated within high-income countries is food insecurity and micronutrient deficiencies: individuals may consume enough calories (high quantity), but not enough essential nutrients (low quality).

Food quality consists of sufficient amounts of both micronutrients and macronutrients that allow for normal growth and development from childhood to adulthood. Diet quality can be assessed through dietary guidelines (i.e. adherence to recommendations), food group variety, or nutrient consumption (Kant, 2004). While there is still debate regarding which foods should be classified as 'healthy' and 'unhealthy', many argue that diets should be

Table 10.1. The four characteristics of a healthy diet.

Characteristic	Components
Quantity	Diets that supply sufficient amounts of calories to meet individual nutrition and health needs
Quality	Diets that offer essential micronutrients and macronutrients to support adequate nutritional development, growth and health through the lifespan
Safety	Diets that contain foods and beverages that are safe to consume, free of additives or toxins harmful to human health
Diversity	Diets that contain a variety of nutrient-dense foods from the six basic food groups (i.e. vegetables and fruits, whole grains and cereals, dairy foods, animal- and plant-based protein foods, fish, and sweets and confectionary items)

evaluated holistically, through overall dietary patterns (Mozaffarian and Ludwig, 2010)[1], or the nutrient density of foods (Drewnowski and Fulgoni, 2014). Diet quality also takes into account 'anti-nutrients' that interfere with nutrient absorption (Fabbri and Crosby, 2016). For example, phytates and oxalates are two particularly harmful anti-nutrients that inhibit iron and zinc absorption (de Pee and Bloem, 2009).

Food safety

Food safety describes the impact and potential hazards that food may cause for human health (HLPE, 2017). Safe foods are those that are free from pathogens, chemicals or contamination that can lead to foodborne illnesses such as *E. coli* or *Salmonella*. In turn, food safety regulations and controls ensure a food product's nutrient density remains intact upon consumption (Di Renzo *et al.*, 2015).

Dietary diversity

Dietary diversity is the consumption of varied combinations of food groups[2] that allows an individual to consume a sufficient quantity and quality of nutrients (FAO, 2011). An important contributor to dietary diversity is *agricultural biodiversity*, or the variety of foods and agricultural species and varieties produced worldwide. Reducing losses in agricultural biodiversity is closely associated with improvements in nutritional status, particularly in low-income settings (Jones, 2017).

To guide food choices, policymakers and researchers have developed four principles of a 'healthy' diet for adults, as shown in Fig. 10.1.

Current Trends of Diets

Globally, unhealthy diets are now the number one risk factor for preventable deaths and losses in disability-adjusted life-years (Forouzanfar *et al.*,

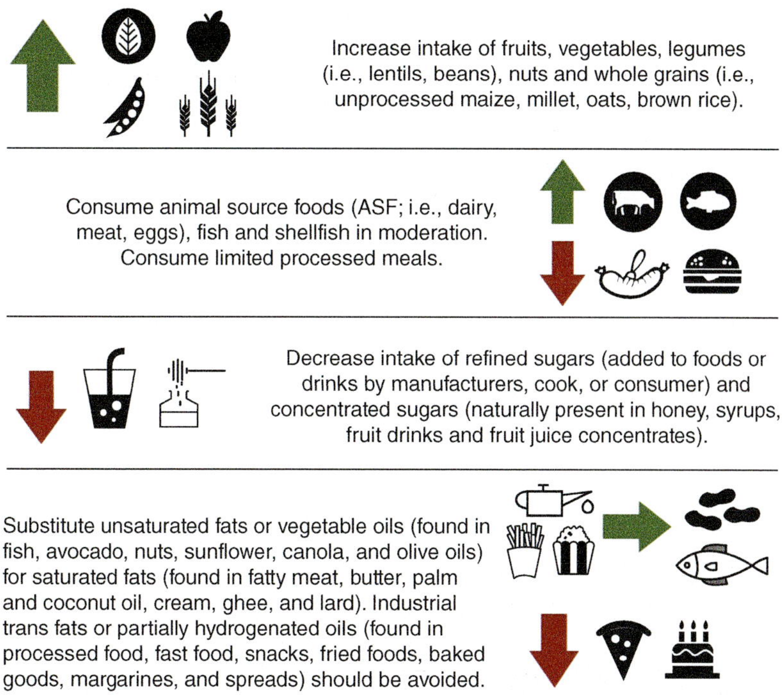

Fig. 10.1. Four recommendations for a healthy diet. Source: created by author, based on Mozaffarian, 2016; Malik *et al.*, 2013; Korat *et al.*, 2014.

2015), surpassing both tobacco smoking and hypertension (HLPE, 2017).

Unhealthy diets are typically contrasted with the four principles of a healthy diet, with nutrient-poor foods replacing nutrient-rich foods (Forouzanfar *et al.*, 2015). Characterized by low consumption of fruits, vegetables, whole grains, nuts and seeds, milk, and seafood, unhealthy diets lack the key micronutrient and macronutrients necessary for the protection against diet-related NCDs. These diets may also contain highly processed foods, are high in trans-fats, sodium and added sugar, as well as red meat, processed meats such as grilled, salted and cured meats, and sugar-sweetened beverages (SSBs) (Monteiro *et al.*, 2015; Baker and Friel, 2014).

Figure 10.2 illustrates changes in food consumption leading to the rise of unhealthy diets and diet-related NCDs; however, the picture is not all negative and there are some positive trends. In most regions, consumption of some nutrient-rich, 'healthy' foods have increased from 1990 to 2013 (Fig. 10.2a). Regional differences include varied consumption of vegetables, whole grains, and seafood. Patterns in the consumption of 'unhealthy' nutrient-poor foods varied worldwide (Fig. 10.2b). While *trans*-fat consumption declined precipitously, consumption of processed meat – linked to both cancer and substantial environmental degradation (Tilman and Clark, 2014; Larsson and Wolk, 2012) – increased in all regions. Red meat consumption declined in all regions except East Asia, where it rose by 40%. SSBs, beverages low in nutrient value – grew in four of the seven studied regions, with the greatest increase in North America.

Access to nutrient-rich animal-source foods (ASFs) remains and will likely continue to be prohibitively expensive or simply unavailable for many people living in low-income countries (Allen, 2012), in which certain populations (i.e. pregnant and lactating women, children,

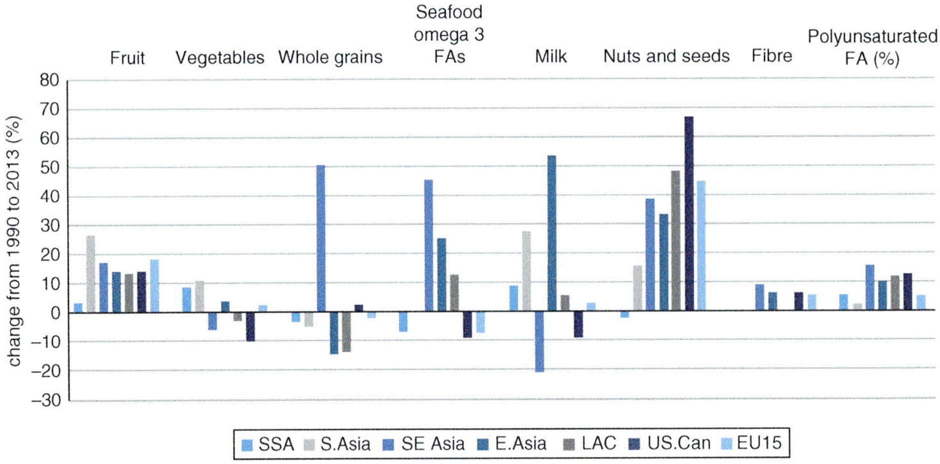

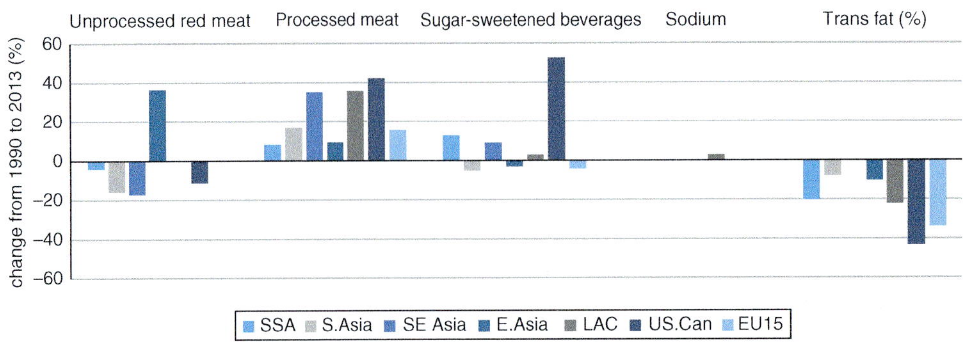

Fig. 10.2. Intake of key foods and dietary components, by region, 1990–2013. (a) Healthy foods. (b) Unhealthy foods. Source: HLPE (2017); GLOPAN (2016).

malnourished people, and the elderly) require key nutrients found in these foods. Diets with low ASF intake can result in iron, zinc, vitamin A, and vitamin B12 deficiencies. While these micronutrients are present in plant-based foods, their bioavailability and nutrient density is, in general, higher or more bioavailable in ASF.

Food Environments: their Influence on Diets

Diets and food choices are bounded by geographic location and cultural context. Diets look substantially different in a low-income area than in high-income settings. One of the key determinants of diet quantity and quality, food safety, dietary diversity and its associated environmental impact are *food environments*, or the physical, economic, political and sociocultural contexts in which consumers engage with the food system to make their decisions about acquiring, preparing and consuming food (HLPE, 2017).

Food environments are the intermediary between the food system and individual food consumption (Swinburn *et al.*, 2014). They are powerful forces, as they have the ability to encourage nutritious diets or enable unhealthy food choices.

For some communities around the world, food environments are primarily the foods they produce or those they purchase from local markets. But for most countries, the food environment is a conglomeration of local, regional and international markets, foods and practices (Hawkes, 2006). Both types of food environments, and the many between the margins of the two, have the capability to be sustainable, supportive of biodiversity and resilient to climate change. But to be healthy and sustainable, food environments must contain four elements: food is available, affordable, truthfully advertised and acceptable to individuals. The following section will discuss these elements in detail, explaining why many food environments around the world are currently unsustainable.

Food availability

Food availability and physical access refer to the adequate supply and distribution of food at the national and/or international levels.[3] Food availability requires a supportive *built environment*, or the presence of adequate infrastructures that support access to 'food entry points' (i.e. markets, food stands, etc.) (HLPE, 2017). Dimensions of the built environment that impact food availability include mobility (i.e. proximity, means of transportation and any disability that inhibits mobility), adequate equipment and kitchen space to cook, and knowledge and skills to prepare foods.

Millions of people throughout the world lack a positive built environment for nutritious food consumption, contributing to unsustainable and unhealthy diets. In LMICs, limited food availability can be caused by geographic constraints and lack of appropriate infrastructure, including refrigeration and other transportation mechanisms for perishable, nutrient-rich foods. In high-income countries (HICs), *food deserts* are geographic areas where residents have restricted or no access to foods due to the absence or low density of food entry points within close proximity (Walker *et al.*, 2010).

Food affordability and availability are often discussed in parallel, for a food environment must provide access to both dimensions for a quality diet. Low food availability and affordability are linked with all three forms of malnutrition (Feng *et al.*, 2010), and are the primary pillars that support unsustainable food environments of all income levels.

Food affordability

Food affordability, or economic access, refers to the relative cost of food as compared to household income and purchasing power (Powel *et al.*, 2013). Worldwide, the relationship between diet quality and affordability is increasingly significant, with many food environments offering an unsustainable dichotomy: expensive foods are nutrient rich, while cheap foods are nutrient poor (de Soysa and de Soysa, 2017).

Economic access to food reflects the relative cost of food compared with a household's income and purchasing power (Powel *et al.*, 2013). People in LMICs tend to spend a greater proportion of their household budget on food, with people in Cameroon and Kenya spending almost half their budgets and people in Nigeria spending even more (USDA ERS, 2016) (see Fig. 10.3).

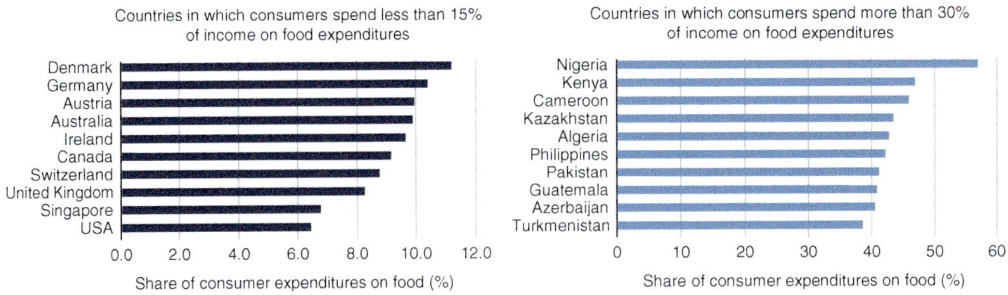

Fig. 10.3. Proportion of household budgets spent on food in different countries, 2015. Source: USDA, 2016.

For millions of the most nutritionally vulnerable throughout the world, nutrient-rich foods such as ASFs, fruits and vegetables remain prohibitively expensive. Their lack of access and thus consumption contributes to high rates of undernutrition and micronutrient deficiencies. Furthermore, high absolute poverty levels in LMICs increases food insecurity and often forces many to skip meals to save money or they are just not affordable or even accessible (FAO, 2013).

Food insecurity is widespread during times of price volatility (HLPE, 2011). This is true particularly in LMICs, in which households spend a greater proportion of household budgets on food than those in HICs. Even within HICs, there is substantial variability between how much the poorest and wealthiest households spend on food (HLPE, 2017). By contrast, nutrient-poor, highly processed foods are often inexpensive and convenient. The affordability of these foods helps explain why they are overconsumed throughout the world, including in LMICs (de Soysa and de Soysa, 2017). Increased globalization and trade liberalization may also contribute to this low cost.

Food advertisements

Food advertisements, or food promotion, incorporate the variety of actors and activities that encourage individuals to purchase and consume food products. Actors include supermarkets, the food industry, advertising firms, advocacy groups and public health agencies; activities include branding and social marketing.

Product placements, billboards, radio, television and internet ads, packaging, labelling, point-of-sale promotions, celebrity sponsorships, branded characters, merchandising, and even signage in markets have direct effects on children's food acceptability, preferences, nutrition knowledge and health status (Cairns, 2013). Television ads are particularly effective, as advertisers use child-oriented persuasion strategies to promote highly processed foods (Kelly, 2010).

Consumers can access information about a food product on its label and any declarations on its packaging. Nutrition labels are effective for both food producers and consumers, as they encourage healthier individual choices and prompt the food industry to reformulate products with more nutritious ingredients (Cowburn and Stockey, 2005; Campos *et al.*, 2011). However, many products carry misleading claims on the health and/or nutrition benefits of foods (e.g. 'heart health', 'high in antioxidants', 'low fat'). Producers often design the product's packaging to ensure these statements are immediately seen by the consumer, who then may or may not evaluate the product's nutrition content. These marketing strategies contribute to unsustainable food environments in which consumers think they know what they are eating, but are continually deceived.

Two interventions that have proven to be effective at enhancing nutrition knowledge among consumers and enhancing the integrity of food environments include easy-to-understand, front-of-pack labelling and menus with nutrition information, such as calories or sodium content (Swartz *et al.*, 2011; Kleef and Dagevos, 2015).

Food acceptability

An acceptable food is one that individuals are willing to purchase and then consume.

Not to be confused with diet quality, food quality describes how foods become acceptable: both the elements of a food (i.e. size, shape, colour, texture, flavour and composition) and its processing (i.e. 'organic', 'cage free', and so on) will determine whether consumers will value and desire a product (HLPE, 2017). Food quality includes both positive (i.e. colourful, fresh) and negative (i.e. off-colour, spoiled) attributes (Giusti *et al.*, 2008).

Food acceptability is interdependent with food safety, defined earlier in this chapter. Food contamination can occur at any point of the food value chain, from pesticide residues to lack of cold-chain storage. The latter is particularly prevalent in LMICs, where perishable foods are stored improperly and become unsafe to eat, increasing the risk of pathogen transmission. Chronic ingestion of aflatoxin, a mycotoxin produced in mould during post-harvest storage, has been linked to increased stunting (Smith *et al.*, 2015).

Food quality and safety are often used in parallel to define food acceptability, as the two strongly impact changes in consumer preferences and affordability. For instance, food safety incidents are robustly associated with a decline in consumer purchases of the affected product (FAO, 2016), and high consumption of low-quality foods are linked to a low-quality diet.

Diet and Environment Links

The four principles of a healthy diet (see Fig. 10.1) reflect a nutrition and human health perspective. But dietary diversity also impacts environmental health, for better or worse. This section evaluates diets and food choices from an environmental perspective.

Food production and consumption practices are substantial strains on natural resources. The food system alters the functions of ecosystems in which people live, grow and consume food, profoundly affecting diets (Leemans and de Groot, 2003). Food production can affect the environment at any point from its transformation from crop to food products found on the shelves of markets. Researchers analyse this 'farm-to-fork' or 'cradle-to-grave' paradigm through life cycle assessments (LCAs) (Nemecek *et al.*, 2016). The stages of a food-based LCA include agricultural conditions, food processing, transportation, packaging, and storage. LCA research has identified the certain commodity impacts on land and water use, eutrophication, pesticide use, nitrogen runoff and greenhouse gas emissions (GHGEs)[4] (Heller *et al.*, 2013; Tilman and Clark, 2014).

Environmental impacts over time vary based on both the stage of production and food group. While the greatest GHGEs occur during the agricultural stage of ASF production, storage accounts for the highest GHGE impact for fruits and vegetables (Drewnowski *et al.*, 2015). ASF production contributes to environmental degradation at a greater rate than nearly all plant-based food products. Figure 10.4 compares the environmental impacts of animal- and plant-based food production. While nutrient-rich beef (last column on the left) is the most water and land-intensive food group, nutrient-poor sugar (first column on the right) has a relatively minor environmental footprint.

A number of recent literature reviews summarized the environmental impacts of dietary patterns (Joyce *et al.*, 2014; Auestad and Fulgoni, 2015; Hallström, *et al.*, 2015; Nelson *et al.*, 2016), including three that explicitly explored health outcomes alongside environmental outcomes (Payne *et al.*, 2016; Perignon *et al.*, 2016). These reviews found that dietary patterns that replace ASF with plant-based alternatives confer the greatest environmental benefits. In their review of 210 scenarios extracted from 63 studies, Aleksandrowicz *et al.* (2016) found that vegan diets were associated with the greatest reductions in GHGEs and in land use, and vegetarian diets with the greatest reductions in water use. Diets that replaced ruminants with other alternatives, such as fish, poultry and pork, also show reduced environmental impacts, although less than plant-based alternatives (Aleksandrowicz *et al.*, 2016; Auestad and Fulgoni, 2015; Hallström, *et al.*, 2015).

While food choices affect the environment, the environment also impacts food choices, and nutritional interventions may have unintended environmental and socioeconomic consequences. Table 10.2 illustrates these and other synergies and trade-offs between the environment, nutrition and sustainability (Pray, 2014; Ingram, 2011).

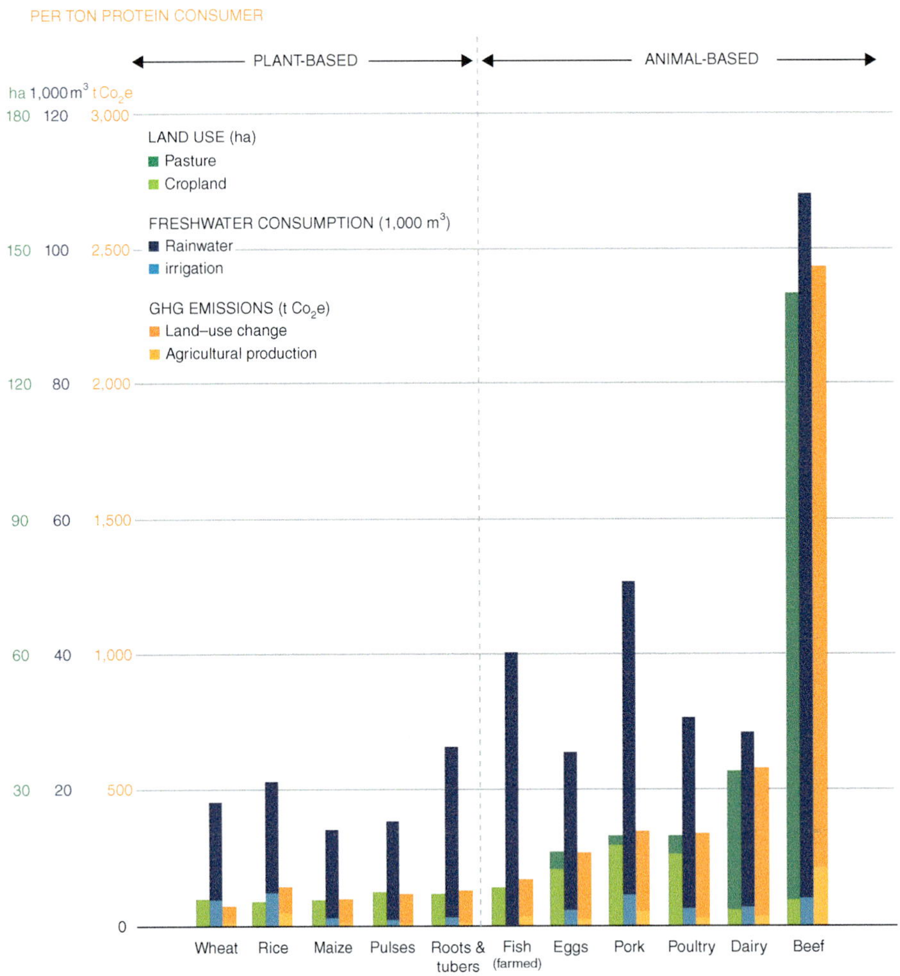

Fig. 10.4. Greenhouse gas emissions, land and water utilization in animal- and plant-based food production. Source: Ranganathan *et al.* (2016).

Sustainable diets are those diets with low environmental impacts that contribute to food and nutrition security and to a healthy life for present and future generations. Sustainable diets are protective and respectful of biodiversity and ecosystems, culturally acceptable, accessible, economically fair and affordable; nutritionally adequate, safe and healthy; while optimizing natural and human resources (Burlingame & Dernini, 2012).

More research is needed to determine the relationship between sustainable diets and health benefits and the metrics to better measure sustainable diets. Currently, substantial evidence indicates sustainable dietary patterns may reduce all-cause mortality and the risks of CVD, colorectal cancer and diabetes (Aleksandrowicz *et al.*, 2016). However, these studies are often statistically insignificant, report heterogeneous outcomes and fail to represent micronutrient deficiencies pervasive in LMICs (Payne *et al.*, 2016).

Future Diet Trends

Currently, enough food is produced throughout the world to feed all people, but malnutrition in its many forms exist in nearly every country worldwide (Popkin *et al.*, 2012). Without immediate and substantial dietary and food-based interventions, scientific projections indicate these trends will continue.

Table 10.2. The environmental, nutritional and sustainability impacts of the six food groups.

Food group	Environmental effects	Nutritional considerations	Other sustainability dimensions
Whole grains and cereals	Fewer GHGs emitted during production, but cultivation has considerable impacts on pesticide use, loss of biodiversity, water use Processing and cooking strain energy systems	*High quantity:* Consumed throughout the world *Processing*: Variable nutrient composition (white, whole grain, whole wheat) *Food type:* Potatoes, rice, bread *Condiments*: Foods simultaneously consumed that add calories (e.g., fats, spreads, sauces)	*Socioeconomics:* Growth in demand for quinoa and other grain products such as fonio, teff, etc., among indigenous and small holder farmers
Animal-source foods (i.e., poultry, beef, pork)	High environmental impacts (GHGs, water, land use, biodiversity), but significant variation between livestock type and system ASF production can enhance efficiency in food systems in which producers grow animal by-products, thereby utilizing lands unsuitable for crop production	*Variable consumption*: Excess in high-income settings; but deficient in low-income settings *Micronutrient content:* High energy and nutrient density *Species type:* Poultry, beef, pork *Cut consumed*: Carcass or processed; lean or fatty meat *Production method*: Grass-fed or grain-fed *Evidence*: Relationship between meat (especially red and processed meats) to negative health outcomes	*Socioeconomics*: Meat production is a source of employment and livelihoods for pastoralists and the world's poor *Labour conditions:* Meat processing is a dangerous profession *Ethics:* Animal welfare concerns *Food security:* Effects of feeding grain to livestock, not people *Food safety:* Zoonoses are leading source of emerging infectious diseases *Culture:* Meat is culturally significant throughout the world
Milk and dairy	High environmental impacts (GHGs, water, land use, biodiversity) Production of dairy products can enhance efficiency in food systems in which producers grow animal by-products, thereby utilizing lands unsuitable for crop production	*Quantity:* Variable worldwide *Type:* Milk, cheese, yogurt, high or low fat *Micronutrient content:* Calcium may protect against heart diseases *Production method:* Intensive or extensive; grain-fed or pasture-fed *Additives*: Added sugar and salt	*Socioeconomics*: Milk and dairy production are sources of employment and livelihoods for pastoralists and the world's poor *Ethics:* Animal welfare concerns *Food safety:* Effect of zoonoses and ease of disease transmission *Culture:* Cheese and yogurt are culturally significant worldwide
Fish	Overfishing has led to the depletion of many species and degradation of wider marine ecosystems Aquaculture is unsustainable and linked to a variety of environmental problems	*Micronutrient content:* Fish is high in omega-3 fatty acids and low in trans-fats *Evidence:* Fish consumption linked to a range of positive health outcomes	*Socioeconomics*: Fish and aquaculture are important sources of food and livelihoods in low-income settings *Long-term viability:* Overfishing has significantly lowered the world's fish population

Continued

Table 10.2. Continued.

Food group	Environmental effects	Nutritional considerations	Other sustainability dimensions
Fruits and vegetables	Variable GHG impact. Products with low GHG emissions are robust, consumed during the season in which they are grown, and transported by sea and land. Products with high GHG emissions are airfreighted, grown in heated greenhouses and reliant on irrigation Trade-offs in water use and GHGs	*Consumption:* Increased consumption throughout the world is recommended by most public health experts *Production:* Organic or non-organic *Seasonality*: Locally grown products vary in season and geographic location *Storage method*: Availability of refrigeration or freezing *Transportation method*: Sea, land or air *Residues*: Pesticides or other anti-nutrients	*Socioeconomics*: Horticulture provides livelihoods for low-income communities *Labour conditions:* Exploitation and low wages rampant
Sugary foods and confectionary items	Production emits low GHGs, but is reliant on land and water use, as well as linkages to pesticides Given absence of nutritional content, the mass production of sugary foods represent a waste of embedded natural resources	*High quantity:* Consumed worldwide *Limited diversity:* Extent to which consumers substitute sugar foods for other food groups, lowering dietary diversity *Empty calories:* Linked to obesity, diet-related NCDs, dental issues	*Socioeconomics*: Processing and production are source of jobs and livelihoods for millions of people worldwide *Fair trade*: Standardized prices *Culture*: Important worldwide

ASF, animal-source food; GHG, greenhouse gas.
Source: Adapted from Garnett (2014).

Through 2030, undernutrition in Asia will only moderately decline, while caloric deficiency will remain stagnant in Africa (GLOPAN, 2016). With studies suggesting climate change will shock cyclical weather patterns, initiating erratic and severe droughts, floods and hurricanes, millions in LMICs will face extreme poverty, food insecurity, insufficient diets, and poor growth and cognitive development (Wheeler and von Braun, 2013; Thornton *et al.*, 2014).

Trends in the production and consumption of all food groups differ worldwide. Fruits, vegetables and pulses will be disrupted by climate change, in both supply chain activities and distribution systems (GLOPAN, 2016). Policies to alter this trajectory include financial support (i.e. crop insurance) and incentives for smallholder farms (Brown-Paul, 2014). Consumption of nutrient-poor, highly processed foods may remain stable in HICs, but will grow substantially in LMICs, particularly in East Asia (GLOPAN, 2016; IFPRI, 2014).

While millions of malnourished individuals in LICs lack access to ASFs, MIC and HICs are expected to overconsume ASFs through 2050. These trends have mixed results on both nutrition and environmental goals, particularly concerning the sustainability of the global ASF supply. LICs will struggle to increase ASF consumption to the levels necessary to reverse micronutrient deficiencies. In contrast, consumption of both unprocessed and processed meats in MIC and HICs will risk obesity, CVDs and other NCDs that strain worldwide and domestic health, food and economic systems. In summary, ASF policy priorities include: reducing consumption of processed meats in HICs, discouraging ASF overconsumption in MICs, and increasing access of all ASFs to the most nutritionally vulnerable in LICs (Henchion *et al.*, 2014; Whitmee *et al.*, 2015).

Conclusion

Improvements in diet quality and the sustainability of the global food system require both individual and collective action. National-level policy commitments (i.e. taxes, subsidies, incentives and regulatory frameworks) can ensure the food industry limits environmental harms throughout the food value chain (Ranganathan *et al.*, 2016). However, *systems*-level approaches are needed, requiring public health, nutrition, agriculture and the food industry to collaborate, work within institutional frameworks and create country-specific points of opportunity (Finley *et al.*, 2017). In HICs and MICs, nutrition education may be an effective intervention to bring awareness to individuals on the nutritional and environmental impacts of their dietary habits. In LICs, social and behavioural change communication use community systems to increase knowledge, attitudes and social norms regarding health, sanitation, nutrition and diet practices (Bhutta *et al.*, 2013).

Without radical shifts towards enhanced sustainability in both food consumption patterns and production mechanisms, the food system will only contribute to rising hunger and obesity rates in the coming decades. But with simultaneous change at individual, community and institutional levels, the global food system could have the capability to deliver culturally-relevant foods that provide sufficient quantities and high-quality diets.

Notes

[1] *Nutrient density* is defined as the proportion of nutrients in foods. Nutrient-dense, or nutrient-rich foods supply relatively more nutrients than calories (i.e. 'healthy'), while nutrient-poor foods are those with higher calories than micronutrients (i.e. 'unhealthy').

[2] Six primary food groups exist: whole grains and cereals; animal-source foods, including poultry, beef, and livestock; milk and dairy; fish; fruits and vegetables; sweets and confectionary foods.

[3] Proximity, defined as close distance to a market, is related to food availability.

[4] Food loss and waste (both in quality and quantity) occur at every point in the food supply chain and thus have been omitted from this chapter.

References

Aleksandrowicz, L., Green, R., Joy, E. J., Smith, P. and Haines, A. (2016) The impacts of dietary change on greenhouse gas emissions, land use, water use and health: A systematic review. *PLoS One* 11(11), e0165797.

Allen, L.H. (2012) Global dietary patterns and diets in childhood: implications for health outcomes. *Annals of Nutrition and Metabolism* 61(1), 29–37.

Auestad, N. and Fulgoni, V.L. (2015) What current literature tells us about sustainable diets: emerging research linking dietary patterns, environmental sustainability and economics. *Advances in Nutrition: An International Review Journal* 6(1), 19–36.

Baker, P. and Friel, S. (2014) Processed foods and the nutrition transition: evidence from Asia. *Obesity Reviews* 15(7), 564–577.

Bhutta, Z.A., Das, J.K., Rizvi, A., Gaffey, M.F., Walker, N., *et al.* (2013) Evidence-based interventions for improvement of maternal and child nutrition: What can be done and at what cost? *The Lancet*, 382(9890), 452–477. DOI:10.1016/S0140-6736(13)60996-4

Brown-Paul, C. (2014) Raising the roof. *Practical Hydroponics and Greenhouses* 143, 38.

Burlingame, B. and Dernini, S. (Eds) (2012) Sustainable Diets and Biodiversity: Directions and solutions for policy, research and action. *Proceedings of the International Scientific Symposium Biodiversity and Sustainable Diets United against Hunger*, 3–5 November 2010, FAO Headquarters, Rome. http://www.fao.org/docrep/016/i3004e/i3004e.pdf (accessed 28 August 2018).

Cairns, G., Angus, K., Hastings, G. and Caraher, M. (2013) Systematic reviews of the evidence on the nature, extent and effects of food marketing to children. A retrospective summary. *Appetite* 62, 209–215.

Campos, S., Doxey, J. and Hammond, D. (2011) Nutrition labels on pre-packaged foods: a systematic review. *Public Health Nutrition* 14(8), 1496–1506.

Chinnakali, P., Upadhyay, R.P., Shokeen, D., Singh, K., Kaur, M., *et al.* (2014) Prevalence of household-level food insecurity and its determinants in an urban resettlement colony in north India. *Journal of Health, Population and Nutrition* 32(2), 227.

Cowburn, G. and Stockley, L. (2005) Consumer understanding and use of nutrition labelling: a systematic review. *Public Health Nutrition* 8(01), 21–28.

de Pee, S. and Bloem, M.W. (2009) Current and potential role of specially formulated foods and food supplements for preventing malnutrition among 6- to 23-month-old children and for treating moderate malnutrition among 6- to 59-month-old children. *Food and Nutrition Bulletin* 30(3), S463. DOI:10.1177/15648265090303S305

de Soysa, I. and de Soysa, A.K. (2017) Do globalization and free markets drive obesity among children and youth? An empirical analysis, 1990–2013. *International Interactions* 1–19.

di Renzo, L., Colica, C., Carraro, A., Cenci Goga, B., Marsella, L.T., *et al.* (2015) Food safety and nutritional quality for the prevention of non-communicable diseases: The nutrient, hazard analysis and critical control point process (NACCP). *Journal of Translational Medicine* 13(1), 128. DOI:10.1186/s12967-015-0484-2

Drewnowski, A. and Fulgoni, V.L. (2014) Nutrient density: Principles and evaluation tools. *The American Journal of Clinical Nutrition* 99(5), 1228S. DOI:10.3945/ajcn.113.073395

Drewnowski, A., Rehm, C.D., Martin, A., Verger, E.O., Voinnesson, M. and Imbert, P. (2015) Energy and nutrient density of foods in relation to their carbon footprint. *The American Journal of Clinical Nutrition* 101(1), 184–191. DOI:10.3945/ajcn.114.092486

Fabbri, A.D.T. and Crosby, G.A. (2016) A review of the impact of preparation and cooking on the nutritional quality of vegetables and legumes. *International Journal of Gastronomy and Food Science* 3, 2–11. DOI:10.1016/j.ijgfs.2015.11.001

FAO (1996) The Rome declaration on world food security. *Population and Development Review* 22(4), 807.

FAO (2011) *Guidelines for Measuring Household and Individual Dietary Diversity*. Food and Agriculture Organization of the United Nations, Rome, Italy.

FAO (2013) *The Food Insecurity Experience Scale: Development of a Global Standard for Monitoring Hunger Worldwide*. Food and Agriculture Organization of the United Nations, Rome, Italy.

FAO (2016) *Influencing Food Environments for Healthy Diets*. Food and Agriculture Organization of the United Nations, Rome, Italy.

Feng, J., Glass, T.A., Curriero, F.C., Stewart, W.F. and Schwartz, B.S. (2010) The built environment and obesity: A systematic review of the epidemiologic evidence. *Health and Place* 16(2), 175–190.

Finley, J.W., Dimick, D., Marshall, E., Nelson, G.C., Mein, J.R. and Gustafson, D.I. (2017) Nutritional sustainability: Aligning priorities in nutrition and public health with agricultural production. *Advances in Nutrition* 8(5), 780–788.

Forouzanfar, M.H., Alexander, L., Anderson, H.R., Bachman, V.F., Biryukov, S., *et al.* (2015) Global, regional, and national comparative risk assessment of 79 behavioural, environmental and occupational, and metabolic risks or clusters of risks in 188 countries, 1990–2013: A systematic analysis for the global burden of disease study 2013. *Lancet* 386(10010), 2287–2323.

Garnett, T. (2014) *What is a Sustainable Healthy Diet? A Discussion Paper*. Food and Climate Research Network, Oxford, UK.

GLOPAN (2016) Food Systems and Diets: Facing the Challenges of the 21st Century. Global Panel on Agriculture and Food Systems, London, UK.

Giusti, A.M., Bignetti, E. and Cannella, C. (2008) Exploring new frontiers in total food quality definition and assessment: From chemical to neurochemical properties. *Food and Bioprocess Technology* 1(2), 130.

Hallström, E., Carlsson-Kanyama, A. and Börjesson, P. (2015) Environmental impact of dietary change: a systematic review. *Journal of Cleaner Production* 91, 1–11.

Hawkes, C. (2006) Uneven dietary development: Linking the policies and processes of globalization with the nutrition transition, obesity and diet-related chronic diseases. *Globalization and Health* 2(1), 4.

Heller, M.C., Keoleian, G.A. and Willett, W.C. (2013) Toward a life cycle-based, diet-level framework for food environmental impact and nutritional quality assessment: A critical review. *Environmental Science and Technology* 47(22), 12632–12647.

Henchion, M., McCarthy, M., Resconi, V.C. and Troy, D. (2014) Meat consumption: Trends and quality matters. *Meat Science* 98(3), 561–568.

HLPE (2011) *Price Volatility and Food Security*: A report of the High-Level Panel of Experts on Food Security and Nutrition of the Committee on World Food Security. Food and Agriculture Organization of the United Nations, Rome, Italy.

HLPE (2017) *Nutrition and Food Systems*: A report of the High-Level Panel of Experts on Food Security and Nutrition of the Committee on World Food Security. Food and Agriculture Organization of the United Nations, Rome, Italy.

IFPRI (2014) *Global Nutrition Report: Actions and Accountability to Accelerate the World's Progress on Nutrition*. International Food Policy Research Institute Washington, DC, USA.

IFPRI (2016) *Global Food Policy Report*. IFPRI, Washington, DC, USA.

Ingram, J. (2011) A food systems approach to researching food security and its interactions with global environmental change. *Food Security* 3(4), 417–431. DOI:10.1007/s12571-011-0149-9

Jones, A.D. (2017) Critical review of the emerging research evidence on agricultural biodiversity, diet diversity, and nutritional status in low-and middle-income countries. *Nutrition Reviews* 75(10), 769–782.

Joyce, A., Hallett, J., Hannelly, T. and Carey, G. (2014) The impact of nutritional choices on global warming and policy implications: examining the link between dietary choices and greenhouse gas emissions. *Energy and Emission Control Technologies* 2, 33–43.

Kant, A.K. (2004) *Dietary patterns and health outcomes*. Elsevier, USA. DOI:10.1016/j.jada.2004.01.010

Kelly, B. *et al*. (2010) Television food advertising to children: a global perspective. *American Journal of Public Health* 100(9), 1730–1736.

Kleef, E. V. and Dagevos, H. (2015) The growing role of front-of-pack nutrition profile labeling: a consumer perspective on key issues and controversies. *Critical Reviews in Food Science and Nutrition* 55(3), 291–303.

Lallukka, T., Laaksonen, M., Rahkonen, O., Roos, E. and Lahelma, E. (2007) Multiple socio-economic circumstances and healthy food habits. *European Journal of Clinical Nutrition* 61(6), 701–710. DOI:10.1038/sj.ejcn.1602583.

Larsson, S. C. and Wolk, A. (2012) Red and processed meat consumption and risk of pancreatic cancer: meta-analysis of prospective studies. *British Journal of Cancer*, 106(3), 603.

Leemans, R. and de Groot, R.S. (2003) *Millennium Ecosystem Assessment: Ecosystems and Human Well-being: A Framework for Assessment*. Island Press, Washington DC, USA.

Monteiro, C.A., Moubarac, J., Cannon, G., Ng, S.W. and Popkin, B. (2013) Ultra-processed products are becoming dominant in the global food system. *Obesity Reviews* 14(S2), 21–28.

Mozaffarian, D. (2016) Dietary and policy priorities for cardiovascular disease, diabetes, and obesity: A comprehensive review. *Circulation,* 133(2), 187–225.

Mozaffarian, D. and Ludwig, D. S. (2010) Dietary guidelines in the 21st century—a time for food. *Journal of the American Medical Association* 304(6), 681–682. DOI: 10.1001/jama.2010.1116.

Nelson, M.E., Hamm, M.W., Hu, F.B., Abrams, S.A. and Griffin, T.S. (2016) Alignment of healthy dietary patterns and environmental sustainability: A systematic review. *Advances in Nutrition* 7(6), 1005–1025.

Nemecek, T., Jungbluth, N., Canalis, L. N. and Schenck, R. (2016) Environmental impacts of food consumption and nutrition: Where are we and what is next? *The International Journal of Life Cycle Assessment* 21(5), 607. DOI: 10.1007/s11367-016-1071-3.

Payne, C.L., Scarborough, P. and Cobiac, L. (2016) Do low-carbon-emission diets lead to higher nutritional quality and positive health outcomes? A systematic review of the literature. *Public Health Nutrition* 19(14), 2654–2661.

Perignon, M., Masset, G., Ferrari, G., Barré, T., Vieux, F., *et al*. (2016) How low can dietary greenhouse gas emissions be reduced without impairing nutritional adequacy, affordability and acceptability of the diet? A modelling study to guide sustainable food choices. *Public Health Nutrition* 19(14), 2662–2674.

Popkin, B.M., Adair, L.S. and Ng, S.W. (2012) Global nutrition transition and the pandemic of obesity in developing countries. *Nutrition Reviews* 70(1), 3–21.

Powell, L.M., Chriqui, J.F., Khan, T., Wada, R. and Chaloupka, F. J. (2013) Assessing the potential effectiveness of food and beverage taxes and subsidies for improving public health: A systematic review of prices, demand and body weight outcomes. *Obesity Reviews* 14(2), 110–128.

Pray, L. (2014) *Sustainable Diets: Food for Healthy People and a Healthy Planet – Workshop Summary*. CAB International, Wallingford, UK.

Ranganathan, J., Vennard, D., Waite, R., Dumas, P., Lipinski, B. and Searchinger, T. (2016) Shifting diets for a sustainable food future. World Research Institute, Washington DC, USA.

Smith, L.E., Prendergast, A.J., Turner, P.C., Mbuya, M.N., Mutasa, K., *et al.* (2015) The potential role of mycotoxins as a contributor to stunting in the SHINE trial. *Clinical Infectious Disease* 61(Suppl 7): S733–737.

Swartz, J.J., Braxton, D. and Viera, A. J. (2011) Calorie menu labeling on quick-service restaurant menus: an updated systematic review of the literature. *International Journal of Behavioral Nutrition and Physical Activity* 8(1), 135.

Swinburn, B., Dominick, C. and Vandevijvere, S. (2014) Benchmarking food environments: experts' assessments of policy gaps and priorities for the New Zealand Government. Faculty of Medical and Health Sciences, School of Population Health, University of Auckland, Auckland, New Zealand.

Thornton, P.K., Ericksen, P.J., Herrero, M. and Challinor, A.J. (2014) Climate variability and vulnerability to climate change: A review. *Global Change Biology* 20(11), 3313–3328.

Tilman, D. and Clark, M. (2014) Global diets link environmental sustainability and human health. *Nature,* 515(7528), 518–522. DOI:10.1038/nature13959

USDA ERS (2016) Calculations based on annual household expenditure data from Euromonitor International. USDA Economic Research Service, Washington, DC, USA.

Vuong, T.N., Gallegos, D. and Ramsey, R. (2015) Household food insecurity, diet, and weight status in a disadvantaged district of Ho Chi Minh City, Vietnam: a cross-sectional study. *BMC Public Health* 15(1), 232.

Walker, R.E., Keane, C.R. and Burke, J.G. (2010) Disparities and access to healthy food in the United States: a review of food deserts literature. *Health and Place* 16(5), 876–884.

Wheeler, T. and Von Braun, J. (2013) Climate change impacts on global food security. *Science* 341(6145), 508–513.

Whitmee, S., Haines, A., Beyrer, C., Boltz, F., Capon, A.G., *et al.* (2015) Safeguarding human health in the Anthropocene epoch: report of the Rockefeller Foundation–Lancet commission on planetary health. *The Lancet*, 386(10007), 1973–2028.

WHO (2015) *Healthy diet fact sheet no. 394*. World Health Organization, Geneva, Switzerland.

11 Highlighting Interlinkages Between Sustainable Diets and Sustainable Food Systems

Alexandre Meybeck and Vincent Gitz

Abstract

Sustainable food systems and sustainable diets are increasingly being called upon as ways to orient action towards the eradication of hunger and malnutrition and the fulfilment of the sustainable development goals. This chapter explores the links between the two notions and how these links can orient policies and consumption choices. To do so, it first considers the relationships between food systems and diets, how food systems condition the availability and accessibility of foods that can be part of a diet, and also how demand determines the foods that are made available and accessible. Diets are thus both the results and the drivers of food systems. A sustainable food system can be defined as a food system that ensures food security and nutrition for all in such a way that the economic, social and environmental bases to generate food security and nutrition of future generations are not compromised. The concept of a sustainable diet combines two totally different perspectives: (i) a nutrition perspective, which is person focused; and (ii) a global sustainability perspective, in all its dimensions – environmental, economic and social. Understanding the links between these two notions can help design policies and incentives to improve the sustainability of food systems and diets, building upon the motivation of various actors, consumers and private actors, which are often related to very different dimensions (health, environment, social and cultural).

Introduction

Sustainable food systems and sustainable diets are increasingly being called upon as ways to orient action towards the eradication of hunger and malnutrition and the fulfilment of sustainable development goals. This chapter explores the links between the two notions and how these links can orient policies and consumption choices. To do so, it first considers the relationships between food systems and diets, how food systems condition the availability and accessibility of foods that can be part of a diet, and also how demand determines the foods that are made available and accessible. It then considers the concepts of a sustainable food system (SFS) and of a sustainable diet in order to analyse their relationships. This highlights some of the links that can help design policies and incentives to improve the sustainability of food systems and diets, building upon the motivations of various actors, consumers and private actors. These are often related to very different dimensions (health, environment, social and cultural), even if consumers also tend to aggregate them in a broader perception of quality.

Food Systems and Diets

Food systems and diets are obviously linked, by food at least. Diets comprise the individual foods that a person consumes, and dietary patterns are the quantities, proportions and combinations of different foods and beverages in diets and the frequency of how they are habitually consumed (Hu, 2002). Dietary patterns interact with food systems, not only as an outcome of existing food systems but also as a driver of change for future food systems (HLPE, 2017). Globally, food systems are focused on food, and diets on its consumption and nutritional outcomes. However, definitions of food systems do differ in the way they integrate, or relate to, diets. Some authors have given a particular space to them, often as linked to outcomes of food systems. For instance, Hammond and Dubé (2012) describe a systems framework for food and nutrition security, and propose a definition of agri-food systems focused on food production and linked to two other systems: the environmental system and the health and disease system. It is the interactions between these three systems that determine outcomes on individuals.

Sobal and colleagues (1998) designed an integrated conceptual model of the food and nutrition system with a focus on nutrition and emphasizing the links between food production, food consumption and nutritional health. They define the food and nutrition system as 'the set of operations and processes involved in transforming raw materials into foods and transforming nutrients into health outcomes, all of which functions as a system within biophysical and sociocultural contexts'. They further identify three subsystems: the producer subsystem, the consumer subsystem and the nutrition subsystem, each flowing into the subsequent one. This model is clearly focused on the linear relationships between production, consumption and nutrition, with a much less comprehensive coverage of the determinants of food systems that are here presented as part of biophysical and sociocultural contexts. It is complemented by the identification of several other systems that interact in many points with the food nutrition system. These include the healthcare, economic, cultural, ecological, governmental and transportation systems, each having its own specific orientations and interacting with others. Such an approach, which puts the consumer at the centre of the system, as an intermediate between food production and nutrition outcomes, gives a particular importance to diets.

Building upon these works and others (Ericksen, 2008; Ericksen *et al.*, 2010; Ingram, 2011; IPCC, 2014), the High Level Panel of Experts on Food Security and Nutrition (HLPE) proposed a comprehensive, descriptive definition:

> A food system gathers all the elements (environment, people, inputs, processes, infrastructures, institutions, etc.) and activities that relate to the production, processing, distribution, preparation and consumption of food, and the outputs of these activities, including socio-economic and environmental outcomes.
>
> (HLPE, 2014)

Here, diets are an integral part of food systems, under consumption, as well as part of their outcomes.

We use here the HLPE definition, and propose, in the light of Sobal *et al.* (1998), to distinguish two subsystems: (i) a production subsystem, from the use of natural resources and inputs to the production, transformation and distribution of food; and (ii) a consumption subsystem (see Fig. 11.1). Consumers select between the foods made accessible to them, physically and economically, by the production system, to compose their diet. Of particular importance is the interface between consumers and the rest of the food systems, which is now often designated under the term 'food environment'. As defined by the HLPE (2017): 'Food environment refers to the physical, economic, political and sociocultural context in which consumers engage with the food system to make their decisions about acquiring, preparing and consuming food'. It consists of both physical 'food entry points' where food is purchased or obtained, and the means to access them and individual determinants of consumer food choices, including income, lifestyle and attitude towards food, which are themselves determined by education, information, culture and social norms. The two subsystems interact. The production subsystem determines the options between which the consumer can choose. On the other hand, the consumption subsystem determines the demand, which, in turn, influences what is produced as well as prices. Each of these subsystems has specific outcomes and impacts. Diets, the eaten part of consumption (the rest becoming waste), determine nutritional outcomes, interacting with health along with other factors. The production system, as influenced by consumption, has, in addition to its main outcome, diets, environmental, economic

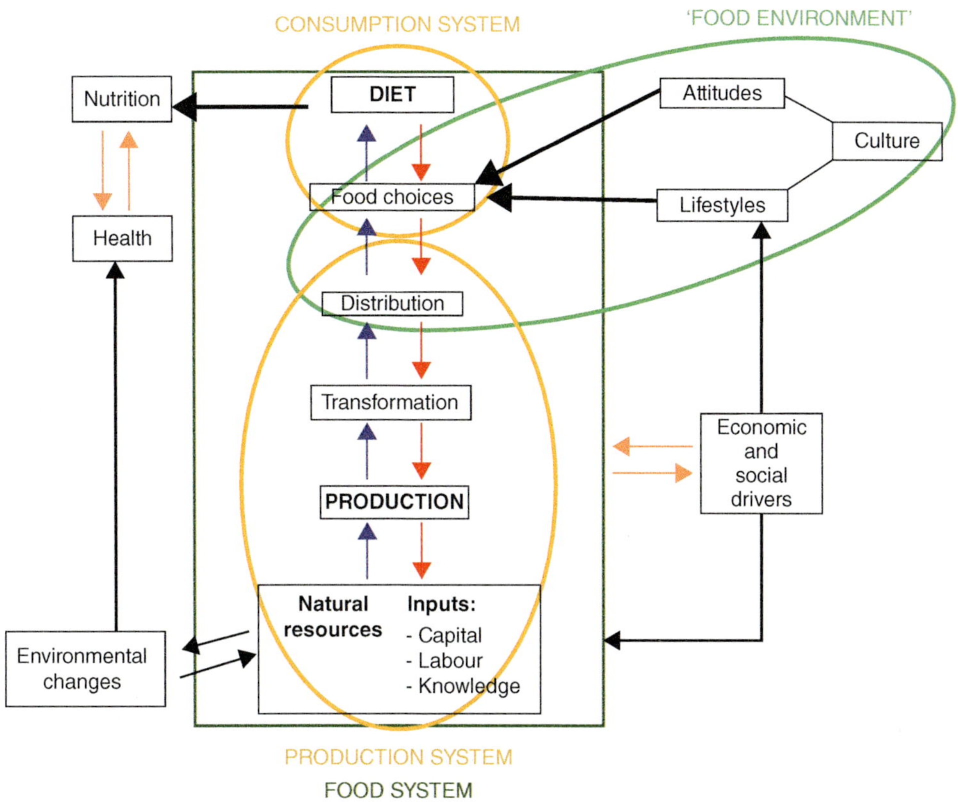

Fig. 11.1. Food system and diet.

and social impacts and outcomes, interacting with other changes and drivers. Some of these outcomes, in turn, influence diets.

Many drivers of food consumption choices are to be found within food systems. As sustainable diets are both an objective and a driver of sustainable food systems, understanding the drivers of food choices is of paramount importance to design ways to improve the sustainability of both diets and food systems. This leads to particular interest for two specific groups of parameters. The first covers economic, social and cultural parameters that, both inside and outside food systems, can drive food consumption choices. The second relates to consumption choices that go beyond diet nutritional composition and take into account characteristics such as quality, origin and mode of production. Such choices can have various impacts on all dimensions of sustainability. Moreover, they can be the expression of attitudes that are also grounding some choices related to diet composition, as shown, for instance, for consumers of organic food (Lairon and Kesse-Guyot, 2015).

Notions and Definitions of Sustainable Food Systems and Sustainable Diets

As defined in 2010, sustainable diets are:

> those diets with low environmental impacts which contribute to food and nutrition security and to healthy life for present and future generations. Sustainable diets are protective and respectful of biodiversity and ecosystems, culturally acceptable, accessible, economically fair and affordable; nutritionally adequate, safe and healthy; while optimizing natural and human resources.
>
> (FAO, 2010)

The concept of sustainable diet combines, in fact, two totally different perspectives: a nutrition perspective that is person focused, and a global sustainability perspective, in all its dimensions – environmental, economic and social, relating to food systems.

The HLPE, in line with the original broad approach of sustainability, has provided a definition of a sustainable food system oriented by its capacity to ensure the positive outcomes of a food system: food security now and for future generations.

> A sustainable food system (SFS) is a food system that ensures food security and nutrition for all in such a way that the economic, social and environmental bases to generate food security and nutrition of future generations are not compromised.
>
> (HLPE, 2014)

The internationally agreed definition of food security dates from the 1996 World Food Summit: 'Food security exists when all people, at all times, have physical and economic access to sufficient, safe and nutritious food that meets their dietary needs and food preferences for an active and healthy life' (WFS, 1996). This definition identifies four dimensions of food security: availability of food, accessibility (economical and physical), utilization (the way it is used and absorbed) and stability of these three dimensions. The HLPE has thus formalized the link between the two concepts of food security and nutrition (FSN) and of sustainable food systems (SFS): the bottom line is that there can be no FSN (short and long term) without SFS. FSN for all, worldwide, and the conditions for their existence over time, could be what ultimately characterizes SFS.

The relationships between diets and food systems, as discussed in the previous section, enable a concrete assessment of the sustainability of diets, intended as their contribution to the sustainability of food systems and ultimately to FSN. The links between the notions could thus be formalized as follows:

> a sustainable diet is a diet that contributes to the good nutritional status and long-term good health of the individual/community, and that contributes to, and is enabled by, sustainable food systems, thus contributing to long-term food security and nutrition.
>
> (Meybeck and Gitz, 2017)

In other words, sustainable diets are both an objective and an essential means, a key driver, to achieve the transformation of food systems, which is needed to achieve FSN. The sustainability of the diet could be characterized by its contribution to the sustainability of the food system.

Relationships Between Sustainability of Diets and of Food Systems

The sustainability of diets and food systems thus largely depend on one another, the sustainability of one being linked to the sustainability of the other. The relationships are not, however, homothetic.

By its very definition, a sustainable food system is the condition *sine qua non* for sustainable diets. Indeed, the whole reflection about sustainability of the food system is fuelled by concerns about its capacity to provide enough and adequate food in the future given, in particular, the growing scarcity of natural resources (Foresight, 2011). The economic sustainability of food systems – and particularly of agriculture – is also put under pressure by the decrease of food prices and by their volatility, both of which discourage investment. These economic threats, and in many countries the growing discrepancy between incomes in the agriculture sector and out of it, and especially between the perspectives of income evolution, translate into social sustainability concerns, with the risk of discouraging young generations from entering agriculture (HLPE, 2013), with immediate consequences for the capacity of the sector to adapt to changes and to innovate, and threatening its capacity to fulfil its function. Such concerns about the sustainability of the system, and about its capacity to provide FSN in the future, have led Berry *et al.* (2015) to propose adding sustainability as a fifth dimension of FSN.

The reverse relationship, between diets and food systems' sustainability, can be understood as part of the broader concept of sustainable consumption and production as highlighted in the outcome document of the Rio conference on sustainable development in 1992 (UNCED, 1992). This concept emphasizes that to increase the sustainability of systems, both production and consumption and supply and demand have to be considered. There are production options; there are consumption choices. Improving sustainability is a matter of both. To a certain extent, and still in many economies, consumption choices are restricted by what production offers. But, globally, with the extent of consumption choices increasing worldwide, there are greater prospects for consumption to drive production, for consumption choices to orient the choices that producers make (which products, how they are made), and to orient 'production' towards the products consumers are more likely to buy. There are thus increasing risks that unsustainable consumption patterns threaten the sustainability of systems and also more opportunities for sustainable consumption patterns and choices to drive towards more sustainable production patterns.

Numerous studies have shown correlations and synergies between healthier diets and a reduced impact on the environment at global (Tilman and Clark, 2014) or national levels (Tukker *et al.*, 2011; Monsivais *et al.*, 2015; Milner *et al.*, 2015). However, the relationships between sustainable diets and food systems are not always that simple, particularly when integrating the environmental impact assessments that are local (e.g. water quality or biodiversity), or economic and social impacts that first often depend on the organization of specific production systems, and second impact differently on consumers and producers. It has, for instance, been noted that generalizing the recommended intake of fish could threaten the sustainability of fisheries (van Dooren *et al.*, 2014).

Price is one of the first determinants of food consumption choices (and of their feasibility). From a consumer perspective, and especially poor net food buyers, the lower food prices are, the better: it facilitates diversified and nutritious diets, and favours capacity to spend on other basic needs. It can thus be a condition for sustainable diets. Gustafson *et al.* (2017) integrate affordability as the main economic indicator of a sustainable food system. But low food prices also impact on the sustainability of food systems. They can have direct negative environmental impacts by not discouraging food waste (FAO, 2011). They reduce investment capacity and thus economic sustainability. By driving the need for low production costs, they also encourage low-cost practices that can be environmentally damaging and drive low income and wages for food producers and workers, with important social impacts. The potentially ambivalent (and at least multi-form) role of prices with respect to sustainability and food security calls for clarification, clearly separating food prices as an indicator of access at consumption level from its use and interpretation inside the food system at large, which requires breaking down the final consumption price into various components to better envision its relationships with economic and social dimensions of sustainability. Such distinctions of level of impacts inside food systems are also particularly important to better understand, conversely, the potential impacts of changing diets on the different stages of food systems, as it will impact prices and economic exchanges (Adinolfi *et al.*, 2015). Food prices need therefore to be also analysed in terms of their impacts on sustainability, with different approaches for diets and for food production, for instance. The contribution of food prices to the various dimensions of sustainability can thus be different when considering only diets or food systems as a whole, particularly when integrating a long-term perspective.

Another difficulty when translating from diet to food system is that a product, while having the same nutritional value, could have different impacts on the food system depending on where it is coming from, how it has been produced, transformed, transported, by whom and how each actor has been remunerated. In other words, the impacts of diets on the sustainability of food systems also depends on the specificities of the food system itself.

Scope and Scales: Fragmented, Combined and Composite Food Systems

The discourse about sustainable diets is often guided by the assumption that diets and food systems are closely linked, and in particular that they share the same spatial limits – with a broad equivalence between consumption and production spaces. Such an equivalence is true at the global level: the global diet summing all the individual diets that compose it as related to the global food system. It was to a great extent true in 'traditional' food systems closely linked to a specific 'traditional' diet, somehow homogeneous, shared by a geographic community, and therefore sharing the same geographical limits as the food production area. To a certain extent, this was true for the Mediterranean diet model, abstracted from the description of traditional diets in the Mediterranean (Meybeck and Gitz, 2017). Such a close association between a diet and a food system, in the same geographic area, is no longer present in most modern diets. Furthermore, within a particular 'system', diet is not the same for everybody, consumption within a food system is not homogeneously distributed and, from a nutrition and health perspective, the individual diets are important, rather than the average.

Such considerations call for questioning the very scope of a food system. It has been noted that 'a household's food system comprises all the food chains it participates in to meet its consumption requirements and dietary preferences, and all the interactions and feedback loops that connect the different parts of these chains' (FAO, 2008). This would lead to considering that food systems are organized around, and by, consumers; each consumer, or group of consumers, being at the centre of a network of relationships with producers. The same could be said of producers, linked to various consumers. Building upon the distinction in a food system between two subsystems (i.e. the consumer and producer subsystems), food systems could be represented as polycentric ensembles of subsystems with more or less strong relationships depending on the intensity and frequency of exchanges (see Fig. 11.2).

Such a representation enables a better understanding and account of the influence of consumer subsystems that can be spatially determined like cities, or groups of consumers sharing the same preferences – what some authors designate as 'food tribes' (Niola, 2015). Such influences can be exerted on local producer subsystems (e.g. through farmers' markets) or more remotely (e.g. by including specific exotic foods, or foods coming from a specific producer system, distinguished by a geographical indication of provenance) or a whole specific food system (e.g. the organic food system, which includes transformation, specific distribution channels, etc.). It thus enables a better understanding of how sustainable diets and food systems can be linked and become common factors of identity for consumers – an identity that is, in itself, a strong and long-term incentive. Such links can give way to symbolic and/or political forms of collective engagement in/for the food system, grounded on a common identity, and that can be linked to a place-based approach to food (Sonnino *et al.*, 2016). This collective sense of identity can play a key role to facilitate change towards more sustainable diets and systems by enabling the coalition of actors that often have different perspectives and drivers. The consumer caring for a certain production system supports its sustainability. Analysing the case of the *Parmigiano Reggiano Terremotato*, Finardi and Menozzi (2014) note that it is the 'social embeddedness' of the product that enabled the avoidance of a market crisis when considerable quantities of it had to be sold following the destruction of storage facilities by the earthquake. In that regard, the link between a diet and a production system can in itself be a key component of their respective sustainability.

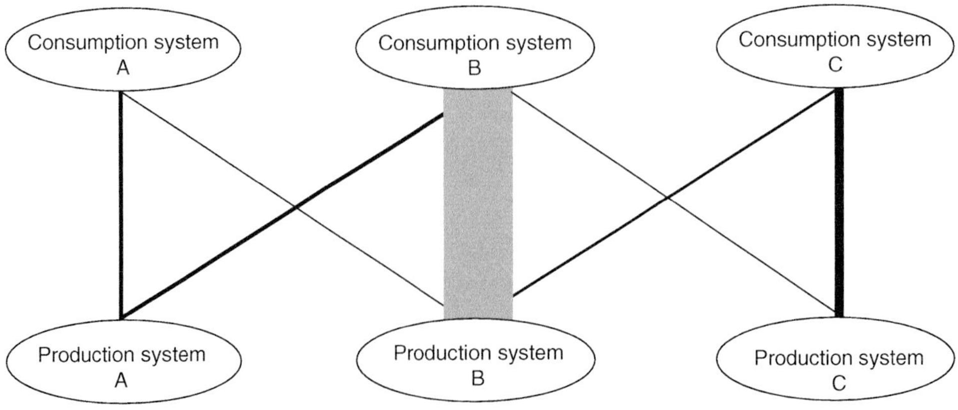

Fig. 11.2. Schematic representation of links between food consumption and production systems. The line thickness of the links is proportionate to the intensity of exchanges between specific systems.

Conclusion

The relationships between diets and food systems are complex, submitted as they are to contradicting tendencies to globalization and specification of diets. Multiple diets interact with multiple food systems; both of which are also influenced by numerous external factors, lifestyles, economic and social changes, and so on. The sustainability of a diet has to be assessed along both nutrition characteristics: contribution to the health of the individual or community, and its contribution to the environmental, economic and social sustainability of the food system(s). Conversely, the sustainability of a food system needs to be assessed both in terms of its capacity of perpetuating itself in the long term and of enabling sustainable diets. In other words, the sustainability of one depends, to a large part, on the influence it has on the sustainability of the other. This calls for identifying the critical links between specific diets and food systems in order to build upon them to improve their respective sustainability. Given the multiplicity of drivers, both internal and external, of food system changes and of actors' motivations for improvement, it is critical to adopt a holistic perspective, accounting not only for nutritional and environmental aspects but also for economic and social aspects, especially as they are often, ultimately, what enables and sustains change. Such a holistic perspective should build upon collective identities, recognize that consumers and other actors are often driven by global understanding of food 'quality', tasty, good for health and the environment, and thrive to accommodate it, while making clear that such objectives require that society recognizes the value of food, including through appropriate economic mechanisms.

References

Adinolfi, F., Capone, R. and El Bilali, H. (2015) *Assessing diets, food supply chains and food systems sustainability: towards a common understanding of economic sustainability*. Proceedings of an International Workshop: Assessing sustainable diets within the sustainability of food systems. Food and Agriculture Organization of the United Nations, Rome, Italy.

Berry, E., Dernini, S., Burlingame, B., Meybeck, A. and Conforti, P. (2015) Food security and sustainability: can one exist without the other? *Public Health Nutrition* 18(13), 2293–2302.

Burlingame, B. and Dernini, S. (Eds) (2012) Sustainable Diets and Biodiversity: Directions and solutions for policy, research and action. *Proceedings of the International Scientific Symposium Biodiversity and Sustainable Diets United against Hunger*, 3–5 November 2010, FAO Headquarters, Rome. http://www.fao.org/docrep/016/i3004e/i3004e.pdf (accessed 28 August 2018).

Ericksen, P.J. (2008) What is the vulnerability of a food system to global environmental change? *Ecology and Society* 13(2), 14.

Ericksen, P.J., Stewart, B., Dixon, J., Barling, D., Loring, P., *et al.* (2010) The value of a food system approach. In: Ingram, J.S.I., Ericksen, P.J. and Liverman, D.M. (eds) *Security and Global Environmental Change*. Earthscan, London, UK, pp. 25–45.

FAO (2008) *Climate change and food security: A Framework Document*. Food and Agriculture Organization of the United Nations, Rome, Italy.

FAO (2011) *Global Food Losses and Food Waste – Extent, Causes and Prevention*. Food and Agriculture Organization of the United Nations, Rome, Italy.

Finardi, C. and Menozzi, D. (2014) PDOs' role in reassuring consumers: the "Parmigiano Reggiano Terremotato" case. In: *Voluntary Standards for Sustainable Food Systems: Challenges and opportunities,* Proceedings of a workshop of the FAO/UNEP program on sustainable food systems. Food and Agriculture Organization of the United Nations, Rome, Italy.

Foresight Report (2011) The Future of Food and farming: Challenges and choices for global sustainability. The Government Office for Science, London. Available at https://assets.publishing.service.gov.uk/government/uploads/system/uploads/attachment_data/file/288329/11-546-future-of-food-and-farming-report.pdf (accessed 27 June 2018).

Gustafson, D., Gutman, A., Leet, W., Drewnowski, A., Fanzo, J. and Ingram, J. (2016) Seven food system metrics of sustainable nutrition security. *Sustainability* 8(3), 196.

Hammond, R.A. and Dubé, L. (2012) A systems science perspective and transdisciplinary models for food and nutrition security. *Proceedings of the National Academy of Sciences of the USA* 109(31), 12356–12363.

HLPE (2013) *Investing in Smallholder Agriculture for Food Security*. A report by the High Level Panel of Experts on Food Security and Nutrition of the Committee on World Food Security. Food and Agriculture Organization of the United Nations, Rome, Italy.

HLPE (2014) *Food Losses and Waste in the Context of Sustainable Food Systems*. A report by the High Level Panel of Experts on Food Security and Nutrition of the Committee on World Food Security. Food and Agriculture Organization of the United Nations, Rome, Italy.

HLPE (2017) *Nutrition and Food Systems*. A report by the High Level Panel of Experts on Food Security and Nutrition of the Committee on World Food Security. Food and Agriculture Organization of the United Nations, Rome, Italy.

Hu, F.B. (2002) Dietary pattern analysis: a new direction in nutritional epidemiology. *Current Opinion in Lipidology* 13(1), 3–9.

Ingram, J. (2011) A food systems approach to researching food security and its interactions with global environmental change. *Food Security* 3, 417–431.

IPCC (2014) *Climate Change 2014: Impacts, Adaptation, and Vulnerability*. Cambridge University Press, Cambridge, UK.

Lairon, D. and Kesse-Guyot, E. (2015) Profiles of organic food consumers, first lessons from the French NutriNet-Santé cohort study: a step towards diet sustainability. Proceedings of an International Workshop: Assessing sustainable diets within the sustainability of food systems. Food and Agriculture Organization of the United Nations, Rome, Italy.

Meybeck, A. and Gitz, V. (2017) Sustainable diets within sustainable food systems. *Proceedings of the Nutrition Society* 76, 1–11.

Milner, J., Greem, R., Dangour, A.D., Haines, A., Chalabi, Z., *et al.* (2015) Health effects of adopting low greenhouse gas emission diets in the UK. *BMJ Open* 5(4), e007364.

Monsivais, P., Scarborough, P., Lloyd, T., Mizdrak, A., Luben, R., *et al.* (2015) Greater accordance with the Dietary Approaches to Stop Hypertension dietary pattern is associated with lower diet-related greenhouse gas production but higher dietary costs in the United Kingdom. *American Journal of Clinical Nutrition* 102(1), 138–145.

Niola, M. (2015) *Homo Dieteticus, Viaggio nelle Tribù Alimentari*. il Mulino, Bolognia, Italy.

Sobal, J., Khanb, L.K. and Bisogni, C. (1998) A conceptual model of the food and nutrition system. *Social Science and Medicine* 47(7), 853–863.

Sonnino, R., Marsden, T. and Moragues-Faus, A. (2016) Relationalities and convergences in food security narratives: towards a place-based approach. *Transactions of the Institute of British Geographers* 41(4), 477–489. DOI: 10.1111/tran.12137

Tilman, D. and Clark, M. (2014) Global diets link environmental sustainability and human health. *Nature* 515, 518–522.

Tukker, A., Goldbohm, R.A., de Koning, A., Verheiden, M., Kleijn, R., *et al.* (2011) Environmental impacts of changes to healthier diets in Europe. *Ecological Economics* 70(10), 1776–1788.

UNCED (1992) Agenda 21. United Nations, New York, USA.

van Dooren, C., Marinussen, M., Blonkb, H., Aiking, H. and Vellinga, P. (2014) Exploring dietary guidelines based on ecological and nutritional values: A comparison of six dietary patterns. *Food Policy* 44, 36–46.

WFS (1996) *Rome Declaration on World Food Security*. Available at http://www.fao.org/docrep/003/w3613e/w3613e00.htm (accessed January 2016).

12 Understanding the Food Environment: the Role of Practice Theory and Policy Implications

Dalia Mattioni, Francesca Galli and Gianluca Brunori

Abstract

The last decade has witnessed an increase in the number of malnourished people worldwide, and particularly of people suffering from overweight and obesity. Research has shown the link between diet quality and the underlying food systems through the intermediation of the food environment. Specifically, a number of studies have analysed the role of the food retail environment and its impact on dietary intake largely by using quantitative geospatial tools – an approach that has been criticized on the grounds of its limited integration of social aspects linked to people's daily paths and lifestyles. This chapter contributes to a better understanding of the food environment by using social practice theory. Social practice theory can help complement the 'objective' measures used to study the retail environment, with more 'subjective' measures linked to its more symbolic and social dimensions by using more qualitative and/or mixed methods. With a view to changing people's food patterns, it is of fundamental importance to understand how food environments shape practices and vice versa, and where change can come about. In some cases, change can be triggered at the level of the material aspects of the food environment, such as the physical outlets where people buy their foods, and sometimes it can be triggered (also) by a change in the meaning attributed to food. This has implications for the types of policies adopted by governments and relevant stakeholders: policies need to be consistent and coherent, and aimed at changing both the material aspects of the food environment as well as the competence people need to make it work and the meaning attached to healthy eating.

Introduction

In the face of rising rates of overweight and obesity worldwide, the work carried out in the past decades by public health academics and practitioners has highlighted the key role played by the food environment in shaping people's food preferences and ultimately their dietary intake. While there are numerous factors influencing dietary intake and eating behaviours, there has been a general recognition of the role of the (global) food system in advancing the nutrition transition worldwide. No doubt the modern food system has increased the amount of dietary diversity in a number of countries, but evidence shows that if in the past 20 years there has been an increase in the consumption of healthy foods, this has been outpaced by the parallel increase in the consumption of unhealthy items (Inamura *et al.*, 2015). More importantly, the current food system keeps influencing individuals to adopt unhealthy nutrition patterns by way of relative prices, behavioural models and marketing stimuli. Taking the cue from several calls for action, governments have put in place a number of food environment policies aimed at tackling the increase in overweight and obese people in their countries – policies that range from taxation on sugary beverages to labelling and zoning. The last decade has also witnessed an increase in research being undertaken on the food environment, particularly around the built (or retail) food environment, with some critiques being levelled both on methodological and on substantive grounds.

Against this backdrop, the aim of this chapter is twofold: first, it wishes to better understand the nature of food environments, and suggests the use of practice theory to provide some conceptual and practical 'strength'; and second, following on from this, it wishes to make some reflections on the implication of this 'revised' notion of the food environment on the types and mix of policies to be implemented to (sustainably) improve diets.

The Food Environment: State of the Art and a Critical Review

Although there is no one agreed upon definition, the food environment can be conceptualized as the collective physical, economic, policy and sociocultural conditions that influence people's food choices. They are the underlying determinants of what people eat, and are made up of the foods that are available, affordable and acceptable to people in their surroundings (Swinburn *et al.*, 2013; IFPRI, 2015). In the short term, food environments influence people's food choices, while in the long term they have an important role to play in affecting people's food preferences and habits (Hawkes *et al.*, 2015). According to a growing literature, food environments have developed in a way that makes it more difficult for people to consume high-quality diets, in spite of rising incomes (Swinburn *et al.*, 2011; Popkin *et al.*, 2011; GLOPAN, 2016).

The food environment concept is closely linked to food security, insofar as the food environment influences availability, access and utilization of food, three of the four pillars of food security (FAO, 2008).[1] Some global trends have strongly affected food environments in the world. In terms of availability, shifts in the global supply system, especially with the rise of transnational food corporations (TFCs) in the past three to four decades have made energy-dense, nutrient-poor processed food products much more available. Technological advances in food processing, such as the extraction of vegetable oils and other processing techniques, have made the production of this type of food possible, profitable and less expensive for consumers (Popkin *et al.*, 2011; Monteiro *et al.*, 2013).

These trends in availability have not come without consequences for affordability: generally speaking, 'unhealthy foods' such as energy-dense, nutrient-poor processed food products are cheaper than 'healthy foods' such as fresh fruits and vegetables. The reasons for this are manifold: among factors that have shaped the cost of food, an important role has been played by the vertical integration of TFCs and global sourcing that has allowed food companies to use cheaper ingredients and take advantage of economies of scale, thus making it possible to cut costs and keep prices – especially of processed packaged foods – low (Hawkes *et al.*, 2009; Gomez and Ricketts, 2013). The topic is, however, the subject of debate as the evidence put forth by various studies on the relationship between healthiness of foods and costs for consumers is still inconclusive (Drewnowski and Darmon, 2005; Rao *et al.*, 2013; Jones *et al.*, 2014) and more research is needed on issues such as the metrics used (e.g. cost per calories versus cost per nutrient), or on the effects of price changes on total diet rather than on the consumption of a specific product (Mozaffarian *et al.*, 2012).

Acceptability, further qualified by some to include desirability and convenience (Herforth and Ahmed, 2015; Turner *et al.*, 2017), has also to a certain extent been shaped by availability and affordability. The vast amount of funds being spent by large companies on advertising and marketing (including the strategic use of shelf space) have all played an important role in shaping acceptability, with an increased use of quick-to-use ultra-processed foods (Hawkes *et al.*, 2009; Cameron *et al.*, 2012; Bereuter and Glickman, 2015). Processed food and modern distribution systems are today, particularly in low- and medium-income countries (LMICs), 'trendy'. One of the implications of this shift has been in terms of loss of knowledge: people, especially young people, are no longer knowledgeable about certain types of (usually traditional) foods, nor do they know how to cook/process them. Another consequence has been a certain level of destructuring of the traditional meal and an increase in snacking and eating out (Mestdag, 2005; Yates and Warde, 2015). While up to 50 years ago snacking was a rare activity among adults, now in countries such as the USA, Canada, Brazil, Mexico and China, up to one-quarter of all calories consumed come in snack form and, as incomes rise, so does the proportion of ultra-processed foods that are used as snacks (Lang *et al.*, 2009; Monteiro *et al.*, 2013).

The effects of the above shifts in the global food system and the global food environment have been mediated in different ways at the local level, reflecting different 'local geographies of consumption' and 'place-based food environments' (Turner *et al.*, 2017). The last decade has seen a burgeoning number of context-specific research work, especially with respect to measures related to the retail setting (Ni Mhurchu *et al.*, 2013; Townsend and Lake, 2017). In this body of literature, the food environment has been investigated mainly with reference to the physical dimension. Indeed, in the last decade, food environment research has been dominated by the use of static geospatial quantitative methods. The evidence produced so far on how retailers are geographically distributed in a given territory, how much of different food types they sell, at what price and their effect on dietary patterns/intake is mixed. A series of systematic reviews carried out mainly in the US, Canada, Australia and some European countries show, for example, that there is moderate evidence of an association between physical distribution patterns of retail outlets and dietary patterns and/or intake (Giskes *et al.*, 2011; Caspi *et al.*, 2012; Black *et al.*, 2014). The strong evidence that has been collected relates to the link between local food environments and inequalities in high income countries, given that there is a higher density of shops selling unhealthy foods in low-income neighbourhoods compared to other areas (Mozaffarian *et al.*, 2012). However, here too, density of shops is not always related to unhealthy dietary patterns.

These weaknesses introduce a series of critiques to the above conceptualization of the food environment. A first set of critiques relates to the heterogeneity of measures and variables used to map out the retail food environment, which may be leading to inconsistent results and conclusions in terms of impact on dietary patterns (Lytle, 2009; Kelly *et al.*, 2011). A second strand of critiques tries to adapt the concept of social environment to people's mobility: they may travel beyond their neighbourhood and consume healthier products (Townsend and Lake, 2017).[2] Some community food environment studies were carried out that consider 'individualized living spaces', that is, the daily paths used by people to buy or eat food beyond their neighbourhood (Black *et al.*, 2014; Zenk *et al.*, 2011; Kestens *et al.*, 2010), or analysis of food shopping routines for in-store food environments (Thompson *et al.*, 2013), or even subjective perceptions of the neighbourhood environment with innovative participatory tools (Díez *et al.*, 2017).

All of the above point to the fact that individual daily paths and lifestyles do not always coincide with (rigid) geographical boundaries (see Fig. 12.1) and that it is important to consider what Turner *et al.* (2017) have labelled as the 'personal food environment' with an aim to reintroduce the 'social' back into a conceptual framework of food environment that has used a rather static geospatial language. Although spelt out in 2007, Cummins' plea for a greater consideration of 'social processes and symbolic relations between individuals and their environment' and of a 'deeper understanding of how "environment" gets into the "body"', is still very valid today (Cummins, 2007).

Food Environment and Social Practice Theory

A third strand of critiques suggests that food environments are in fact made up not only of material aspects, such as the geographical distribution of food outlets for example, but also of symbolic and social dimensions, and that an understanding of all three dimensions is important to help make food environments work for high-quality diets. This approach points to the need to complement the 'objective' measures used to study the retail environment, such as geographical information systems and Google Earth, with more 'subjective' measures linked to people's purchasing practices, perceptions, motivations and taste by using qualitative approaches or a mix of both (Lytle, 2009; Kelly *et al.*, 2011; Caspi *et al.*, 2012; Turner *et al.*, 2017). In this context, we believe that theory

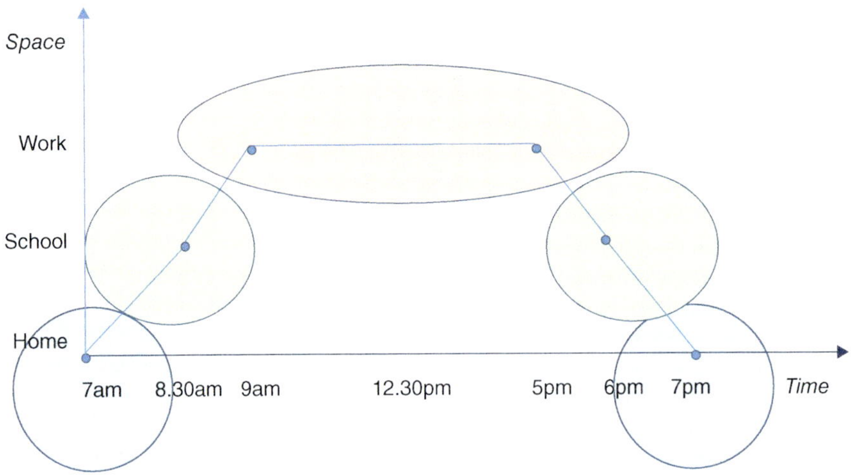

Fig. 12.1. Daily time–space life path of a mother. As Susan goes about her daily life (blue line in the graph), she encounters different food retail environments (coloured ovals) where she buys her food. She takes her children to school, and because of convenience and good prices, she shops at a nearby supermarket just after having dropped them off. She then goes to work and her eating habits are influenced by what she finds at work (canteen, or restaurants around her work area). On the way back, her children are hungry after having finished school and a small convenience store close to school sells snacks that her children love and she feels that she has been a 'good mother'. She may thus actually never shop in the neighbourhood where she lives (blue circles). Source: diagram adapted from Giddens (1984, p. 115).

of practice can be a useful theoretical framework to reintroduce the social back into the concept of food environment as it has so far been used. It can help better understand how individuals intersect with, shape, and at the same time 'embody' the food environment, and it can explain how food fits into people's daily lives and routines, how it is valued and perceived, and what are the formal and informal conventions that govern food practices in specific contexts.

A social practice can be considered as a 'routinized type of behavior', an entity made up of various interdependent elements: 'forms of bodily activities, forms of mental activities, "things" and their use, a background knowledge in the form of understanding, know-how, states of emotion and motivational knowledge' (Reckwitz, 2002, p. 249). What holds together the various elements of a practice are: understandings (of what to do and say); procedures (explicit rules, principles, precepts and instructions); and engagements (projects, purposes, beliefs, emotions) (Schatzki, 1996; Warde, 2005). A practical illustration of this definition is provided by Shove:

> skateboarding consists of a complex amalgam of skateboards and street spaces along with the bodily competencies required to ride the board and to use the affordances of the street to turn tricks; the rules and norms that define the practice of skateboarding; its meaning to practitioners and to outsiders ... and so on. As such skateboarding exists as a recognizable conjunction of *elements*, consequently figuring as an *entity*.
>
> (Shove, 2012, p. 7)

The shared tacit knowledge and 'embodied know-how' that underlie practices is central in understanding first, and using later, practice theory. In contrast with voluntarist and rational notions of action, conduct within practices is governed by convention, tradition, routine and what Giddens (1984) refers to as 'practical knowledge' as opposed to discursive knowledge. Codifications therefore that often do not require 'much reflection or conscious awareness on the part of the bearers ... notwithstanding a capacity for reflective monitoring of performance' (Warde, 2005).

We have noted above that practices are made up of elements. Different practice theorists have developed different typologies of elements that make up practices (cf. Evans *et al.*, 2012). In this paper, we will use the typology used by Shove *et al.* (2012) for its stated intent of 'conceptualizing stability and change' which fits in well with the objective of this chapter. In *The Dynamics of Social Practice*, Shove *et al.* distinguish three types of elements:

- materials: including things, technologies, tangible physical entities, and the stuff that objects are made of;
- competences: skills, know-how and technique; and
- meanings: symbolic meanings, ideas and aspirations.

Practices come to being when elements are linked up in specific configurations, and stability over time is ensured when 'connections between defining elements [are] renewed time and again' through regular enactment (Shove *et al.*, 2012). As the individual is seen as a 'carrier' of a practice, in fact of many different practices ('bundles' of practices), practices can share some elements, and change in one practice can occur as the result of a change in an element of another practice (see Fig. 12.2). Change is therefore a reconfiguration of the elements that make up practices.

Just like practices, food environments are also made up of material, symbolic and social dimensions. Both are subject to 'duality' (Giddens, 1984): food environments provide rules and resources, opportunities and constraints for individual action, they shape individual action and at the same time they are shaped by actors' attitudes, choices and behaviour; practices, as routinized individual activities that can be justified in terms of given rules and resources, mirror this dualism in their being at the same time moulded by people's daily actions but also living a trajectory of their own that shapes people's way of 'doing' and 'knowing'. In this paper, we consider food environments primarily as the 'structural' guise of how people 'do' food – the physical outlets people buy their food at, the prices, the labels, the way a product is formulated, and so forth – and social practices as more closely linked to people's actions or 'agency' given their nature of routinized individual activities as outlined above. Social practice theory can help understand how structural aspects (e.g. the retail environment) are connected to the meaning and competence needed to make a specific food practice persist or change over time.

How Does Change Come About?

In the challenging context of aiming at changing people's food patterns and how they eat, it is of fundamental importance to understand how

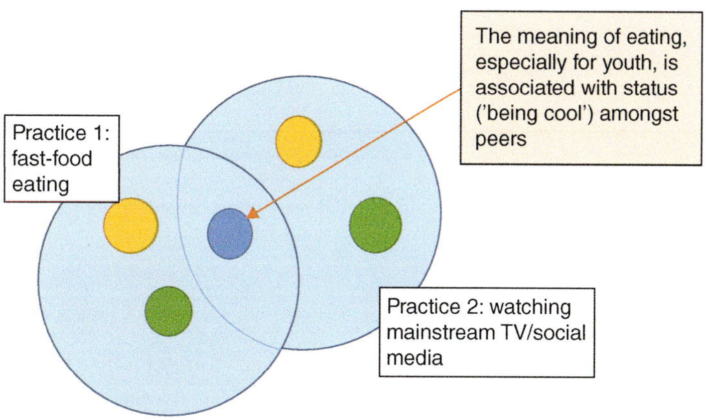

Fig. 12.2. An illustration of how unhealthy eating is linked to other practices: many young individuals today are carriers of these two practices (large light blue circles): fast-food eating and watching mainstream TV/social media. The two practices share the 'meaning' element (small blue circle), a 'new' meaning given to eating out that has allowed the new practice of 'fast-food eating' to emerge.

food environments shape practices (rather than individual behaviour) and vice versa, in a mutually and dynamic reinforcing configuration. As elements of a practice interact and co-evolve, novelty can come from any quarter (Shove *et al.*, 2012).

As seen above, ample evidence exists on the way that the food environment influences people's food practices,[3] and of how these, in turn, strengthen the existing food environment. A school child attending a school where meals are catered for by a fast-food company will not have many options but to eat in an unhealthy way in that particular setting. Exposure at such a young age to that type of food may have an influence on his/her aesthetic judgement of taste and lead to similar types of choice also outside of school, thus reinforcing a specific type of food retail environment. A change in the 'meaning' element of food, however, may trigger a change in the whole practice, as is illustrated by the intervention of food movements and alternative food networks (AFN) in a number of countries, and this in turn can influence the broader food environment (see Fig. 12.3). The GAS (Gruppo di Acquisto Solidale [Solidarity Purchase Group]) movements in Italy, for example, started off as a movement of (predominantly urban) consumers acting according to their values based on solidarity with farmers, environmental care and a desire to eat in a healthier way, reflecting the three 'interlocking cares' illustrated by Dowler *et al.* (2009) in their study of AFNs in the UK: a care for the environment and the local economy, a care about transparency and integrity in the food system, and a care for health and wholeness. This shift in values is what triggered the co-production of a new way of purchasing/supplying food, which in turn requires new competences in terms of cooking and conserving food, organizing the collective purchase and collection of goods and new understandings on shifting tastes within the household (Brunori *et al.*, 2012; Fonte, 2013). Linkages were thus forged and stabilized through routine enactment between values/meaning, competences and new material infrastructure.

This change in practices can, in turn, influence the broader food environment in several ways: through a 'graduation process' whereby consumers bring their new skills, values and expectations about 'good food' into more conventional arenas thus shifting demand (Kneafsey *et al.*, 2008), or by influencing the way conventional players behave (Brunori *et al.*, 2012), or even still by introducing new food conventions into a food system thus creating the grounds for new food 'assemblages' to be created along more sustainable and healthier lines (Brunori *et al.*, 2017). In other words, practices can enrich the diversity or pool of food languages, thus allowing for a greater variety of food environments to be shaped and adjusted according to the different contexts.

Change can also be effected at the level of food environments first through structural interventions. There are several examples of government action aimed at doing so, such as making health-promoting labelling mandatory (Chile), introducing fast-food zoning around schools (USA, South Korea) and regulating food marketing and advertising to children (Norway).

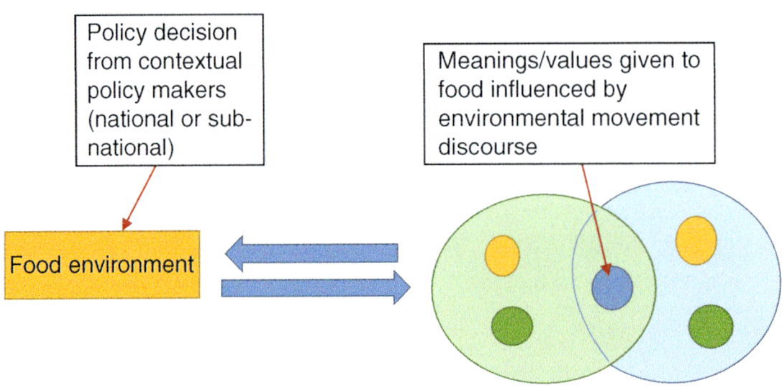

Fig. 12.3. An illustration of the reciprocal nature of food environment and food practices: change (arrows) can be triggered in any part of the diagram depending on the socioeconomic and institutional context and depending on other practices that the food practices may be 'bundled' with.

Other examples have had the explicit objective of changing (low income) people's food practices by making sure that more farmers' markets (FM) open in low-income areas, and more importantly, by equipping them with machines that would allow them to accept federal food benefit vouchers (SNAP), as has occurred in the USA. Here, the practice of shopping at FMs with SNAP subsidies has thus become a 'normal activity' thanks to the establishment of material elements (the physical location of FMs and the machines to 'read' the vouchers), the evolving meaning about food access (as a result of national and local efforts aimed at food justice and health), and improved competences (new skills on how to use the new machines, cooking classes) (Cohen and Ilieva, 2015). Following on from the example of school meals, the introduction of a healthy meal at school coupled with an educational programme could change the 'meaning' and 'competence' elements of the child's eating practice, thus possibly changing his/her food practices and 'bringing' the new practice home (Morgan and Sonnino, 2008; Giuca, 2016).

Policy Implications

A number of high-level and global reports have been published in recent years that show the rising importance of affecting a change in the food environment as a major avenue towards diminishing worryingly high levels of malnutrition worldwide (Swinburn *et al.*, 2013; IFPRI, 2015; GLOPAN, 2016; FAO, 2017; HLPE, 2017). All converge on a number of policy areas where action is needed to foster a healthier food environment: (i) nutrition labelling; (ii) food provisioning (or food offered in specific settings, such as schools); (iii) economic incentives/disincentives; (iv) food composition (or nutritional quality of foods produced by food processors); (v) food promotion (advertising and marketing); (vi) food retailing; (vii) food systems (including: agricultural production; storage; processing; retail). Some policy frameworks – such as the NOURISHING[4] framework – also include (viii) behaviour change communication as a key complementary area of work, and this includes: informing people about food and nutrition through public awareness, nutrition advice and counselling in health care settings, and giving nutrition education and skills.

With respect to the sustainable diets, a mix of actions coming from each of these policy areas will need to be set in motion by policymakers – both national and local – to transition towards a sustainable diet. Some material elements have already been put in place in several countries that point in the right direction, such as public procurement of local and organic food in schools, and the creation of FMs in multiple cities (Morgan and Sonnino, 2008), and although the evidence is not solid enough on the extent to which this has led people to eat in a healthier way, there are studies that show that those who buy at FMs tend to consume more fresh fruits and vegetables in their overall diet (McCormack *et al.*, 2010; Dannefer *et al.*, 2016; Pitts *et al.*, 2013; Minaker *et al.*, 2016).

In these cases, the material itself has contributed to changing competences and meanings around healthy eating through a form of 'experiential learning' that through performance has become the practical knowledge that underlies social practices. What will be needed, however, is an emphasis not only on the material aspects of diets as a practice – on labelling, for example, to improve information – but concurrently on those aspects related to competence and meaning so as to 'bind' the three elements together.

South Korea is an example where, in the face of rising obesity levels, a number of actions were taken by the government to change the country's food environment: green food zones around schools to protect children from unhealthy food items, healthy school feeding, but also a consistent effort aimed at keeping the traditional Korean diet alive, such as social campaigning, investing in cooking classes, organizing public events and other information activities (Lee *et al.*, 2002).

'Policy', however, is not only the domain of government: civil society and the private sector too can forge policy from their respective quarters (Lang *et al.*, 2009). As seen above, it was the work of local-level, grassroots initiatives, such as the GAS in Italy or the AMAPs (Association pour le maintien d'une agriculture paysanne [Association for Maintaining Small Scale Family Farming]) in France, which encouraged a shift in food practices. Translated in governance terms, what this means is a need to 'further support, fund and link these movements, coalitions and networks together if change is to come about' (HLPE, 2017, p. 117). Meanings and competences change when routinized food practices are exposed to 'ethical problematization' (Barnett *et al.*, 2010), and a food system that opens spaces of deliberation accelerates this change. Indeed, ethical problematization encourages private business to anticipate emerging societal concerns by adapting their marketing strategies. In a number of countries such as Costa Rica, Germany and Japan, voluntary standards have been adopted on salt, fat and sugar reduction, and market narrative has changed accordingly. In some countries, the private sector has also tried to increase the acceptability of foods by producing packaged foods that are nutritious, and can make it more convenient for people to cook and eat healthy meals, for example by selling bagged, cut up or shredded vegetables that can easily be added to salads or cooked (HLPE, 2017). In Mediterranean cultures, for example, where eating vegetables is still considered acceptable, but where problems of time make it more and more difficult for people to do so, a joint action by government (national and local), grassroots and private actors needs to be taken to shift the material obstacles and to strengthen the narrative and competence around healthy eating.

From the perspective of sustainable diets, the question related to food environment is: what needs to change in the elements of the current food practices and the connections between them, in order to move them towards healthier and more sustainable grounds? The point made here is that, as governments begin to design specific policies aimed at shaping food environments intended to help people eat better, it will be important to ensure governance arrangements that include a mix of consistent and coherent policies, that is, 'integrated policy strategies' (Howlett and Rayner, 2007) aimed at changing the material aspects of the food environment, concurrently with the competence people need to make it work and the meaning attached to healthy eating.

Notes

[1] The four pillars of food security are: (i) physical availability of food; (ii) economic and physical access to food; (iii) food utilization; and (iv) stability of the other three dimensions over time (FAO, 2008).

[2] It is important to note that food environment research in low- and middle-income countries (LMICs) has been sparse and is only now beginning. A seminal review carried out in 2013 on food value chains and nutrition points to a good supply of cheap fresh food in poor urban areas due to the predominance of traditional short supply chains in many LMICs. At the same time, the growth of modern retailing is leading to an increase of cheap, ultra-processed foods possibly leading to over-nutrition (Gomez and Ricketts, 2013). More research on food retail environments is warranted in LMICs as the nutrition transition progresses.

[3] For the purposes of this chapter, a food practice (or 'eating') has been taken to be a compound practice that can be divided up into four distinct integrative practices: purchase/supply of food, cooking, organization of the meal and the aesthetic judgement of taste (Warde, 2015).

[4] Available at http://www.wcrf.org/int/policy/our-policy-work/our-policy-framework-promote-healthy-diets-reduce-obesity (accessed 15 October 2017).

References

Barnett, C., Cloke, P., Clarke, N. and Malpass, A. (2010) *Globalizing Responsibility: The Political Rationalities of Ethical Consumption*. John Wiley & Sons, Chichester, UK.

Bereuter, D. and Glickman, D. (2015) *Healthy Food For a Healthy World: Leveraging Agriculture and Food to Improve Global Nutrition*. The Chicago Council on Global Affairs, Chicago, Illinois, USA.

Black C., Moon, G. and Baird, J. (2014) Dietary inequalities: what is the evidence for the effect of the neighbourhood food environment? *Health and Place* 27, 229–242.

Brunori, G., Rossi, A. and Guidi, F. (2012) On the new social relations around and beyond food. Analysing consumers' role and action in Gruppi di Acquisto Solidali (Solidarity Purchasing Groups). *Sociologia Ruralis* 52, 1.

Brunori, G., Galli, F. and Grando, S. (2017) Sustainable agri-food systems: a reflection on assemblages and diversity. *Systèmes alimentaires/Food Systems* 1, 21–39.

Cameron, A., Thornton, L., McNaughton, S.A. and Crawford, D. (2012) Variation in supermarket exposure to energy-dense snack foods by socio-economic position. *Public Health Nutrition* 16(7), 1178–1185.

Caspi, C., Sorensen, G., Subramanian, S.V. and Kawachi, I. (2012) The local food environment and diet: a systematic review. *Health and Place* 18(5), 1172–1187.

Cohen, N. and Ilieva, R. (2015) Transitioning the food system: A strategic practice management approach for cities. *Environmental Innovation and Societal Transitions* 17, 199–217.

Cummins, S. (2007) Neighbourhood food environment and diet-time for a new conceptual model? *Preventive Medicine* 44, 196–197.

Dannefer, R., Bryan, E., Osborne, A. and Sacks, R. (2016) Evaluation of the Farmers' Markets for Kids programme. *Public Health Nutrition* 19(18), 3397–3405.

Díez, J., Conde, P., Sandin, M., Urtasun, M., López, R., *et al.* (2017) Understanding the local food environment: A participatory photovoice project in a low-income area in Madrid, Spain. *Health and Place* 43, 95–103.

Dowler, E., Kneafsey, M., Cox, R. and Holloway, L. (2009) 'Doing food differently': reconnecting biological and social relationships through care for food. *The Sociological Review* 57, 200–221.

Drewnowski, A. and Darmon, N. (2005) Food choices and diet costs: an economic analysis. *The Journal of Nutrition* 135(4), 900–904.

Evans, D. (2011) Blaming the consumer – once again: the social and material contexts of everyday food waste practices in some English households. *Critical Public Health* 21(4), 429–440.

FAO (2008) *An Introduction to the Basic Concepts of Food Security*. Food and Agriculture Organization of the United Nations, Rome, Italy.

FAO (2017) *The Future of Food and Agriculture: Trends and Challenges*. Food and Agriculture Organization of the United Nations, Rome, Italy.

Fonte, M. (2013) Food consumption as social practice: Solidarity Purchase Groups in Rome, Italy. *Journal of Rural Studies* 32, 230–239.

Giddens, A. (1984) *The Constitution of Society*. Polity Press, Cambridge, UK.

Giskes, K., van Lenthe, F., Avendano-Pabon, M. and Brug, J. (2011) A systematic review of environmental factors and obesogenic dietary intakes among adults: are we getting closer to understanding obesogenic environments? *Obesity Reviews* 12, 95–106.

Giuca, S. (2016) I programmi della Pac per la fornitura di prodotti agricoli alle scuole. Available at https://agriregionieuropa.univpm.it/it/content/article/31/46/i-programmi-della-pac-la-fornitura-di-prodotti-agricoli-alle-scuole (accessed 18 October 2017).

GLOPAN (2016) *How can Agriculture and Food System Policies improve Nutrition*? Technical Brief. GLOPAN, London, UK.

Gomez, M. and Ricketts, K. (2013) Food value chain transformations in developing countries: selected hypotheses on nutritional implications. *Food Policy* 42, 139–150.

Hawkes, C., Chopra, M. and Friel, S. (2009) Globalization, trade and the nutrition transition. In: Labonte, R., Schrecker, T., Packer, C. and Runnels V. (eds) *Globalization and Health: Pathways, Evidence and Policy*. Routledge, New York, USA, pp. 235–262.

Hawkes, C., Smith, T., Jewell, J., Wardle, J., Hammond, R., *et al.* (2015) Smart policies for obesity prevention. *The Lancet* 385, 2410–2421.

Herforth, A. and Ahmed, S. (2015) The food environment, its effects on dietary consumption, and potential for measurement within agriculture–nutrition interventions. *Food Security* 7(3), 505–520.

HLPE (2017) *Nutrition and Food Systems*. A report by the High Level Panel of Experts on Food Security and Nutrition of the Committee on World Food Security. Food and Agriculture Organization of the United Nations, Rome, Italy.

Howlett, M. and Rayner, J. (2007) Design principles for policy mixes: cohesion and coherence in 'new governance arrangements'. *Policy and Society* 26, 1–18.

IFPRI (2015) *Global Nutrition Report*. International Food Policy Research Institute, Washington, DC, USA.

Inamura, F., Micha, R., Kathibzadeh, S., Fahimi, S., Powels, J. and Mozaffarian, D. (2015) Dietary quality among men and women in 187 countries in 1990 and 2010: a systematic assessment. *The Lancet* 3, 132–142.

Jones, N.R., Conklin, A.I., Suhrcke, M. and Monsivais, P. (2014) The growing price gap between more and less healthy foods: analysis of a novel longitudinal UK dataset. *PLoS One* 9(10), e109343.

Kelly, B., Flood, V. and Yeatman, H. (2011) Measuring local food environments: an overview of available methods and measures. *Health and Place* 17, 1284–1293.

Kestens, Y., Lebel, A., Daniel, M., Theriault, M. and Pampalon, R. (2010) Using experienced activity spaces to measure foodscape exposure. *Health and Place* 16, 1094–1103.

Kneafsey, M., Cox, R., Holloway, L., Dowler, E., Venn, L. and Tuomainen, H. (2008) *Reconnecting Consumers, Producers and Food: Exploring Alternatives*. Berg Publishers, Oxford, UK.

Lang, T., Barling, D. and Caraher, M. (2009) *Food Policy*. Oxford University Press, Oxford, UK.

Lee, M., Popkin, B. and Kim S. (2002) The unique aspects of the nutrition transition in South Korea: the retention of healthful elements in their traditional diet. *Public Health Nutrition* 5(1A), 197–203.

Lytle, A. (2009) Measuring the food environment: state of the science. *American Journal of Preventive Medicine* 36, 134–144.

McCormack, L.A., Laska, M.N., Larson, N.I. and Story, M. (2010). Review of the nutritional implications of farmers' markets and community gardens: a call for evaluation and research efforts. *Journal of the American Dietetic Association* 110(3), 399–408.

Mestdag, I. (2005) Disappearance of the traditional meal: temporal, social and spatial destructuration. *Appetite* 45(1), 62–74.

Minaker, L., Olstad, D., Thompson, M., Raine, K., Fisher, P. and Lawrence F. (2016) Associations between frequency of food shopping at different store types and diet and weight outcomes: Findings from the NEWPATH study. *Public Health Nutrition* 19(12), 2268–2277.

Monteiro, C., Moubarac, J., Cannon, G., Ng, S. and Popkin, B. (2013) Ultra-processed products are becoming dominant in the global food system. *Obesity Reviews* 14(2), 21–28.

Morgan, K. and Sonnino, R. (2008) *The School Food Revolution*. Earthscan, London, UK.

Mozaffarian, D., Afshin, A., Benowitz, N., Bittner, V., Daniels, S., *et al.* (2012) Population approaches to improve diet, physical activity, and smoking habits: a scientific statement from the American Heart Association. *Circulation* 126(12), 1514–1563.

Ni Mhurchu, C., Vandevijvere, S., Waterlander, W., Thornton, L.E., Kelly, B., *et al.* (2013) Monitoring the availability of healthy and unhealthy foods and non-alcoholic beverages in community and consumer retail food environments globally. *Obesity Reviews* 14(3), 108–119.

Pitts, S., Wu, Q., McGuirt, J.T., Crawford, T.W, Keyserling, T.C. and Ammerman A.S. (2013) Associations between access to farmers' markets and supermarkets, shopping patterns, fruit and vegetable consumption and health indicators among women of reproductive age in eastern North Carolina, U.S.A. *Public Health Nutrition* 16(11), 1944–1952.

Popkin, B., Adair, L. and Ng, S. (2011) Global nutrition transition and the pandemic of obesity in developing countries. *Nutrition Reviews* 7(1), 3–21.

Rao, M., Afshin, A., Singh, G. and Mozaffarian, D. (2013) Do healthier foods and diet patterns cost more than less healthy options? A systematic review and meta-analysis. *BMJ Open*, 3(12), e004277.

Reckwitz, A. (2002) Towards a theory of social practices: a development in culturalist theorizing. *European Journal of Social Theory* 5(2), 243–263.

Schatzki, T. (1996) *Social Practices: A Wittgensteinian Approach to Human Activity and the Social*. Cambridge University Press, Cambridge, UK.

Shove, E., Pantzar, M. and Watson, M. (2012) *The Dynamics of Social Practice*. Sage, London, UK.

Swinburn, B., Sacks, G., Hall, K., McPherson, K., Finegood, D., *et al.* (2011) The global obesity pandemic: shaped by global drivers and local environments. *The Lancet* 378, 804–814.

Swinburn, B., Sacks, G., Vandevijvere, S., Kumanyika, S., Lobstein, T., *et al.* (2013) INFORMAS (International Network for Food and Obesity/non-communicable diseases Research, Monitoring and Action Support): overview and key principles. *Obesity Reviews* 14(2), 1–12.

Thompson, C., Cummins, S., Brown, T. and Kyle, R. (2013) Understanding interactions with the food environment: an exploration of supermarket food shopping routines in deprived neighbourhoods. *Health and Place* 19C, 116–123.

Townsend, T. and Lake, A. (2017) Obesogenic environments: current evidence of exploring the built and food environments. *Perspectives in Public Health* 137(1), 38–44.

Turner, C., Kadiyala, S., Aggarwal, A., Coates, J., Drewnowski, A., *et al.* (2017) *Concepts and Methods for Food Environment Research in Low and Middle Income Countries*. Agriculture, Nutrition and Health Academy Food Environments Working Group (ANH-FEWG). Innovative Methods and Metrics for Agriculture and Nutrition Actions (IMMANA) Programme. London, UK.

Warde, A. (2005) Consumption and theories of practice. *Journal of Consumer Culture* 5(2), 131–153.

Warde, A. (2015) On the sociology of eating. *Revue d'Etudes en Agriculture et Environnement* 96(1), 7–15.

Yates, L. and Warde, A. (2015). The evolving content of meals in Great Britain. Results of a survey in 2012 in comparison with the 1950s. *Appetite* 84, 299–308.

Zenk, S.N., Schulz, A.J., Matthews, S.A., Odoms-Young, A., Wilbur, J., *et al.* (2011) Activity space environment and dietary and physical activity behaviors: a pilot study. *Health and Place* 17, 1150–1161.

13 Sustainable Diets: Social and Cultural Perspectives

F. Xavier Medina and Alicia Aguilar

Abstract

The incorporation of sustainability issues into the international agri-food and nutritional agenda has been increasingly discussed over the last decades. In this framework, anthropological concerns with food and nutrition have increased greatly in the last five decades, and the development has been across the subdisciplines of anthropology and in conjunction with other academic disciplines. Nevertheless, social and cultural aspects related to food are, even today, frequently neglected, regarded as secondary or less important in comparison to other 'main' subjects like health or economy. In this sense, the aim of this chapter is to focus on the social and cultural perspective of food and its intrinsic relationship with diets, territories and sustainability, highlighting this point of view as an essential part of a very complex panorama, helping to have a more comprehensive and less partial view of the situation.

Introduction: What is a Sustainable Diet?

The incorporation of sustainability issues into the international agri-food and nutritional agenda has been increasingly discussed over the last decades. The concept of sustainable diets acknowledges the interdependencies of food production and consumption with food requirements and nutrient recommendations (Dernini *et al.*, 2016), and at the same time, expresses the notion that food (including production, distribution and consumption, social and cultural aspects, health, or economy, among others) cannot work separately from that of the *ecosystem*.

As the Food and Agriculture Organization (FAO) has highlighted in today's most accepted definition of *sustainable diets*, established after the International Scientific Symposium on Biodiversity and Sustainable Diets United Against Hunger: 'Sustainable diets are protective and respectful of biodiversity and ecosystems, culturally acceptable, accessible, economically fair and affordable; nutritionally adequate, safe and healthy; while optimizing natural and human resources' (FAO, 2011; Burlingame and Dernini, 2011). After this main definition, some authors, like Jonston *et al.* (2014) adds that sustainable diets must be also *culturally sensitive and acceptable*.

As in other different fields and items, cultural aspects have traditionally been neglected, observed only as subservient or complementary to other, more important items. Even after the declaration by UNESCO of the Mediterranean diet as intangible cultural heritage of humanity, definitions of *diet, sustainable diets*, or the *Mediterranean diet* continue to relegate to the background those aspects more closely linked to culture, even the more open groups and those drafted by supranational institutions. In this sense, sustainable diets are protective and respectful of biodiversity and ecosystems, *culturally acceptable,*[1] accessible, economically fair and affordable; nutritionally adequate, safe and healthy; while optimizing natural and human resources (FAO, 2011; Burlingame and Dernini, 2011). Jonston *et al.* (2014) add that food must be *culturally sensitive and acceptable*.

But more than 'acceptable', food (but not only) must be culturally 'coherent' (Medina, 2015, 2017). The Cambridge Dictionary (online edition) defines 'acceptable' as: 'satisfactory and able to be agreed to or approved of', or simply 'just good enough, but not very good'. In this sense, if something must be agreed or approved, usually it is because it is not taking part of the system itself, and must be *accepted* from the outside. Or, as the second of the meanings cited above explains very well, it is 'just good enough, but not very good'.

On the other hand, 'coherent' is defined as 'it is clear and carefully considered, and each part of it connects or follows in a natural or reasonable way'. In this sense, many things may be *acceptable*, but very few are *coherent*. From a local and sustainable point of view, betting on the cultural 'acceptability' of a food can open up too much the spectrum of what is acceptable as edible. But its cultural coherence within a system appeals to other aspects that have nothing to do with what is simply acceptable.

Regarding this last point, it is necessary to remark that any discussion about food also involves the consequences of social and cultural constructions. Food and eating behaviours, in general, fall within the framework of the societies that produce and recreate them, and therefore within specific sociocultural systems (Medina, 1996).

Anthropological concerns with food and nutrition have increased greatly in the last five decades, and the development has been across the subdisciplines of anthropology and in conjunction with other academic disciplines. Nevertheless, social and cultural aspects related to food are, even today, frequently neglected, regarded as secondary or less important in comparison to other 'main' subjects like health or economy. In this sense, the aim of this chapter is to focus on the social and cultural perspective of food and its relationship with diets, territories and sustainability.

Food, Culture, Ecology and Sustainability: an overview from Social Anthropology[2]

As British social anthropologist Mary Douglas (1979, p. 145) pointed out:

> the choice of food is undoubtedly, of all human activities, riding in a most disconcerting way on the dividing line between nature and culture. Food is linked to the satisfaction of the needs of the body, but also, to a large extent, those of society.

But although food and nutrition correspond to a primary human need, anthropologists rarely focused on aspects of food until the 1930s. Collaboration between anthropologists and nutritionists was first achieved by Richards and Widdowson (1936), and was developed further in the British Colonial Office and in the French *Organisation de Recherche sur l'Alimentation et la Nutrition Africaines* (ORANA). Interest in anthropological perspectives on food grew during the Second World War and the post-war conditions (Guthe and Mead, 1945) and among members of the culture and personality school (Kardiner *et al.*, 1945).

Social and cultural anthropologists have pointed out after their fieldwork with different traditional societies around the world, that human action over the environment and, after that, food as the main necessity to live, have always been as sustainable as possible. In the words of the American anthropologist Margaret Mead (1975, p. 74): 'at any given moment in history, the interrelationships between food and the technical and social habits of those who use it constitute the indispensable structure within which these people gets its nutrition and its relationship with the environment'.

After the 1950s, dissatisfaction with the rigid structures and theories of cultural change stimulated the transition to an ecological perspective. In the 1930s, Julian Steward (1955) studied the Shoshone tribes of the Great Basin, hunter-gatherers that he defined as heavily dependent on the pinon nut tree. After his fieldwork research, he illustrated the direct relationship between resource base and population density after he observed that lower population densities exist in areas where the tree is sparsely distributed. He also observed this relationship in regard to water availability and management, and theorized on how cross-cultural regularities exist due to the presence of similar environments.

Steward's cultural ecology is at the base of ecological anthropology, which provides a materialist explanation on the relationship between society, culture and environment. As Salzman and Attwood (1996, p. 169) pointed out a couple of decades ago, human societies have ongoing contact with and impact upon their territory, plants and animals in their vicinities, but also upon the climate, and all those elements of their environment have at the same time a reciprocal impact on humans. In this sense, and following the same authors, we know that ecological anthropology investigates the ways that a society shapes its environment and the subsequent manners in which this relationship culturally forms (and transforms) social, economic, or political life (Salzman and Attwood, 1996).

In the 1960s and 1970s, the importance of sustainability and using food as a basis for analyses (frequently as a main subject) was growing. In this framework, Rappaport (1968) and Lee (1979) created a distinction between their 'human ecology' and the 'cultural ecology' of anthropologists such as Harris and Ross (1987), that could be directly described as 'cultural materialists' and turns their social anthropology into a Darwinian direction. The study by Rappaport on the Tsembanga Maring in New Guinea also represents a turning of the former culturalist approach into a biological approach.

In the area of applied nutrition, information from anthropologists was sought and this brought an anthropological contribution to the understanding of food scarcity and malnutrition. Agencies of the United Nations, such as the International Children's Centre, World Health Organization and FAO included anthropological perspectives in their programmes on food and nutrition, and in 1966 a Food Habits Section was created within the FAO. In 1964, the International Biological Programme included food and nutrition as a component in its Human Adaptability Section, allowing collaboration at a worldwide level between specialists from the biological and sociocultural sciences, including anthropologists.

We must also bear in mind that, with the modernization of agriculture and globalization of foods that took place in the second half of the twentieth century, concepts such as sustainable diets or human ecology have been neglected in favour of intensification and industrialization of agricultural systems. The recent growing concern over food safety has motivated a renewed interest in organic foods (after the pioneering work by Herrin and Gussow [1989]) and locally produced and sustainable foods.

In recent times, social anthropologists like Kottak (1999) or Townsend (2009) developed actual visions on anthropological ecology. The 'new ecological anthropology' proposed by Kottak is located at the intersection of global, national, regional and local systems, studying the outcome of the interaction of multiple levels and multiple factors. It blends theoretical and empirical research with applied, policy-directed, and critical work going to an 'engaged' anthropology; it is otherwise attuned to the political aspects and implications of ecological processes. Carefully laying out a critique of previous ecologies by way of announcing newer approaches, his article insists on the need to recognize the importance of culture mediations in ecological processes rather than treating culture as epiphenomenal and as a mere adaptive tool (Kottak, 1999, p. 23). On the other hand, Townsend (2009) developed the difference between ecological anthropology and environmental anthropology, going firmly to the analysis of the relationships between humans and their environment across space and time to build the basis of a political ecology that allows a specific activism on the field.

Food and Cultural Heritage

Cultural heritage is alive, composed of selected elements deemed to belong to a particular culture at the expense of other elements, to serve particular interests. Although it is part of a social contract (it should be perceived by the majority of the population as their own), it is very often the establishment that proposes, promotes and/or recognizes heritage.

Aspects of intangible cultural heritage such as food have only recently been officially recognized as such, legitimizing its importance to our identities. It is not until the recent turn of the last century, with the 'UNESCO Proclamation of Masterpieces of the Oral and Intangible Heritage of Humanity' international distinction, and subsequently with the UNESCO declaration of the first spaces that constituted intangible heritage of humanity, that the 'official' concept of heritage can be considered to have begun to take an interest in areas beyond the purely monumental and environmental fields, broadening its scope to more ethno-anthropological and less tangible aspects.

An important point to emphasize here is that gastronomic heritage, and therefore human consumption in general, falls within this emerging intangible heritage. If the UNESCO heritage declaration of 2005 had an outstanding characteristic (from the point of view of this study) it was that for the first time, a country like Mexico was presenting its culinary art at a national level in order for it to be declared world heritage.

The candidacy was rejected, but Mexico announced that it would submit it once again, as an essential part of its culture. In order to highlight the heritage value of its cuisine, in July 2008 the Mexican candidacy convened an international academic meeting in the city of Campeche (Mexico) entitled: 'The cuisine as cultural heritage, criteria and definitions'. Its aim was to provide UNESCO with a series of recommendations that would lead to greater awareness of food/gastronomic candidacies (known as the Declaration of Campeche, 2008). A similar initiative took place a year later in Barcelona, at the request of the candidacy of the Mediterranean diet (the Barcelona Declaration, 2009).

However, the zenith of UNESCO recognition for food candidacies did not arrive until November 2010, when the three proposals submitted at that time were declared intangible cultural heritage: traditional Mexican cuisine, the gastronomic meal of the French (le *Répas gastronomique des Français*) and the Mediterranean diet. Food had been recognized by UNESCO as intangible cultural heritage of humanity for the first time.

Nevertheless, it is true that both beforehand and afterwards, various initiatives and candidacies related (albeit tangentially) to the food sphere have been submitted (or are in the process of being submitted) to UNESCO. These include the jurisdiction and the landscape of the vineyards of Saint-Emilion (1999) in France, the wine region of the Alto Douro (2001) and the wine-producing landscape of the island of Pico (2004) in Portugal, the cultural landscape and the vine region of Tokaj in Hungary (2004), and the agave landscape and the ancient industrial facilities of tequila in Mexico (2006), to name just a few related to cultural landscapes.

Regarding this last point, we must also have in mind that cultural heritage is today considered a highly effective tourist resource, and the official recognition of 'food culture' (production landscapes, food, dishes, wines and drinks, tourist routes, industries...) is taking place in the context of tourism and its benefits (or not) for local development. And this aspect is significantly affecting the management of the heritage, and the conception of heritage itself.

Given the scale of the heritage area proposed in the candidatures approved by UNESCO, it is clear that the governments and institutions that led it have assumed a major responsibility in its safeguard as a whole complex, which we are not sure they have understood to its full extent. It is no simple challenge: the food culture is a complex one, which cannot be defined by merely listing the foodstuffs within it, or the culinary preparations and its most distinctive rituals.

Many important challenges were assumed when submitting a candidacy, due to the urgent need to safeguard (and disseminate) the heritage values which, according to the dossiers, affect peoples and their cultures, their cultural spaces, identity and intercultural dialogue, knowledge and creativity. The initiative now lies with the governments that promoted it, who must not shirk a responsibility that demands urgent and necessary action to find new ways to value something as human and as important as food.[3]

The Cultural Imperative and Holistic Perspectives

A local food system always embodies territory, natural resources and landscapes, biodiversity, production, distribution and consumption, all related to established social and cultural functions and values, jobs and occupations, organization of time, as well as nutritional health and welfare aspects.

Every food system in its own biosocial context is an outstanding resource for the achievement of an effective sustainable development. In this sense, we must always bear in mind that food systems are always a significant part of an interdependent social and cultural body, and must never be considered a separate element in itself, as has often been observed, especially from different perspectives and disciplines (other than social anthropology). For example, in the globalized world like the one we live in today, we have many dietary patterns that can be considered to be healthy, but they can vary substantially, for example, in terms of their resource cost or their environmental impact (in terms of extensive production, import and export, etc.).

A food system is a complex network of interdependent cultural aspects, and we must remember that all the links in the chain must be identified, valorized and, if it is the case, officially recognized and protected (Medina, 2015), from production to the dish, including distribution, sales, cooking techniques, food and consumer behaviour, etc. Every food system (and affecting every singular local diet) must be considered as a complex web of relationships and interests. A very integrated social articulation that *makes the world turn.*

Conclusion

Today, many aspects such as health, economy or food consumption are still considered separately from agricultural or fisheries production, economics (sales, import-export, etc.) or the maintenance of traditional structures of distribution or sale. As the anthropological studies showed us many decades ago, we need to recognize the importance of culture mediations in ecological processes rather than treating culture as epiphenomenal and as a mere adaptive tool. Culture must be understood as the main framework to understand and to negotiate, but also to act on the most pressing needs and concepts from a holistic and open perspective.

From this point of view, while good nutrition should be a goal of agriculture, it is imperative that concerns of sustainability are not lost in the process. An adequate attention to cultural aspects can help to have a less partial and interested view of the situation.

Notes

[1] The authors' emphasis in italic.

[2] We follow here briefly some aspects that were already raised in a previous publication focused on the history of food anthropology (Garine *et al.*, 2009).

[3] Regarding the specific case of the candidature of the Mediterranean diet to UNESCO and its consequences and results, see also González Turmo and Medina (2012).

References

Burlingame, B. and Dernini, S. (2011) Sustainable diets: the Mediterranean diet as an example. *Public Health Nutrition* 14(12A), 2285–2287.

Dernini, S., Berry, E., Serra Majem, L., La Vecchia, C., Capone, R., *et al.* (2016) Med Diet 4.0. The Mediterranean diet with four sustainable benefits. *Public Health Nutrition* 20(7), 1322–1330.

Douglas, M. (1979), Les structures du culinaire. *Communications* 31.

FAO (2011) International Scientific Symposium. Biodiversity and Sustainable Diets United Against Hunger. Rome: FAO. Available at http://www.fao.org/ag/humannutrition/28506-efe4aed57af34e2dbb8dc578d465df8b.pdf (accessed 10 April, 2017).

Garine, I., Macbeth, H., Ávila, R., Duhart, F., Garine, V., *et al.* (2009) Nutrition and the Anthropology of Food. In: Nas, P.J.M. and Zhang, J. (eds) *Anthropology Now: Essays by the Scientific Commissions of the International Union of Anthropological and Ethnological Sciences (IUAES) and History of the IUAES.* Intellectual Property Rights Publishing House, Beijing, China, pp. 117–148.

González Turmo, I. and Medina, F. X. (2012) Défis et responsabilités suite à la déclaration de la diète méditerranéenne comme patrimoine culturel immatériel de l'humanité (Unesco). *Revue d'Ethnoécologie* 2. DOI: 10.4000/ethnoecology.957

Guthe, C. and Mead, M. (1945) Manual for the study of food habits. *National Academy of Sciences' Bulletin of National Research Council* 111.

Harris, M. and Ross, E.B. (1987) *Food and Evolution: Towards a Theory of Human Food Habits.* Temple University Press, Philadelphia, Pennsylvania, USA.

Herrin, M. and Gussow, J.D. (1989) Designing a Sustainable Regional Diet. *Journal of Nutritional Education* 21(6), 270–275.

Jonston, J.L., Fanzo, J.C. and Cogill, B. (2014) Understanding sustainable diets: a descriptive analysis of the determinants and processes that influence diets and their impact on health, food security and environmental sustainability. *Advances in Nutrition* 5, 418–429.

Kardiner, A., Linton, R., Du Bois, C. and West, J. (1945) *Psychological Frontiers of Society*. Columbia University Press, New York, USA.

Kottak, C. (1999) The new ecological anthropology. *American Anthropologist* 101(1), 23–35.

Lee, R.B. (1979) *The !Kung San: Men, Women and Work in a Foraging Society.* Cambridge Univesity Press, Cambridge, UK.

Mead, M. (1975) *La Antropología y el Mundo Contemporáneo.* Siglo XX, Buenos Aires, Argentina.

Medina, F.X. (1996) Alimentación, dieta y comportamientos alimentarios en el context mediterráneo. In: Medina, F.X. (ed.) *La Alimentación Mediterránea. Historia, Cultura, nutrición.* Icaria, Barcelona, Spain.

Medina, F.X. (2015) Assessing sustainable diets in the context of sustainable food systems: socio-cultural dimensions. In: Meybeck, A., Redfern, S., Paoletti, F. and Strassner, P. (eds) *Assessing Sustainable Diets within the Sustainable Food Systems. Mediterranean Diet, Organic Food: New Challenges.* FAO, CREA e International Research Network for Food Quality and Health, Rome, Italy.

Medina, F.X. (2017) Five relevant points on food, culture and Mediterranean diet. In: Meybeck, A., Redfern, S., Hachem, F., Capone, R. and Dernini, S. (eds) *Development of Voluntary Guidelines for the Sustainability of the Mediterranean Diet in the Mediterranean Region*. Food and Agriculture Organization of the United Nations, Rome, Italy, pp. 83–87.

Rappaport, R. (1968) *Pigs for the Ancestors: Ritual in the Ecology of a New Guinea People.* Yale University Press, New Haven, Connecticut, USA.

Salzman, Ph.C. and Attwood, D.W. (1996) Ecological anthropology. In: Barnard, A. and Spencer, J. (eds) *Encyclopedia of Social and Cultural Anthropology*. Routledge, London, UK, pp. 169–172.

Steward, J. (1955) *Theory of Culture Change: The Methodology of Multilinear Evolution*. University of Illinois Press, Urbana, Illinois, USA.

Townsend, P. (2009) Environmental anthropology: from pigs to policies. Waveland Press, Prospect Heights, Illinois, USA.

14 Nutritional Indicators to Assess the Sustainability of the Mediterranean Diet

Lorenzo M. Donini, Sandro Dernini, Denis Lairon, Lluis Serra-Majem and Marie-Josèphe Amiot-Carlin

Abstract

There is increasing evidence of the multiple effects of diets on public health nutrition, society and environment. Sustainability and food security are closely inter-related. The traditional Mediterranean diet (MD) is recognized as a healthier dietary pattern with a lower environmental impact. As a case study, the MD may guide innovative inter-sectorial efforts to counteract the degradation of ecosystems and loss of biodiversity and homogeneity of diets due to globalization, through the improvement of sustainable healthy dietary patterns. This chapter defines a suite of the most appropriate nutrition and health indicators for assessing the sustainability of diets based on the MD. Thirteen nutrition indicators of sustainability were identified in five areas: *biochemical characteristics of food* (A1. Vegetable/animal protein consumption ratios; A2. Average dietary energy adequacy; A3. Dietary energy density score; A4. Nutrient density of diet and foods); *food quality* (A5. Fruit and vegetable consumption/intakes; A6. Dietary diversity score); *environment* (A7. Food biodiversity composition and consumption; A8. Local/regional foods and seasonality; A9. Organic/eco-friendly production and consumption); *lifestyle* (A10. Physical activity/physical inactivity prevalence; A11. Adherence to the Mediterranean dietary pattern); and *clinical aspects* (A12. Diet-related morbidity/mortality statistics; A13. Nutritional anthropometry). These proposed nutrition indicators will be a useful methodological framework for designing health, education and agricultural policies in order to conserve the traditional diets of the Mediterranean area as a common cultural heritage and lifestyle and also to enhance the sustainability of diets in general.

There is increasing evidence of the multiple effects and cost of diets on public health nutrition, society and environment (O'Kane, 2012; Clonan and Holdsworth, 2012; Heller *et al.*, 2013; Tilman and Clark, 2014). The sustainable diets concept (FAO/Bioversity, 2012) highlights the role of sustainable consumption as a driver of sustainable production, for food systems' transformation toward more sustainable food consumption and production patterns that are among the most important drivers of environmental pressures (Kearney, 2010; Reisch *et al.*, 2013). Food systems need to grow within the framework of finite and often reduced funds and need to make use of natural resources and skills in a sustainable manner to conserve the fragile ecosystem balance, while with food globalization and the increased industrialization of agricultural systems, the concepts of sustainable diets and agri-food systems had been neglected.

The traditional Mediterranean diet (MD) has been studied in-depth and is recognized as a healthier dietary pattern characterized by a lower environmental impact (Gussow, 1995; Duchin, 2005; Baroni *et al.*, 2007; Sáez-ciheam *et al.*, 2013). However, despite these well-documented health benefits and the low environmental impact of the MD, current surveys show a decline in its adherence in Northern, Southern and Eastern Mediterranean countries, because of multifactorial influences – lifestyle changes, globalization of food markets and economic and sociocultural factors (Alexandratos, 2006; Belahsen and Rguibi, 2006; Garcia-Closas *et al.*, 2006; Da Silva *et al.*, 2009; Vareiro *et al.*, 2009; Bach-Faig *et al.*, 2011; León-Muñoz *et al.*, 2012; Belahsen, 2014; Roccaldo *et al.*, 2014; Bonaccio *et al.*, 2014).

During several recent international seminars, four main thematic areas of sustainability have been identified: (i) nutrition, health and lifestyle; (ii) environment including agribiodiversity; (iii) economy; and (iv) society and culture (FAO/CIHEAM, 2012). The assessment and development of sustainable diet models requires awareness among consumers, producers and governments that agriculture, food, nutrition, health, culture, environment and sustainability are strongly interdependent.

In particular, in the context of sustainable consumption and production, indicators are necessary to monitor time trends to see if a society's consumption and production patterns lead to more socially equitable and environmentally sustainable development. They are also necessary to evaluate the impact of dietary patterns on long-term health status and, in particular, on the pathogenesis and incidence of non-communicable chronic diseases.

An international working group was informally developed in 2011 with the contribution of different national and international institutions to define the nutritional and health indicators relevant to assessing the sustainability of diets, with a focus on the MD. To select the most effective indicators, standardized criteria were considered (Watson *et al.*, 2010) and a set of nutrition indicators of sustainability was identified (Table 14.1). The indicators were attributed to five different domains related to nutritional aspects of the diet: *biochemical quality of food* (A1. Vegetable/animal protein consumption ratios; A2. Average dietary energy adequacy; A3. Dietary energy density score; A4. Nutrient density of diet); *food quality* (A5. Fruit and vegetable consumption/intakes; A6. Dietary diversity score); *environment* (A7. Food biodiversity composition and consumption; A8. Rate of Local/regional foods and seasonality; A9. Rate of eco-friendly food production and/or consumption); *lifestyle* (A10. Physical activity/Physical inactivity prevalence; A11. Adherence to the Mediterranean dietary pattern); and *clinical aspects* (A12. Diet-related morbidity/mortality statistics; A13. Nutritional Anthropometry) (Donini *et al.*, 2016). The proposed nutrition indicators will be useful for further developing a methodological framework for designing policies in order, not only to conserve and preserve the traditional diet such as the MD as a common cultural heritage and lifestyle, but also to enhance the sustainability of dietary models. The MD, in its various national forms, may be used as a case study: a model to describe, understand and improve the sustainability of current food consumption because of the high and increasing pressure on its fragile natural resources exacerbated by the changes of Mediterranean food consumption patterns (Vareiro *et al.*, 2009; De Marco *et al.*, 2014).

A medium-term research and action framework needs to be implemented to analyse the sustainability of the diets in the Mediterranean area (FAO/CIHEAM, 2012; Dernini *et al.*, 2013). The use of the selected indicators and their validation may represent a first step of a 'pilot sustainability laboratory' aimed at the definition of a validated procedure that will help governments and policy makers to formulate sustainability-sensitive policies in the promotion of sustainable food systems development in different areas.

It has, however, to be considered that the choice of the indicators is indeed a compromise between what is desirable and what is practical and available in which countries. In one sense these indicators represent an 'ideal' list; it remains to be seen how useful they are for practical application.

Table 14.1. Nutritional indicators of sustainability.

		Definition and methodology	Data sources	Limitations of the indicator
A1	**Plant and animal protein consumption ratios**	Ratio of the relative intakes of protein from plant and animal sources	FAOSTAT food balance sheets and commodity balances	It reflects the domestic availability of foods, not consumption or production *per se*
A2	**Average dietary energy adequacy**	Dietary energy supply (kcal/capita/day): average supply available for each individual in the total population Average dietary energy requirement (ADER) (kcal/capita/day): the amount of food energy needed to balance energy expenditure in order to maintain body size, body composition and a level of necessary and desirable physical activity consistent with long-term good health	http://www.fao.org/economic/ess/ess-fs/ess-fadata/en/#.VPhu3y7K1i0. National individual dietary surveys, household budget surveys, food balance sheets and ADER from national energy requirements, FAO human energy requirements	Being expensive and labour intensive, surveys are undertaken only in a limited number of countries, often at regional or local level or in specific population groups; furthermore, it is difficult to accomplish comparability at the international level, because the assessment methods are variable, self-reported and consequently subject to considerable measurement errors
A3	**Dietary energy density score**	Dietary energy density (kcal/g) calculated by dividing total dietary energy by the edible weight of foods and caloric beverages consumed	National individual dietary surveys, household budget surveys (HBS), FAO food balance sheets	Data obtained from food balance sheets do not reflect the effective food intake, because they relate to the food quantities theoretically available for consumption
A4	**Nutrient density of diet and foods**	Mean adequacy ratio based on the mean percentage of the recommended intakes for 29 key needed nutrients, alone or in combination with the mean excess ratio for nutrients to be limited Nutrient density scores referring to either 100 g, 100 kcal/kJ or cost/kg or L of a given food	National individual dietary surveys, household budget surveys, FAO food balance sheets	(i) lack of accurate and quantitative dietary intake data and food composition databases; (ii) comparisons between countries are limited by possibly different daily recommended intakes (energy, nutrients and fibre); (iii) comparisons between studies need the use of the same nutrients and total number of nutrients
A5	**Fruit and vegetable consumption/intakes**	Measure of the consumption (supply, availability, intake) of fruit and vegetables (g/capita/day), including pulses, nuts and seeds	National individual dietary surveys, household budget surveys, FAO food balance sheets	Data obtained from FBS do not reflect the effective food intake, because they relate to the food quantities available to the consumer (but not necessarily consumed)

Continued

Table 14.1. Continued.

		Definition and methodology	Data sources	Limitations of the indicator
A6	**Dietary diversity score**	Dietary diversity scores are defined as the number of food groups consumed over a reference period: • individual dietary diversity score: used as proxy of the nutritional quality of individual diet has for aim to assess the adequacy of nutrient intake • dietary diversity score at the household level (HDDS) is used as proxy of the socio-economic level of the household and intends to reflect the economic ability of a household to consume a variety of foods Dietary variety score corresponds to the number of foods consumed among a list of foods US Healthy Food Diversity index is a tool for the simultaneous measurement of dietary variety, quality and proportionality at individual level	Specific questionnaires to be administered	The issue of the number and the choice of these food groups has not yet been resolved
A7	**Food biodiversity composition and consumption**	Food composition: a count of the number of foods: • at variety/cultivar/breed level for common foods • at species level for wild/indigenous/underutilized foods with at least one value for component found in published and unpublished sources Food consumption: the taxonomic diversity of foods reported in food consumption/dietary intake surveys. Data collected and reported include: • the study instrument (e.g., diet history, food frequency) with details (scope, date, number and description of subjects, geographical/ethnic coverage; reference, total number of studies examined) • the qualifying biodiverse foods reported (number of foods, food lists) the number of surveys with at least one reported food counting for biodiversity	FAO/INFOODS compile data and report periodically. For food composition, data are obtained by searching peer-reviewed journals using the search engines Scopus and Science Direct, and through a call for data conducted via INFOODS (International Network of Food Data Systems). These data are then compiled in a Biodiversity Food Composition Database	The development and reporting on the indicators are recent, and only two to three time points are available. The usefulness of the indicators should be assessed in the future, and judged against market survey data as well as nutritional outcomes

A8	**Local/regional foods and seasonality**	The distance between consumer purchase location and producing area; it is usually considered that it should be at maximum 150 km (around 100 miles) The number of intermediates between producer and consumer with zero when direct from producer, one when one intermediate is present (one can be considered as a cut point for discrimination) The consumer choice: • directly to local/regional producers (on-farm, farmers' market/shop, food baskets made of local foods) as a share of total food purchases • share of fresh vegetables or fruits consumed coming from open field or unheated greenhouse cultivation The duration between fruit harvest (known or estimated from agriculture statistics of the concerned growing location or country) and purchase of fresh fruit, as a direct reflect of distance from seasonal production (and cold storage duration)	Dedicated studies where such specific questions are addressed	The parameters to use are still under debate and need further testing. The present availability of data can be restricted to a limited number of studies
A9	**Organic/eco-friendly production and consumption**	The % of consumers buying organic foods and the frequency of consumption The organic food consumption in % of total food amount or money per capita (e.g. Bionutrinet cohort survey in France – http://bionutrinet.etude-nutrinet-sante.fr) The % of the organic market volume The % of land use under organic certification	In most industrialized countries, data on the organic market volume as well as the market shares are available as well as recorded. Detailed data for specific food types can be available too During some consumer cohort surveys or in national consumption surveys, individual data are collected on organic food consumption (e.g. Germany, France)	In some countries organic production can be marginal only or data on organic production or consumption are not available at national or regional level

Continued

Table 14.1. Continued.

		Definition and methodology	Data sources	Limitations of the indicator
A10	**Physical activity/ physical inactivity prevalence**	Attributable DALYs (disability adjusted life years) from physical inactivity Physical activity questionnaires (e.g. WHO Global Physical Activity Questionnaire [GPAQ]; International Physical Activity Questionnaire [IPAQ])	National surveys, WHO Global Infobase	It is difficult to use questionnaires that are comparable across cultures. All the questionnaires dealing with physical activity present some limitations, in particular considering the shorter-forms and the versions to be used without personal interview. Moreover, data on population-based physical inactivity may be limited in some countries
A11	**Adherence to the Mediterranean dietary pattern**	Mediterranean Diet Score (MDS)	The MD indexes were estimated in their majority from information collected through detailed food frequency questionnaires or repeated measures of 24 hour recall dietary questionnaires	Usually cut-off points used in most scores are sample-dependent making the interpretation of any identified association of this pattern with health outcomes difficult to generalize Since many MD indexes exist, a natural question is whether some work better than others with respect to capturing the adherence to MD, as well as, to identifying associations of this diet with a specific health outcome
A12	**Diet-related morbidity/ mortality statistics**	The prevalence of individuals having physician-diagnosed obesity, cardiovascular diseases (coronary heart disease, stroke, hypertension), type II diabetes, osteoporosis, neurodegenerative diseases, obesity-related cancers The DALY is a measure of overall disease burden expressed as years lost due to illness, disability or early death associated with nutrition related factors: high blood pressure, high cholesterol (total and low density lipids), high blood sugar (insulin resistance and or diabetes)	National surveys, WHO world health statistics	Some pathologies can be undiagnosed or underreported in some countries. Data may not be available for the same age groups

A13	**Nutritional anthropometry**	Undernutrition: prevalence of individuals having a BMI <18.5 kg/m^2 calculated from self-reported weight and height Overweight or obesity: prevalence of individuals having a BMI ≥25.0 kg/m^2 calculated from self-reported weight and height and/or waist circumference > 88 cm in women and 102 cm in men	WHO Global Database, data locally available through National surveys	Individuals tend to overestimate their height and underestimate their weight, leading to underestimation of BMI and of the prevalence of overweight and obesity. Moreover, anthropometric measurements have to be performed by skilled personnel according to a standardized procedure

References

Alexandratos, N. (2006) The Mediterranean diet in a world context. *Public Health Nutrition* 9(1A), 111–117.

Bach-Faig, A., Berry, E.M., Lairon, D., Reguant, J., Trichopoulou, A., *et al.* (2011) Mediterranean diet pyramid today. Science and cultural updates. *Public Health Nutrition* 14, 2274–2284.

Baroni, L., Cenci, L., Tettamanti, M. and Berati, M. (2007) Evaluating the environmental impact of various dietary patterns combined with different food production systems. *European Journal of Clinical Nutrition* 61, 279–286.

Belahsen, R. (2014) Nutrition transition and food sustainability. *Proceedings of the Nutrition Society* 73(3), 385–388. DOI: 10.1017/S0029665114000135

Belahsen, R. and Rguibi, M. (2006) Population health and Mediterranean diet in southern Mediterranean countries. *Public Health Nutrition* 9(8A), 1130–1135.

Bonaccio, M., Di Castelnuovo, A., Bonanni, A., Costanzo, S., De Lucia, F., *et al.* (2014) Decline of the Mediterranean diet at a time of economic crisis. Results from the Moli-sani study. *Nutrition, Metabolism and Cardiovascular Diseases* 24(8), 853–860. DOI: 10.1016/j.numecd.2014.02.014

Clonan, A. and Holdsworth, M. (2012) The challenges of eating a healthy and sustainable diet. *American Journal of Clinical Nutrition* 96(3), 459–460. DOI: 10.3945/ajcn.112.044487

Da Silva, R., Bach-Faig, A., Raidó Quintana, B., Buckland, G., Vaz de Almeida, M.D. and Serra-Majem, L. (2009) Worldwide variation of adherence to the Mediterranean diet, in 1961–1965 and 2000–2003. *Public Health Nutrition* 12, 1676–1684.

De Marco, A., Velardi, M., Camporeale, C., Screpanti, A. and Vitale, M. (2014) The adherence of the diet to Mediterranean principle and its impacts on human and environmental health. *International Journal of Environmental Protection and Policy* 2(2), 64–75.

Dernini, S., Meybeck, A., Burlingame, B., Gitz, G., Lacirignola, C., *et al.* (2013) Developing a methodological approach for assessing the sustainability of diets: The Mediterranean diet as a case study. *New Medit* 12(3), 28–36.

Donini, L.M., Dernini, S., Lairon, D., Serra-Majem, L., Amiot, M.J., *et al.* (2016) A consensus proposal for nutritional indicators to assess the sustainability of a healthy diet: The Mediterranean diet as a case study. *Frontiers in Nutrition* 3, 37. DOI: 10.3389/fnut.2016.00037

Duchin, F. (2005) Sustainable consumption of food: a framework for analysing scenarios about changes in diets. *Journal of Industrial Ecology* 9, 99–114.

FAO/Bioversity (2012) Sustainable diets and bioversity. Directions and solutions for policy, research and action. Available at www.fao.org/docrep/016/i3004e/i3004e00.htm (accessed 01 October 2018).

FAO/CIHEAM (2012) Towards the development of guidelines for improving the sustainability of diets and food consumption patterns in the Mediterranean area. Available at www.fao.org/docrep/016/ap101e/ap101e.pdf (accessed 01 October 2018).

Garcia-Closas, R., Berenguer, A. and González, C.A. (2006) Changes in food supply in Mediterranean countries from 1961 to 2001. *Public Health Nutrition* 9(1), 53–60.

Gussow, J.D. (1995) Mediterranean diets: are they environmentally responsible? *American Journal of Clinical Nutrition* 61, 1383S–1389S.

Heller, M.C., Keoleian, G.A. and Willett, W.C. (2013) Toward a life cycle-based, diet-level framework for food environmental impact and nutritional quality assessment: a critical review. *Environmental Science and Technology* 47(22), 12632–12647. DOI: 10.1021/es4025113

Kearney, J. (2010) Food consumption trends and drivers. *Philosphical Transactions of the Royal Society of London: Series B Biological Sciences* 365(1554), 2793–2807. DOI: 10.1098/rstb.2010.0149

León-Muñoz, L.M., Guallar-Castillón, P., Graciani, A., López-García, E., Mesas, A.E., *et al.* (2012) Adherence to the Mediterranean diet pattern has declined in Spanish adults. *Journal of Nutrition* 142(10), 1843–1850.

O'Kane, G. (2012) What is the real cost of our food? Implications for the environment, society and public health nutrition. *Public Health Nutrition* 15(2), 268–276. DOI: 10.1017/S136898001100142X

Reisch, L., Eberle, U. and Lorek, S. (2013) Sustainable food consumption: An overview of contemporary issues and policies. *Sustainability: Science, Practice and Policy* 9(2), 7–25.

Roccaldo, R., Censi, L., D'Addezio, L., Toti, E., Martone, D., *et al.* (2014) Adherence to the Mediterranean diet in Italian school children (The ZOOM8 Study). *International Journal of Food Science and Nutrition* 65(5), 621–628. DOI: 10.3109/09637486.2013.873887

Sáez-ciheam, S., Obrador, B., Bach-Faig, A. and Serra-Majem, L.L. (2013) Environmental footprints of Mediterranean versus Western dietary patterns: beyond the health benefits of the Mediterranean diet. *Environmental Health* 12, 118.

Tilman, D. and Clark, M. (2014) Global diets link environmental sustainability and human health. *Nature* 515, 518–522.

Vareiro, D., Bach-Faig, A., Raidó Quintana, B., Bertomeu, I., Buckland, G., *et al.* (2009) Availability of Mediterranean and non-Mediterranean foods during the last four decades: comparison of several geographical areas. *Public Health Nutrition* 12(9A), 1667–1675. DOI: 10.1017/S136898000999053X

Watson, D., Lorenz, U., St Hansen, M., Szlezak, J., Zoboli, R., *et al.* (2010) European Environment Agency (EEA); towards a set of indicators on sustainable consumption and production (SCP) for EEA reporting. European Topic Centre on Sustainable Consumption and Production (ETC/SCP) working paper 1/2010, Copenhagen, Denmark.

15 Assessing the Environmental Impact of Diets

Corné van Dooren

Abstract

At the global level, the planetary boundaries approach addresses the current global environmental state and helps to prioritize the most pressing issues related to the agri-food system as a driver. These issues are climate change, nitrogen and phosphorus cycle disruption, land-use change and freshwater use. At the national level, the footprints approach is used to identify indicators. This footprint family includes ecological, land, carbon, energy and water footprints. At the product level, life cycle assessment includes eleven pressure indicators. We conclude that greenhouse gas emissions (GHGEs) and land use fulfil the selection criteria and address most of the environmental impact of diets well. In the future, these indicators should be supplemented with an indicator addressing the nitrogen and phosphorous efficiency of food products. The function of food is to deliver required nutrients to the human body, not only filling (volume) or fuel (kcal). In order to find an appropriate unit, we analysed and evaluated existing nutrient density scores, quantifying the amounts of essential nutrients per gram or kcal. We propose the nutrient density unit – at least for solid foods – since it reflects the food's function of supplying the essential macronutrients within human metabolic energy needs. Greenhouse gas emissions and land use are the most frequently used indicators in diet studies. Some examples (i.e. the Netherlands) of those studies are given. Low GHGE intensity per 100 gram correlated with positive nutritional characteristics of food products. This is true for low energy density, and high nutrient density, expressed as the well-established NRF9.3 index. This index was improved to include the contribution of food products to GHGEs. GHGEs of product groups correlate more strongly with the proposed sustainable nutrient-rich foods index (SNRF). This index summarizes six distinctive nutrients (three which should be encouraged and three limited), as well as (metabolic) energy density. Including such an index on food product labels could assist consumers in making better informed food choices.

Introduction

It was more than 30 years ago that Joan Dye Gussow formulated her first dietary guidelines for sustainable diets, with reference to ecosystems and environmental sustainability (Gussow and Clancy, 1986; Gussow, 1999). At the time, these guidelines were not given much attention by nutrition professionals in either the health or agriculture sectors.

In the last decade, the Dutch (Health Council, 2011), British (Reddy *et al.*, 2009), Swedish (Livsmedelsverket, 2009), Finnish (Steering Group, 2010) and Belgian (FRDO, 2011) governments have appointed committees to give policy advice on 'sustainable' diets. In 2008, the Dutch Steering Committee on Technology Assessment advised the government to increase awareness of food quality among a wider audience. This would combine sustainable food and healthy eating with the Dutch 'Wheel of Five' (nutritional education model). According to the committee, websites need to offer consumers information on health, environment, climate, animal welfare and social values (van der Weijden, 2008). In fact, as early as 1996, calls were made to combine environmental scores with health scores (Aiking *et al.*, 1996).

Although it has been 30 years since Brundtland issued the call for sustainable economic development (Brundtland, 1987), the focus on sustainable dietary habits is more recent. It was only in 2010 that the Food and Agriculture Organization (FAO) of the United Nations attained sufficient consensus to define sustainable diets:

> Sustainable Diets are those diets with low environmental impacts which contribute to food and nutrition security and to healthy life for present and future generations. Sustainable diets are protective and respectful of biodiversity and ecosystems, culturally acceptable, accessible, economically fair and affordable; nutritionally adequate, safe and healthy; while optimizing natural and human resources.
>
> (FAO, 2010)

A broad scientific consensus has now emerged that the food system is a key element in the challenge of global environmental sustainability (Godfray *et al.*, 2010; Foley *et al.*, 2011). Based on the FAO's definition, it is clear that the challenge facing scientists is to precisely outline, for all citizens, what constitutes a healthy, acceptable and affordable diet that also has a low environmental impact.

The Need for Indicators

The search for simple metrics to measure that lower environmental impact and interpret the human impact on our environment is nothing new. The challenge of finding ways to maintain the carrying capacity of the global ecosystem has resulted in concepts of boundaries based on the idea of keeping safe distances from threshold values of control variables (Rockström *et al.*, 2009; Heijungs *et al.*, 2014; Fang and Heijungs, 2015). Disproportional human use of the Earth's resources transgresses planetary boundaries. This raises the question whether it is possible to decrease the human impact on the environment to stay within those boundaries. Rockström *et al.*'s frequently cited study (Rockström *et al.*, 2009) is an important attempt to simplify the complexity of our environment, by using a number of control variables to evaluate the effects of anthropogenic activities on the Earth's ecological system.

Human activities have a complex impact on the Earth's system, but the effects are becoming increasingly clear (Steffen *et al.*, 2007). Food production and consumption contribute significantly to these effects (Vringer *et al.*, 2010). On the other hand, human behaviour can also reduce these effects through informed choices in diets, and selection of food products.

The issue is complex, and communicating about it and developing appropriate policies demands simplification. Indicators are a useful way to simplify how the environmental system is described (Smeets and Weterings, 1999). Proper indicators can provide understandable and readily interpretable information, which can be the basis for informed choices and effective policy responses. These indicators need to reflect the crucial factors that cause diets to put pressure on the environment (Smeets and Weterings, 1999), but there is always a trade-off between completeness and simplicity.

The relation between these different types of indicators can be illustrated via the DPSIR framework for reporting on environmental issues (Fig. 15.1): drivers, pressures, state (changes), impacts and responses. This framework of the European Environmental Agency (Smeets and Weterings, 1999; Turner *et al.*, 2004) is seen as providing a structure to present the indicators to policy makers. This makes it possible to provide feedback on environmental quality and on the impact of political choices. The DPSIR framework illustrates a chain of causal links starting with 'driving forces' (economic sectors, human activities) via 'pressures' (resource extraction, emissions, waste) to the changes in 'state' (physical, chemical and biological) of, and 'impacts' on ecosystems, human health, resources, and functions, eventually leading to societal 'responses' (prioritization, target setting) (Kristensen, 2004).

Top-Down or Bottom-Up Approach

The environmental impacts of diets can be assessed at different levels of abstraction: global (supra), national (macro) and product (micro).

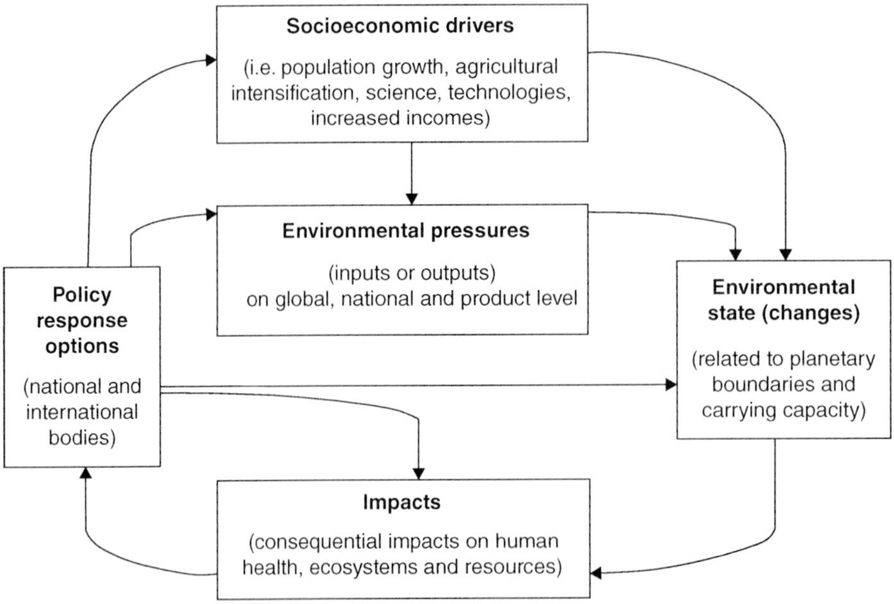

Fig. 15.1. The DPSIR framework for reporting on environmental issues. DPSIR is the abbreviation of the causality chain: Drivers, Pressures, State (changes), Impacts, and Responses (van Dooren *et al.*, 2018).

Classification into levels is necessary because each level needs different methodologies and indicators specifically developed for this application. Appropriate methodologies have been empirically selected at sub-global scales: micro-scale assessments (e.g. food products, packaging) are often subject to bottom-up life cycle assessment (LCA), whereas macro-scale assessments (e.g. national, regional) generally depend on top-down input–output analysis (Peters, 2010; Fang *et al.*, 2014). These two approaches can also be applied to diets in order to assess their ecological sustainability:

- global approach at the supra level: Global agri-food system related to the planetary boundaries (PB);
- top-down approach at the macro level: National footprint of the diet per capita related to PBs based on input and/or output indicators;
- bottom-up approach at the micro level: aggregation of LCA data of food products into diets.

Application of the different approaches depends on the functional unit chosen, the level of abstraction needed, the scale of assessment and the geographic scope (Schader *et al.*, 2014).

At each level, specific methods are available to assess the environmental impacts of diets. The highest level of abstraction is the global one, useful to compare impacts of the diet of humans to the global environmental space. This environmental space is defined by 'PB', primarily covering the (change in) state of the environment and its constituent parts. At the macro level, diets are compared between nations or in relation to national diets. This assessment approach employs macro-scale indicators, called 'footprints'. Footprints are an attempt to evaluate the impacts of anthropogenic activities (pressures) on the biosphere (state, in terms of soil use, ecosystems, air and water quality) (Kristensen, 2004). The lowest level is the product scale. At this level, diets are compared between consumers at the product level (micro level; section Application of the Life Cycle Assessment Methodology). This assessment is purely focussed on measuring pressures of the production of food products, both inputs and outputs, better known as the 'LCA approach'. Each of these assessment methods is based on its own specific indicators to determine the environmental sustainability of diets, which will be described in subsequent sections.

Environmental Footprints: the Footprint Family

The footprint concept is the most elaborate concept, providing macro-scale indicators of the impacts of anthropogenic drivers (i.e. on areas, regions, countries), evaluating ecological sustainability. One of the first footprint concepts, ecological footprint (EF) analysis, was originally introduced and advocated to evaluate the effects of anthropogenic activities on the biosphere (Rees, 1992). It compiles the inputs of biological resources in a specific area and the outputs of carbon emissions compared to the 'biocapacity' of the ecosystem, as a reference (Wackernagel and Rees, 1996). This biocapacity parameter for the available bio-productive area of the Earth is one of the attempts to quantify the Earth's carrying capacity. Likewise, it is in fact a planetary boundary (Rees, 1992; Wackernagel and Rees, 1996) with a global reference value of 1.8 gha/cap/y (global hectares per person per year) (GFN, 2010). The capacity for the production of food is half of this: 0.9 gha/cap/y (van Dooren and Bosschaert, 2013). Biocapacity is an eleventh, combined boundary. Nevertheless, the EF is the most frequently applied footprint, has a reference value and is relevant for diets. Indeed, the EF is an example that footprints can be applied to both products and diets (van Dooren and Bosschaert, 2013). A footprint is intended for easy communication of results, including a supply chain or a full life cycle perspective, which permits it to be applied at the macro national, economy-wide level.

A review of the existing literature indicates the top four most studied and applied footprints (Fang and Heijungs, 2015, Fang *et al.*, 2014): (i) ecological; (ii) carbon; (iii) water; and (iv) energy footprint; followed by nitrogen, biodiversity, land, and phosphorus (Fang *et al.*, 2014). All these footprints are related to diets. Several researchers have illustrated how they transformed PBs to national and per capita footprints (Frischknecht *et al.*, 2016), but the footprints have been applied to diets in no more than a dozen studies (Jones *et al.*, 2016). On the one hand, a European project combined the top three into a footprint family, and then integrated this into an environmentally extended multiregional input–output model (Galli *et al.*, 2012, 2013). On the other hand, Fang *et al.* (2014) confirmed that national data are available for the top four footprints, and that they are methodologically standardized, globally comparable, and generically applicable. Additionally, Mason and Zeitoun (2013) argued that these four footprints should be used because they are closely related to four global concerns over threats to human society: namely food security, energy security, climate security and water security. According to Fang *et al.* (2014), the ecological, energy, carbon and water footprints can be regarded as complementary as each footprint focuses on an important environmental issue. In fact, the EF is an aggregated footprint of land input and carbon output (Fig. 15.2).

The advantage is that this footprint family can provide policy makers with a more complete picture of environmental complexity and better reflect the essence of sustainability than single footprints. Data for the footprint family are available at national and international levels (Fang *et al.*, 2014).

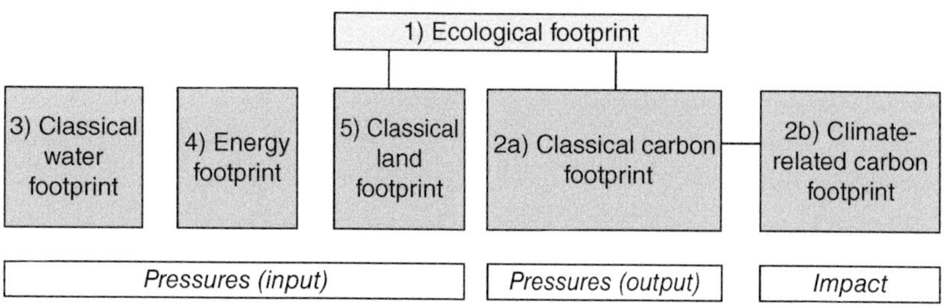

Fig. 15.2. The footprint family with five footprints classified as impact or pressures indicators. The ecological footprint combines the classical land and carbon footprints (van Dooren *et al.*, 2017b).

Application of the Life Cycle Assessment Methodology

The LCA or life cycle impact assessment is based on pressure-related indicators of the needs of individual citizens. This approach is complementary to the other approaches, because it acts on the micro scale. In fact, LCA studies products (as 'functional units', generally 1 kg of product), rather than considering a system at the macro level, economy-wide. In Europe, the LCA method has been standardized, broadly accepted, and further developed over the last 20 years (Hayashi *et al.*, 2006; JRC, 2011). This LCA technique is interesting, because it is frequently used to assess the environmental characteristics of an agricultural product, process or service, as well as the potential impacts associated with it, through all stages of its life cycle (Wegener Sleeswijk *et al.*, 1996). Most European institutions use midpoint ('pressure') indicators within LCAs (Hayashi *et al.*, 2006).

The ReCiPe method (Goedkoop *et al.*, 2013) was the first LCA method to also assess the impacts of products, using endpoints (Hayashi *et al.*, 2006). Distinguishing the boundary between (change in) state indicators and impacts will depend on the end values of the DPSIR system. LCA seems to be human-centred, viewing ecosystem changes and loss of species as a (change in) state. Putting the ReCiPe method into this perspective, human health could be considered in this framework as an impact, ecosystems and resources as a (change in) state or as impact indicators. This method calculates the environmental impacts in terms of endpoints, namely damage to human health, damage to ecosystem diversity (biodiversity), as well as damage to resource availability (see Fig. 15.3). The frequently applied LCA method described by Goedkoop *et al.* (2013) and JRC (2011) consists of 16 midpoint pressure categories and their indicators, related to the endpoints: four of them are input oriented and twelve are output oriented. The individual midpoint indicators are used to calculate a weighted score – 'ReCiPe score' – based on the ReCiPe method (Goedkoop *et al.*, 2013; Sevenster *et al.*, 2010).

Approximately half of the studies assessing environmental impacts of diets examine climate impact as greenhouse gas emissions (GHGEs) (Auestad and Fulgoni, 2015). Jones *et al.* (2016) demonstrated that the LCA approach is the most commonly applied method (50% of the studies) to measure the environmental impacts of products and diets. They performed a systematic literature review of the measurement of sustainable diets, including 113 studies. Within these studies, the GHGEs of diets is by far the most commonly measured component, followed by a frequent application of land, energy and water use (Jones *et al.*, 2016). Of the sixteen LCA indicators, only four are frequently applied.

In practice, most studies on the environmental impacts of food products and diets only look at one to three categories of the LCA (excluding water use):

- climate change or GHGE (expressed in CO_2-equivalents) (63% of the studies);
- agricultural land occupation or land use (LU, defined as land occupation plus land transformation) (28%);
- fossil fuel depletion or fossil energy use (FEU) (24%) (Jones *et al.*, 2016).

Choice of the Functional Unit

The functional unit is a measure of the function of the studied system and it provides a reference to which the inputs and outputs can be related. Diets are by definition useful as functional units (on different scales) while the agri-food system is the set of processes generating the functionality. At the product level, this smallest functional unit is generally either one portion or one kilogram of product (van Dooren, 2016). As choice of a functional unit, most studies focus on the total personal diet that consists of several portions of different food products per day (averaged over a period of at least one week). A national diet consists of the sum of the individual diets. Whereas this is a bottom-up approach, a top-down approach would define the national diet as agricultural production, minus export and food losses, plus imports. At the highest, global level, the unit is the global diet, which is simply the total daily output of the worldwide agri-food system which humans can consume. A national diet or global diet can be reduced to the level of an (average) individual diet, by dividing it by the number of inhabitants (nationally or globally).

According to Heller et al. (2013), however, it is desirable to quantitatively link the environmental

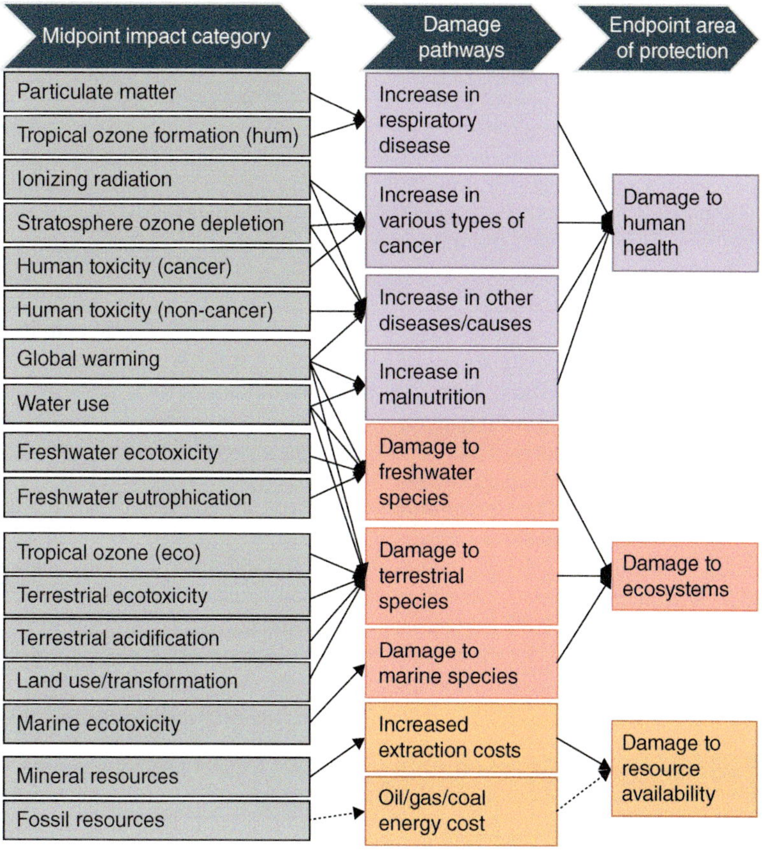

Fig. 15.3. Overview of the structure of ReCiPe. Source: www.rivm.nl/en/Topics/L/Life_Cycle_Assessment_LCA/ReCiPe.

impact of dietary patterns to their nutritional function. Nutritional quality indexes are a potential approach, but need further refining. In fact, the 'function' of food is to deliver required nutrients and energy to the human body, not mass, volume, or portions. More studies have concluded that neither 100 g nor 100 kcal are the best functional units to identify which foods to include in a sustainable diet (Masset *et al.*, 2015; Drewnowski *et al.*, 2015). The function of food is more than filling (g or mL) or fuelling (kcal), it also provides essential nutrients.

In order to find an appropriate unit, van Dooren (2016) analysed and evaluated existing nutrient density scores, quantifying the amounts of essential nutrients per gram or kcal. The conclusion is that these models have a common structure: they include macronutrients to encourage (protein, dietary fibre, and sometimes essential fatty acids) and macronutrients to limit (salt, added sugar and saturated fatty acids), generally accomplished with one or more micronutrients (vitamins, minerals). An index with macronutrients per kcal is sufficient to predict the total nutrient density. This resulted in the formulation of the nutrient density unit (NDU; see equation 1), reflecting total protein, essential fatty acids, and dietary fibre, per energy density. These elements of the NDU correlate significantly with all other essential nutrients. The limiting macronutrients were left out because they can result in a negative unit, which cannot be addressed in LCAs. van Dooren proposes the NDU – at least for solid foods– since it reflects the food's 'function' of supplying the essential macronutrients within human metabolic energy needs (van Dooren, 2016).

Higher NDU represents a higher nutritional functionality, based on delivery of protein, essential fatty acids and dietary fibre per kcal. The NDU is based on the nutrient content per 100 g product. Although products with plant proteins are lower in environmental impact than animal proteins (van Dooren *et al.*, 2017), both sources contribute to the total dietary reference intake of protein. For products without calories or essential nutrients, it is necessary to choose a virtual value: their NDU is set at 0.01. (See Equation 1 at the bottom of page).

The application of the proposed NDU is illustrated with the example of protein-supplying products. Table 15.1 compares the climate impact of six protein-supplying food products, based on different functional units. In this example, the NDU ranks the product from low to high: milk (0.99), eggs (1.48), nuts (1.52), pork (2.05), salmon (2.31) and pulses (2.87). Greenhouse gas emissions per NDU are lowest for pulses and highest for pork. Although pulses have higher GHGEs per 10 g protein than eggs, they have lower GHGEs per NDU. Table 15.1 illustrates that the impacts of milk (NDU = 1), nuts and salmon depend strongly on the functional unit chosen.

Recent Reviews on Diet Assessments

The review by Jones *et al.* (2016) found that although there was substantial heterogeneity in the components of sustainability measured, the estimated GHGEs of various dietary patterns were by far the most commonly measured ($n = 71$ studies) (Jones *et al.*, 2016). The review by Hallström *et al.* (2015) included 14 peer-reviewed journal articles assessing the GHGEs and land-use demands of 49 dietary scenarios. The results suggest that dietary change in affluent areas could play an important role in reaching environmental goals, with the potential to reduce GHGEs and land-use demands by up to 50% when compared with current diets. For healthy diets, the figure is 0–35%; for vegan and vegetarian diets, it is 25–55% and 20–35%, respectively. The choice of functional unit, system boundaries and methods for scenario development, and accounting for uncertainties are methodological aspects thus far identified to have a major influence on the quality and results of dietary scenario analysis (Hallström *et al.*, 2015).

Aleksandrowicz *et al.* (2016) found fourteen common sustainable dietary patterns across the reviewed studies that in average could lead to water use reductions of 50% and reductions of GHGEs and LU as high as 70%. Of the 210 dietary patterns, 197 demonstrated reduced environmental impacts when switching from baseline to alternative dietary patterns, while 13 showed either an increase in impact or no impact. The median changes in GHGEs, LU, and water use across all sustainable diet types were −22%, −28% and −18%, respectively. The largest

Table 15.1. Greenhouse gas emissions (g CO_2eq) of six protein-rich food products measured with five different functional units: 100 g, portion size, 100 kcal, g protein and the nutrient density unit (NDU).

			Greenhouse gas emissions (g CO_2eq)				
Product	Portion size (g)	NDU	per 100 g	per portion	per 100 kcal	per 10 g protein	per NDU
Pulses, brown beans (canned)	75	2.87	250	188	225	352	**87**
Milk, semi-skimmed	250	0.99	**108**	270	235	318	110
Nuts, mixed, salted	25	1.52	229	**57**	**36**	**102**	150
Egg, chicken, boiled	50	1.48	282	141	207	229	190
Salmon, aquaculture, prep.	130	2.31	485	631	220	192	210
Pork, raw, 5–14% fat	100	2.05	709	709	449	336	345

$$\text{NDU} = \frac{\left(\frac{\text{g essential fatty acids}}{12.4\text{ g}}\right)+\left(\frac{\text{g protein}}{50\text{ g}}\right)+\left(\frac{\text{g fibre}}{25\text{ g}}\right)}{3\times\left(\frac{\text{kcal energy}}{2000\text{ kcal}}\right)} \qquad \text{(Eq. 1)}$$

environmental benefits across indicators were seen in those diets that most reduced the amount of animal-based foods, such as vegan, vegetarian and pescatarian diets. Studies modelling the health impacts of shifts from typical Western diets to sustainable dietary patterns showed modest health gains from reductions in mortality rates and associated risks (Aleksandrowicz *et al.*, 2016). According to sixteen studies analysed by Payne *et al.* (2016), some trends related to nutritional quality can be observed. In particular, reduced saturated fat and salt are often associated with reduced GHGEs due to lower usage of animal products. Yet, these diets are also often high in sugar and low in essential micronutrients. Of 151 dietary comparisons across all studies, 79 showed reduced levels of nutrients when shifting to lower-GHGE diets, including 27 for saturated fat, 35 for salt and 17 for sugar (Payne *et al.*, 2016). While Joyce *et al.* (2014) generated inconsistent results, they did demonstrate that higher levels of plant and plant-based foods are generally associated with both positive health outcomes and lower diet-related GHGEs (Payne *et al.*, 2016). Hallström *et al.* (2015) produced similarly mixed results. In line with Payne and colleagues, van Dooren *et al.* (2017) demonstrated that six nutrients (plant protein, dietary fibre, essential fatty acids, salt, added sugars and saturated fat) and metabolic energy intake explains the correlation between GHGEs and nutritional quality. Most of the studies mentioned in the reviews focused on developed countries, particularly in Europe (see Macdiarmid *et al.*, 2011; van Dooren *et al.*, 2014).

An Example from the Netherlands

The Netherlands National Institute for Public Health and the Environment (RIVM), together with Netherlands Nutrition Centre, recently performed a study with the objective to determine the differences in environmental impact and nutrient content of the current Dutch diet and four healthy diets aimed at lowering GHGE (van de Kamp *et al.*, 2018). GHGE (as proxy for environmental impact) and nutrient content of the current Dutch diet and four diets adhering to the Dutch food-based dietary guidelines (Wheel of Five) were compared in a scenario study. Scenarios included a healthy diet with or without meat, and the same diets in which only foods with relatively low GHGEs are chosen. For the current diet, data from the Dutch National Food Consumption Survey 2007–2010 were used (van Rossum *et al.*, 2011). GHGEs (in kg CO_2-equivalents) were based on LCAs. Results are reported for men and women aged 19–30 years and 31–50 years.

The effect on GHGEs of changing the current Dutch diet to a diet according to the Wheel of Five (corresponding with the current diet as close as possible), ranged from −13% for men aged 31–50 years to +5% for women aged 19–30 years. Replacing meat in this diet and/or consuming only foods with relatively low GHGEs resulted in average GHGE reductions varying from 30% to 49% (Table 15.2). In the scenarios in which only foods with relatively low GHGEs are consumed, fewer dietary reference intakes (DRIs) were met than in the other healthy diet scenarios. However, in all healthy diet scenarios the number of DRIs being met was equal to or higher than that in the current diet.

The conclusion was that diets adhering to food-based dietary guidelines did not substantially reduce GHGEs compared with the current Dutch diet, when these diets stayed as close to the current diet as possible. Omitting meat from these healthy diets or consuming only foods with relatively low associated GHGEs both resulted in GHGE reductions of around one-third. These findings may be used to expand food-based dietary guidelines with information on how to reduce the environmental impact of healthy diets (van de Kamp *et al.*, 2018).

Table 15.2. Reduction of greenhouse gas emissions of different options within the Wheel of Five, compared to the current Dutch diet (Brink *et al.*, 2016; van de Kamp *et al.*, 2017).

	Wheel of Five (male/female)	Wheel of Five: most sustainable choices (male/female)
With meat (500 g/week)	−13%/+5%	−30%/−34%
Pesco-vegetarian	−35%/−37%	−47%/−49%

been possible to develop a nutrient density index that quantifies this relationship. The researchers did this on the basis of 403 products that are commonly eaten in the Netherlands. They calculated the energy density, the nutrient density (nutrient-rich foods index) and GHGEs. Low GHGE per 100 g product proved to be correlated

Sustainable Nutrient-Rich Foods Index

Another study attempted to create an overview of the relationship between the climate impact of foods and their nutritional characteristics (van Dooren *et al.*, 2017). On this basis it has

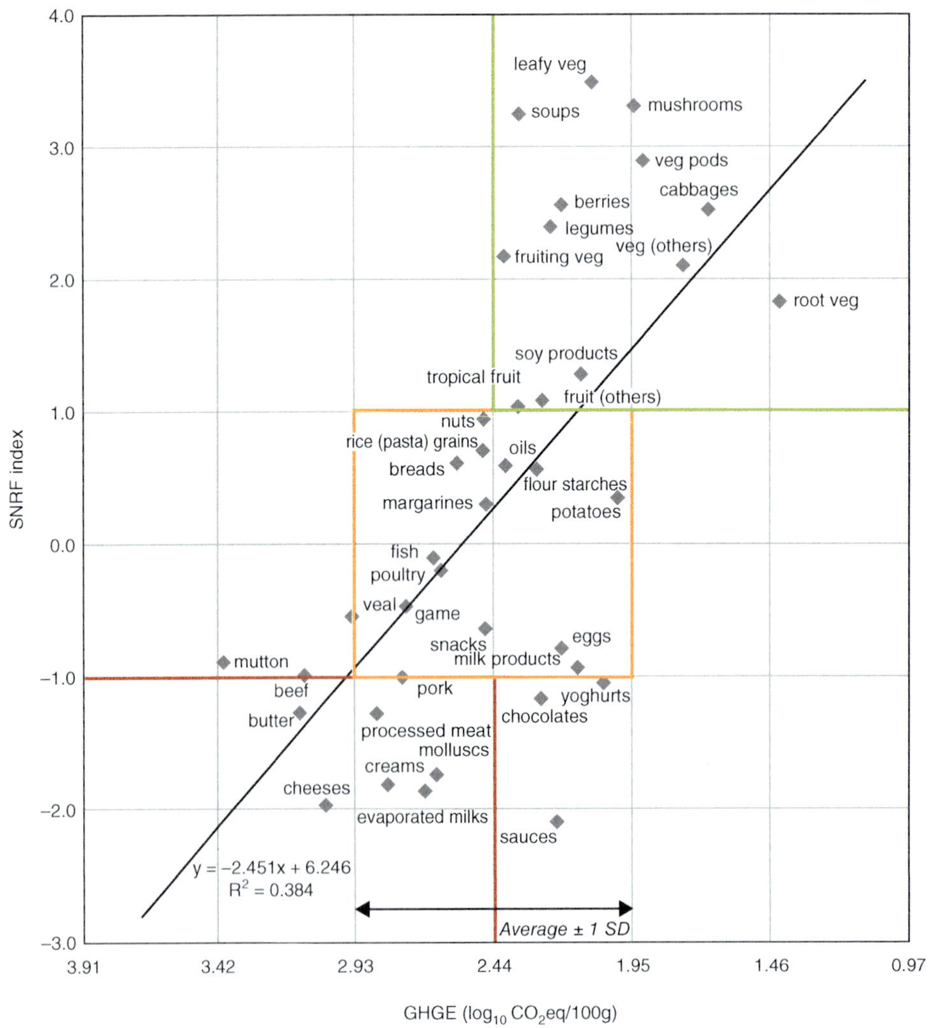

Fig. 15.4. The sustainable nutrient-rich foods index is a novel proposed nutrient density index, based on six distinctive nutrients (three which should be encouraged and three which should be limited), combined with (metabolic) energy density. By combining health-related nutritional characteristics and greenhouse gas emissions of foods, we can create three general groups: red, indicating foods with a negative nutrient profile and high climate impact; amber, indicating foods with a moderate nutrient profile and medium climate impact; and green, indicating a positive nutrient profile and low climate impact. The borders of the amber group are defined by the average GHGE ± 1 SD (x-axis 2.44 ± 0.49) (van Dooren *et al.*, 2017).

with positive nutritional qualities. These products have a low energy density and a high nutrient density, expressed in the commonly used NRF9.3 index of Drewnowski (2009). They enhanced this index by incorporating the climate impact of the products. This new proposed sustainable nutrient-rich foods (SNRF) index correlates better with the GHGE. The SNRF contains six distinct macronutrients in relation to the energy density: three nutrients to be encouraged (plant-based protein, essential fatty acids and dietary fibre) and three that should be limited (salt, saturated fat and added sugar). (See equation 2 below).

Based on the findings, three product groups can be defined. The red group contains foods with a negative nutritional profile and a high climate impact. The amber group contains foods with an average nutritional profile and a moderate climate impact. The green group contains foods with a positive nutritional profile and a low climate impact. For example, the green group contains vegetables, fruit, mushrooms, legumes and soya products (Fig. 15.4). The proposed SNRF index can assist in rating food products. An index of this type can be used for labelling or for education designed to help consumers make choices that are both healthy and sustainable.

Conclusion

Assessing the environmental impact of diets necessitates methods of measuring the environmental sustainability on global, national and product level. GHGEs and LU appeared to be applicable and representative indicators of the environmental sustainability of diets. From the perspective of a healthy diet, comparing the environmental impacts of products by using metabolic energy content (kilocalorie), or even more accurately, nutrient density (defined as NDU) as a functional unit is a more practicable approach than using weight (kilogram). Environmental quality (especially impacts on climate change and land use) could be connected to nutritional quality at diet and product levels. This synergy is described by an index on the basis of seven nutritional quality indicators. The SNRF index can assist in the rating of food products. For consumers, such rating helps in adhering to diets that are both healthy and sustainable.

$$\text{SNRF} = \frac{\left(\frac{\text{g EFA}}{12.4\text{ g}} - \frac{\text{g SFA}}{20\text{ g}}\right) + \left(\frac{\text{g plant protein}}{50\text{ g}} - \frac{\text{g sodium}}{2.4\text{ g}}\right) + \left(\frac{\text{g fibre}}{25\text{ g}} - \frac{\text{g added sugars}}{50\text{ g}}\right)}{3 \times \left(\frac{\text{kcal energy}}{2000\text{ kcal}}\right)} \quad \text{(Eq. 2)}$$

References

Aiking, H., Dorland, C. and ten Hoor, F. (1996) *Food Product, Environment and Health* (In Dutch). NRLO, The Hague, Netherlands.

Aleksandrowicz, L., Green, R., Joy, E.J.M., Smith, P. and Haines, A. (2016) The impacts of dietary change on greenhouse gas emissions, land use, water use, and health: a systematic review. *PLoS One* 11, e0165797.

Auestad, N. and Fulgoni, V.L. (2015) What current literature tells us about sustainable diets: emerging research linking dietary patterns, environmental sustainability, and economics. *Advances in Nutrition* 6, 19–36.

Brink, L., Postma-Smeets, A., Stafleu, A., van Dooren, C. and Wolvers, D. (2016) *Guidelines Wheel of Five* (in Dutch). Voedingscentrum, The Hague, Netherlands.

Brundtland, G.H. (1987) *Our Common Future, United Nations World Commission on Environment and Development*. Oxford University Press, Oxford, UK.

Drewnowski, A., Rehm, C.D., Martin, A., Verger, E.O., Voinnesson, M. and Imbert, P. (2015) Energy and nutrient density of foods in relation to their carbon footprint. *The American Journal of Clinical Nutrition* 1, 184–191.

Fang, K. and Heijungs, R. (2015) Investigating the inventory and characterization aspects of footprinting methods: lessons for the classification and integration of footprints. *Journal of Cleaner Production* 108(Pt A), 1028–1036.

Fang, K., Heijungs, R. and de Snoo, G.R. (2014) Theoretical exploration for the combination of the ecological, energy, carbon, and water footprints: Overview of a footprint family. *Ecological Indicators* 36, 508–518.

FAO (2010) Biodiversity and sustainable diets united against hunger. International Scientific Symposium, Food and Agriculture Organization of the United Nations, Rome, Italy, 3–5 November 2010.

Foley, J.A., Ramankutty, N., Brauman, K.A., Cassidy, E.S., Gerber, J.S., *et al.* (2011) Solutions for a cultivated planet. *Nature* 478, 337–342.

FRDO (2011) *Advice on Animal and Vegetable Proteins* (in Flemish). Federale Raad voor Duurzame Ontwikkeling, Bruxelles, Belgium.

Frischknecht, R., Stolz, P. and Tschümperlin, L. (2016) National environmental footprints and planetary boundaries: from methodology to policy implementation 59th LCA forum, Swiss Federal Institute of Technology, Zürich, June 12, 2015. *The International Journal of Life Cycle Assessment* 21, 601–605.

Galli, A., Wiedmann, T., Ercin, E., Knoblauch, D., Ewing, B. and Giljum, S. (2012) Integrating ecological, carbon and water footprint into a "footprint family" of indicators: definition and role in tracking human pressure on the planet. *Ecological Indicators* 16, 100–112.

Galli, A., Weinzettel, J., Cranston, G. and Ercin, E. (2013) A footprint family extended MRIO model to support Europe's transition to a one planet economy. *Science of the Total Environment* 462, 813–818.

GFN (2010) *National Footprint Accounts Data Tables*. Global Footprint Network, Oakland, California, USA.

Godfray, H.C.J., Beddington, J.R., Crute, I.R., Haddad, L., Lawrence, D., *et al.* (2010) Food security: the challenge of feeding 9 billion people. *Science* 327, 812–818.

Goedkoop, M.J., Heijungs, R., Huijbregts, M., de Schryver, A., Struijs, J. and van Zelm, R. (2013) ReCiPe 2013, A life cycle impact assessment method which comprises harmonised category indicators at the midpoint and the endpoint level; First edition Report I: Characterisation. Version 1.08. Pré Consultants, Amersfoort, Netherlands.

Gussow, J.D. (1999) Dietary guidelines for sustainability: twelve years later. *Journal of Nutrition Education* 31, 194–200.

Gussow, J.D. and Clancy, K. (1986) Dietary guidelines for sustainability. *Journal of Nutrition Education* 18, 1–5.

Hallström, E., Carlsson-Kanyama, A. and Börjesson, P. (2015) Environmental impact of dietary change: a systematic review. *Journal of Cleaner Production* 91, 1–11.

Hayashi, K., Gaillard, G. and Nemecek, T. (2006) Life cycle assessment of agricultural production systems: current issues and future perspectives. Available at http://www.fftc.agnet.org/library.php?func=view&id=20110721140039 (accessed 28 June 2018).

Health Council (2011) *Guidelines for a Healthy Diet: The Ecological Perspective*. Gezondheidsraad, The Hague, Netherlands.

Heijungs, R., de Koning, A. and Guinée, J. B. (2014) Maximizing affluence within the planetary boundaries. *The International Journal of Life Cycle Assessment* 19, 1331–1335.

Heller, M.C., Keoleian, G.A. and Willett, W.C. (2013) Toward a life cycle-based, diet-level framework for food environmental impact and nutritional quality assessment: a critical review. *Environmental Science and Technology* 47, 12632–12647.

Jones, A.D., Hoey, L., Blesh, J., Miller, L., Green, A. and Shapiro, L.F. (2016) A systematic review of the measurement of sustainable diets. *Advances in Nutrition* 7, 641–664.

Joyce, A, Hallett J, Hannelly T and Carey G. (2014) The impact of nutritional choices on global warming and policy implications: examining the link between dietary choices and greenhouse gas emissions. *Energy and Emission Control Technologies* 2014, 33–43.

JRC (2011) *ILCD Handbook: Recommendations for Life Cycle Impact Assessment in the European Context*. Joint Research Centre – European Commission, Luxemburg.

Kristensen, P. (2004) *The DPSIR Framework*. European Environmental Agency, Copenhagen, Denmark.

Livsmedelsverket (2009) *The National Food Administration's Environmentally Effective Food Choices*. Livsmedelverket, National Food Administration Sweden, Stockholm, Sweden.

Macdiarmid, J., Kyle, J., Horgan, G., Loe, J., Fyfe, C., *et al.* (2011) *Livewell: A Balance of Healthy and Sustainable Food Choices*. WWF, Rowett Institute of Nutrition and Health, Aberdeen, UK.

Mason, M. and Zeitoun, M. (2013) Questioning environmental security. *The Geographical Journal* 179, 294–297.

Masset, G., Vieux, F. and Darmon, N. (2015) Which functional unit to identify sustainable foods? *Public Health Nutrition* 18, 2488–2497.

Payne, C.L., Scarborough, P. and Cobiac, L. (2016) Do low-carbon-emission diets lead to higher nutritional quality and positive health outcomes? A systematic review of the literature. *Public Health Nutrition* 15, 1–8.

Peters, G.P. (2010) Carbon footprints and embodied carbon at multiple scales. *Current Opinion in Environmental Sustainability* 2, 245–250.

Reddy, S., Lang, T. and Dibb, S. (2009) *Setting the Table: Advice to Government on Priority Elements of Sustainable Diets*. Sustainable Development Commission, London, UK.

Rees, W.E. (1992) Ecological footprints and appropriated carrying capacity: what urban economics leaves out. *Environment and Urbanization* 4, 121–130.

Rockström, J., Steffen, W., Noone, K., Persson, A., Chapin, F. S., *et al.* (2009) A safe operating space for humanity. *Nature* 461, 472–475.

Schader, C., Grenz, J., Meier, M.S. and Stolze, M. (2014) Scope and precision of sustainability assessment approaches to food systems. *Ecology and Society* 19(3), 42.

Sevenster, M. N., Blonk, H. and van der Flier, S. (2010) *Milieuanalyses Voedsel en Voedselverliezen. Ten Behoeve van Prioritaire Stromen Ketengericht Afvalbeleid*. CE Delft, Blonk Milieu Advies, Delft, Netherlands.

Smeets, E. and Weterings, R. (1999) *Environmental Indicators: Typology and Overview*. European Environmental Agency, Copenhagen, Denmark.

Steering Group (2010) *Food for Tomorrow, Proposal for Finland's National Food Strategy*. Ministry of Agriculture and Forestry, Helsinki, Finland.

Steffen, W., Crutzen, J. and McNeill, J.R. (2007) The Anthropocene: are humans now overwhelming the great forces of Nature? *Ambio* 36, 614–21.

Turner, K., Georgiou, S., Clark, R., Brouwer, R. and Burke, J. (2004) Economic Valuation of Water Resources in Agriculture. FAO Water Reports 27. Food and Agriculture Organization of the United Nations, Rome, Italy.

van de Kamp, M. E., van Dooren, C., Hollander, A., Geurts, M., Brink, E. J., *et al.* (2018) Healthy diets with reduced environmental impact? The greenhouse gas emissions of various diets adhering to the Dutch food-based dietary guidelines. *Food Research International* 104, 14–24.

van der Weijden, W. (2008) *Food Quality: values for your Money* (in Dutch). Stuurgroep Technology Assessment, Culemborg, Netherlands.

van Dooren, C. (2016) Proposing the Nutrient Density Unit as the Functional Unit in LCAs of Foods. International Conference on Life Cycle Assessment of Food 2016, 2016 UCD Institute of Food and Health, Dublin, Ireland.

van Dooren, C. and Bosschaert, T. (2013) Developing and disseminating a footprint tool to raise awareness about healthy and environmentally conscious food choices. *Sustainability* 9, 70–82.

van Dooren, C., Marinussen, M., Blonk, H., Aiking, H. and Vellinga, P. (2014) Exploring dietary guidelines based on ecological and nutritional values: A comparison of six dietary patterns. *Food Policy* 44, 36–46.

van Dooren, C., Douma, A., Aiking, H. and Vellinga, P. (2017) Proposing a novel index reflecting both climate impact and nutritional impact of food products. *Ecological Economics* 131, 389–398.

van Dooren, C., Aiking, H. and Vellinga, P. (2018) In search of indicators to assess the environmental impact of diets. *International Journal of Life Cycle Assessment* 23(6), 1297–1314.

van Rossum, C.T.M., Fransen, H. P., Verkaik-Kloosterman, J., Buurma-Rethans, E. J. M. and Ocke, M. C. (2011) *Dutch National Food Consumption Survey 2007–2010: Diet of Children and Adults aged 7 to 69 years*. RIVM, Bilthoven, Netherlands.

Vringer, K., Benders, R., Wilting, H., Brink, C., Drissen, E., *et al.* (2010) A hybrid multi-region method (HMR) for assessing the environmental impact of private consumption. *Ecological Economics* 69, 2510–2516.

Wackernagel, M. and Rees, W. (1996) *Our Ecological Footprint: Reducing Human Impact on the Earth*. New Society Publishers, Gabriola Island, Canada.

Wegener Sleeswijk, A., Kleijn, R., Meeusen–van Onna, M.J.G., Leneman, H., Sengers, H.H.W.J.M., *et al.* (1996) Application of LCA to agricultural products. 1. Core methodological issues. 2. Supplement to the LCA Guide. 3. Methodological background. Centre of Environmental Science (CML), Leiden University, Leiden, Netherlands.

16 Sustainable Diets and Food-based Dietary Guidelines

Rebekah Jones, Christopher Vogliano and Barbara Burlingame

Abstract

Food-based dietary guidelines (FBDGs) have been developed by countries around the world as simple policy instruments to promote better diets for individuals and populations. The guidance historically has been based on country-specific, diet-related morbidity and mortality. As the environmental impacts of food consumption and production push planetary boundaries, the case for inclusion of elements of environmental sustainability into FBDGs becomes compelling. Issues addressed include biodiversity, plant-based diets, meat and dairy consumption and production, sustainable fish consumption, processed foods, local, seasonal and organic production, standards of ethical treatment for livestock, waste and lifestyle behaviours. Examples from official FBDGs are presented, along with examples of quasi-official guidelines. Challenges and failures are also discussed, related to lack of political support and vested interests. With consideration given to all the international agreements signed by nations related to both nutrition and environmental sustainability, the logical integration should yield country-specific *sustainable* FBDGs.

Introduction

Food-based dietary guidelines (FBDGs) have a long history in providing basic and simple reference standards for healthy eating. They are recommendations and advice given to the public on foods, food groups and dietary patterns to encourage adequate nutrient intakes, promote overall health and prevent chronic diseases. In addition, they are often used as a basis for health and agricultural policies and nutrition education programmes.

Ancient Greek and Roman philosophers can be credited with developing some of the earliest iterations of FBDGs. Plato's writings from the 5th and 4th century BC detail the elements of a healthy diet and the importance of moderation, with recommendations that still appear in modern-day versions of FBDGs. Over the next two millennia, and particularly after the invention of the printing press in the 15th century, physicians and philosophers alike were expounding their dietary advice. Again, moderation was recommended by most; and food safety, more than chronic disease, was an over-arching theme (Albala, 2002). Sustainability issues were not typically featured.

At the end of the 19th century, one of the earliest university-level nutrition programmes was developed at the Massachusetts Institute of Technology (MIT) by Ellen Swallow Richards, a chemist and MIT's first female instructor. She is credited with introducing the word 'ecology' into the English language and establishing the curriculum for 'human ecology', with nutrition encompassing health, agriculture and environmental sciences. Her many publications presented dietary guidelines integrated with guidelines for environmental sustainability. For Richards (1904), one of the pioneers of sustainable diets, there was no separation between human health and environmental health (Burlingame, 2014).

One of the earliest visual representations of FBDGs comes from the early 20th century in the USA, during the period of World War I. Fig. 16.1 shows a poster from the US Food Administration, the predecessor of the US Department of Agriculture, which lists several simple recommendations for the general public for acquiring and consuming food. The focus was mainly on food waste, with recommendations that have been absent for decades but are now re-appearing in very recent versions of sustainable FBDGs, e.g. use less meat, buy local foods and minimize food waste. However, for most of the 20th century, as agriculture became more industrialized and diets became more reliant on processed and convenience foods, the environmental impact of food consumption was not addressed in the study of nutrition, or in FBDGs.

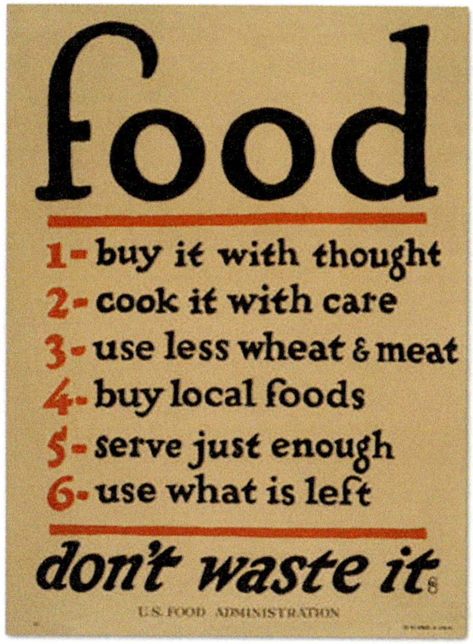

Fig. 16.1. Poster from the US Food Administration, 1917.

The topic of environmental sustainability was to re-emerge briefly in the 1980s. Gussow and Clancy (1986) published a paper, 'Dietary guidelines for sustainability', in which they argued:

> ... information on the relationship between human health and food choices is not a sufficient basis for nutrition education. In our time, educated consumers need to make food choices that not only enhance their own health but also contribute to the protection of our natural resources. Therefore, the content of nutrition education needs to be broadened and enriched not solely by medical knowledge, but also by information arising from disciplines such as economics, agriculture, and environmental science.

Unfortunately, this call for action was largely ignored by the nutrition community. FBDGs were being developed by governments and presented to the public, with little or no consideration of sustainability issues.

The era of sustainable development, which was hastened by the 1987 United Nations report, 'Our Common Future', also known as the Brundtland Report, brought environmental sustainability into sharp focus for the United Nations and its member nations. The sustainable development framework recognized both present and future generations, and the equal importance of people and planet. But again, for the most part, FBDGs were not building on this foundation. By the end of the 20th century, more than 100 countries had FBDGs presented as published food guides, posters and infographics, often depicting food pyramids, food plates and other symbolic representations. They provided a framework for nutrition education programmes and informed policies in health and agriculture, but sustainability recommendations were, for the most part, absent.

However, in the first decade of the 21st century, with heightened awareness of environmental issues, a more holistic view was being applied to FBDGs. Maturing from other fields of study, the mutually dependent relationships between nutrition, human health and planetary health were being recognized and brought into nutrition policies and programmes, and the first modern examples of sustainable FBDGs were produced.

The Case for Sustainability in Food-based Dietary Guidelines

The Brundtland Report (Brundtland, 1987) defined sustainability as sustainable development, and sustainable development as 'development that meets the needs of the present without compromising the ability of future generations to meet their own needs'. This concept necessarily featured in the processes leading up to the concept of sustainable diets, as did the outcomes of many other processes and international instruments (see Chapter 29).

The Convention on Biological Diversity (CBD) considered that human nutrition had a key role in the conservation and sustainable use of food biodiversity. In its COP 8 Decision VIII/23, Cross-cutting Initiative on Biodiversity for Food and Nutrition (CBD, 2006), a rationale was put forward and a framework was proposed, along with operational objectives and activities. The case for integrating issues of biodiversity specifically, and environmental sustainability generally, into human nutrition is well made in the rationale of Decision VIII, reproduced below:

- Biodiversity is essential for food security and nutrition.
- Environmental integrity is critical for maintaining and building positive options for human well-being.
- Existing knowledge warrants immediate action to promote the sustainable use of biodiversity in food security and nutrition programmes.
- Such action would counteract the simplification of diets, agricultural systems and ecosystems, and the erosion of food cultures.
- Considering the difficulty in precisely identifying optimal diets, a diversity of foods from plants and animals remains the preferred choice for human health.
- Traditional food systems provide positive synergies between human and ecosystem health, and culture offers an essential context for mediating positive dietary choices.
- An interdisciplinary initiative on biodiversity for food and nutrition, based on the ecosystem approach that makes the most of locally available biodiversity and will assist countries and stakeholders in achieving the Millennium Development Goals.
- Without urgent action that directly engages the environmental, agricultural, nutrition and health communities, biodiversity and the positive options offered by domesticated and wild biodiversity for addressing food security, nutrient deficiencies, and the emerging burden of non-communicable disease, will be lost.

One of the activities featured in this COP VIII decision specifically addresses FBDGs: 'Integrate biodiversity concerns into nutrition instruments, *inter alia*, food-based dietary guidelines' (CBD, 2006).

In 2010, the Food and Agriculture Organization (FAO) and Biodiversity International convened The International Scientific Symposium 'Biodiversity and Sustainable Diets: United Against Hunger', largely in response to the Cross-cutting Initiative on Biodiversity for Food and Nutrition. This symposium leveraged the expertise of key stakeholders from civil society, government and the private sector to develop the evidence base to provide directions and solutions for policy, research and action (Burlingame and Dernini, 2012). During the final session of the Symposium a consensus Platform for Action was presented and endorsed. Article five in the platform states the following:

> Food-based dietary guidelines and policies should give due consideration to sustainability when setting goals aimed at healthy nutrition. A guidance document on how to develop such guidelines and policies at national level could be elaborated by FAO, in collaboration with Bioversity International and other partners.

This recommendation was addressed and achieved in 2016 with the publication of 'Plates, pyramids, and planet' (Fischer and Garnett, 2016).

Countries' FBDGs and Successful Inclusion of Sustainable Characteristics

Erve *et al.* (2017) looked at 226 countries for the presence of official FBDGs. A total of 103 countries had FBDGs, of which 11 countries were found to share one set. This resulted in the evaluation of 93 international FBDGs. Most countries in Europe and North America have official FBDGs, while the lowest numbers are found in Africa (12%). No official FBDGs were found for 123 countries, although the numbers are growing year by year, particularly in low- and middle-income countries.

Societies around the world have vastly different dietary patterns, which are influenced by culture, accessibility, income levels and climate. Nevertheless, countries show remarkable similarities in their FBDGs. To date, the vast majority of national FBDGs, many of which are old and require updating, do not address sustainability *per se*. However, even when the guidelines do not directly specify sustainable dietary patterns, the recommendations are consistent with some of the recognized characteristics of sustainable diets. Overwhelming evidence indicates that diets higher in plant-based foods (e.g. vegetables, fruits, legumes, seeds, nuts, whole grains) and lower in animal-based foods (particularly red meat) help promote health while reducing adversely associated environmental impacts. In addition, maintaining energy balance and the inclusion of a wide diversity of foods are considered compatible characteristics of a sustainable diet and are common to nearly all FBDGs (Tilman and Clark, 2014; Melina *et al.*, 2016). These recommendations are common to nearly all FBDGs.

Only a handful of countries have guidelines that specifically promote diets and food systems that are both healthy and sustainable (EUPHA, 2017). During the past decade, several revised and newly developed FBDG have successfully incorporated sustainability considerations, while others have attempted their inclusion but were met with opposition from industry and other groups.

Table 16.1 shows some of the characteristics and rationale for sustainable FBDGs, and Table 16.2 lists thirteen countries and the elements of sustainable diets explicit in the SFBDGs.

The countries displayed in Table 16.2 were chosen based on resources available from the FAO as a website that lists and catalogues FBDGs, summarizing the main messages and how guidelines have been developed. All the information is provided and approved by the member countries.

Other SFBDS

In addition to official guidelines, a range of other guidelines exist, including 'quasi-official' guidelines and guidelines from civil society organizations and the private sector. These are defined as advice produced by institutions, some of which are recognized or accredited by governments, but whose recommendations do not constitute official policy. Most are founded on good scientific evidence and illustrate interesting or helpful approaches to integrating sustainability and nutritional advice.

One of the most popular and widely disseminated is the double food and environmental pyramid developed by the BCFN Foundation (Fig. 16.2). The double pyramid model illustrates the relationships between foods and food groups and their ecological footprint. The broad base and lower levels of the food pyramid represent food to be consumed generously. Beside it, the inverted environmental pyramid depicts how these same foods produce the lowest environmental impact. Conversely, the top of the food pyramid depicts food that should be consumed in small quantities and infrequently, alongside the breadth of the inverted pyramid depicting the high environmental impact of these foods. Similarly, the Mediterranean Diet Pyramid makes specific mention of biodiversity and environmental sustainability (IFMed, 2015; Dernini and Berry, 2015).

Highlights and Challenges

Historically, challenges have faced both the development and implementation of FBDGs. These challenges persist and have been amplified as the aim of FBDGs expands to encompass not only human health, but also planetary health.

Countries are faced with the challenges of addressing the diversity that exists in the complex social, economic and political interactions between people and their food. This complexity relates to long-standing cultures and traditions, along with more recent changes related to industrialization (e.g. ultra-processed foods), populations (e.g. migration) and the environment (e.g. climate).

Although there are no guidelines that perfectly encompasses every characteristic of a sustainable diet, several countries have successfully incorporated many aspects. Examples of both the challenges and successes of this implementation are discussed below.

Multi-sectoral, multi-disciplinary approach

Systemic change is required to address the most pressing sustainability issues. Achieving this requires interdisciplinary collaboration from academia, government and industry stakeholders. In many countries, the development of an FBDG falls under the management of ministries of health and the process is therefore driven

Table 16.1. Sustainable diet characteristics and rationale.

Characteristic	Rationale
Promotes diet diversity of whole foods	Diets that include a diversity of whole foods are linked to improved health outcomes by ensuring dietary adequacy, increased food security, a reduced intake of toxicant and protection against chronic diseases (Kant *et al.*, 1993; La Vecchia *et al.*, 1997; Michels and Wolk, 2002; Foote *et al.*, 2004; Jansen *et al.*, 2004; Steyn *et al.*, 2006; Arimond *et al.*, 2010; Vandevijvere *et al.*, 2010). Agricultural biodiversity can help achieve nutrition security by supplying a wide range of nutrients, including phytonutrients, vitamins, and minerals (Toledo and Burlingame, 2006)
Promotes plant-based diets	Eating a predominantly whole food, plant-based diet is consistent with improved health outcomes and a reduced risk for many chronic diseases, certain cancers, obesity, and diabetes mellitus type 2. Additionally, plant-based diets require significantly fewer agricultural inputs such as (energy, petroleum, fertilizers, pesticides, herbicides and water) and emit far fewer greenhouse gas emissions than meat-heavy diets (Tilman and Clark, 2014; Melina *et al.*, 2016)
Reduce/limit red meat consumption	Consuming high quantities of red meat is linked to numerous adverse health effects (Pan *et al.*, 2012; Zelber-Sagi *et al.*, 2018). Meat from ruminant animals such as cattle and lamb is documented as being the largest food-based emitters of greenhouse gas emissions (Scarborough *et al.*, 2014)
Reduce/limit processed meat	Processed meat is linked to certain types of cancer, cardiovascular disease, and diabetes mellitus type 2 when consumed in excess quantities (Chan *et al.*, 2011; Boada *et al.*, 2016)
Eat dairy products and alternatives in moderation	Dairy products can be a part of a healthy diet when consumed in moderation, however dairy is a significant source of greenhouse gas emissions (FAO, 2010). Some guidelines encourage the reduction of dairy, particularly if sweetened with excess amounts of sugar
Encourage sustainable seafood consumption	Seafood contains many essential nutrients including protein, calcium and omega 3 fatty acids. However, the demand for seafood has led to many species of fish becoming threatened due to overfishing. Choosing sustainable seafood options can reduce the ecological impact of overfishing (Jackson *et al.*, 2001). Certain fish varieties are high in the neurotoxin methylmercury and should be consumed in moderation (Sheehan *et al.*, 2014)
Limit ultra-processed foods high in fat and sugar	Ultra-processed foods have been stripped of the majority of their nutrients, and are often high in fat, sugar, and sodium. These foods are a significant source of calories for many around the world, yet do little to contribute to nutrition security (da Costa Louzada *et al.*, 2015; Monteiro *et al.*, 2010)
Water conservation and promotion	Many areas around the world are experiencing water security issues. Some dietary guidelines discuss the benefits of minimizing water in cooking and food production. Where water is safe to drink from the tap, many guidelines encourage tap water over bottled water, encouraging the reduction of plastic bottles
Promotes buying local foods	While the definition of 'local' varies considerably, local foods are shown to improve farmer–consumer relationships, increase revenue for small farmers, and encourage consumption of a wider diversity of foods (Brown and Miller, 2008)
Standards for the ethical treatment of animals	Animal welfare practices and standards vary widely from country to country. Dietary guidelines are beginning to promote the ethical treatment of animals as a part of a sustainable food system
Promotes reduction of food waste	Food waste occurs primarily in the developed world, with the majority occurring in the retail and consumer sectors. Reducing food waste can significantly reduce greenhouse gases while improving food security (Gustavsson *et al.*, 2011)

and directed by health sector professionals and governing bodies. Some of the early reports and manuals produced by the World Health Organization (WHO) and FAO for the preparation and use of FBDGs specified that the primary consideration and purpose was to address the epidemiology of diet-related health problems, including non-communicable diseases and other forms of

Table 16.2. Thirteen countries and the elements of sustainable diets.

Country (year)	Visual representation	Promotes dietary biodiversity	Promotes plant-based diets	Moderate red meat consumption	Limit processed meat	Moderate dairy consumption	Encourages sustainable seafood consumption	Limit processed/ ultra-processed foods	Water conservation in cooking	Promotes buying local foods/ seasonal/most sustainable options	Encourages food and packaging waste reduction	Behavioural e.g. exercise/ cooking etc.	Animal welfare?/ species under threat	Directly links diets to sustainability
Australia (2013)	Plate				X	X		X						
Brazil (2014)	NA	X	X	X	X	X		X		X				X
Canada (2007)	Rainbow				X									
China (2016)	Pagoda							X			X			
Estonia (2006)	Pyramid		X	X	X			X		X	X	X	X	X
France (2011)	NA							X		X				
Germany (2013)	Circle/disc		X	X			X				X	X	X	X
Netherlands (2015)	Wheel of Five		X	X	X	X	X	X		X	X			X
Qatar (2015)	Oyster		X	X	X		X	X	X	X	X			X
Sierra Leone (2016)	Plate				X			X		X				
Sweden (2015)	Traffic light		X	X	X	X	X	X		X			X	X
UK (2016)	Plate			X	X	X	X	X						X
USA (2016)	Plate													

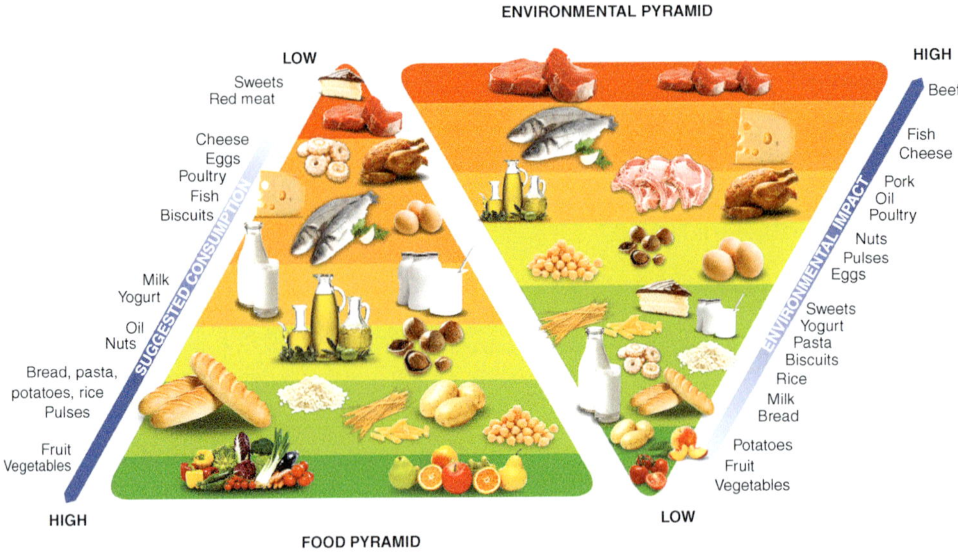

Fig. 16.2. Barilla double pyramid, reproduced with permission. Source: US Food Administration (1917).

malnutrition. In a 100-page publication from an FAO and WHO consultation on FBDG preparation, the evidence presented refers to health and disease, with only a single paragraph on agriculture and environmental sciences (WHO, 1998).

However, some countries have begun to incorporate a wider range of expertise and representation in their consultation and development processes. For example, the 2014 Brazilian FBDGs have included representation from the education, social welfare and agriculture sectors, as well as the public. This strategy ensures the broader societal and environmental issues are addressed and included. It also ensures the characteristics of a sustainable diet are understood by those it is targeting by including them in the translation process.

Political lobbying

The scientific committees in both the US and Australia have attempted to include environmental considerations in their respective FBDGs (Merrigan *et al.*, 2015). However, due to a lack of government endorsement and negative response and resistance towards their implementation from a range of sectors and bodies, the most recent revisions of the guidelines do not include sustainability characteristics.

During the development of the Australian guidelines, a public media campaign strongly opposed incorporation of sustainability in the guidelines. Advocated by the food industry, farmers and fisheries groups, the main argument raised was that the environmental concerns were out of the mandate of the dietary guidelines. Consequently, sustainability guidance is only included in an appendix in the final version of the Australian Dietary Guidelines (NHMRC, 2013).

Sustainability guidance is not included at all in the US guidelines. The inclusion of environmental sustainability into the 2015 US Dietary Guidelines was heavily opposed by the food industry. In a joint statement to the press, the US Secretaries of Agriculture and Health and Human Services stated: 'we do not believe that the 2015 DGAs (Dietary Guidelines for Americans) are the appropriate vehicle for this important policy conversation about sustainability' and that the purpose of dietary guidelines was simply to educate the population about weight control and chronic disease prevention (USDA, 2016).

Although not as dominant as the lobbying in the US, the debate regarding sustainability guidance in Swedish FBDGs continued for ten years before implementation. The biggest opposition came from LRF Dairy Sweden, who claimed that it was too early for advice based on both human and planet health via FBDGs due to the immaturity of research regarding the environmental impact from food production and the importance of milk as a food that naturally contains many important nutrients. Although initially critical, the dairy organizations eventually did express their support for the ambition to incorporate environmental sustainability in the Swedish FBDGs advice once they became engaged in the development.

Diet variety

Historically, most FBDGs promote a diverse diet. Through pyramids, circles, discs or rainbows, FBDGs emphasize the importance of consuming a variety of foods across all food groups to achieve adequate nutrient intakes. For example, the US guidelines (USDA, 2015a) make use of broad statements including a 'Focus on variety, nutrient density, and amount. To meet nutrient needs within calorie limits, choose a variety of nutrient-dense foods across and within all food groups in recommended amounts'. Many guidelines then go on to give an explanation or examples of how to achieve a diverse diet within such food groups, for example, by selecting a variety of vegetables from subgroups such as by colour (dark green, red and orange). However, most guidelines end their recommendations at this level of detail. With few exceptions, no explicit link is generally made between diet variety, biodiversity and sustainability. One exception is the Brazilian guidelines, in which the rationale for consuming a diverse diet is linked directly to sustainability:

> Choosing diets based on a variety of foods of plant origin with sparing amounts of foods of animal origin implies the choice of a food system that is relatively equitable, and less stressful to the physical environment, for animals and biodiversity in general.
>
> (Ministry of Health of Brazil, 2015)

Another example is the Mediterranean Diet Pyramid (IFMed, 2015) where biodiversity is linked to environmental sustainability.

Plant-based foods

Some guidelines explicitly promote, and link, the consumption of a plant-based diet to health and sustainability. For example, Germany's guidelines say to 'Choose mainly plant-based foods. They have a health promoting effect and foster a sustainable diet' (The German Nutrition Society, 2013).

In addition, the name of the 'protein' food group in the UK guidelines has been updated to include plant-based proteins such as beans and pulses, along with the traditionally included fish, eggs, meat and other proteins.

In contrast, the US 2015–2020 dietary guidelines use generalized statements without an environmental sustainability justification. For example, they recommend shifting dietary patterns to consume more vegetables in place of foods high in calories, saturated fats or sodium. However, these recommendations are solely reinforcing the diet–disease relationship, despite the recommendation by US Dietary Guideline Advisory Committee (DGAC) to make the explicit statement that 'a dietary pattern that is higher in plant-based foods, and lower in animal-based foods is more health promoting and is associated with lesser environmental impact than is the current average U.S. diet' (USDA, 2015b).

Seasonal and local food

Many guidelines promote the purchasing of foods that are in season from local farms to support the farmers, the local economy and the environment. The Estonian recommendations include 'eating ecologically', stating that foods should be regional or local in origin, and be seasonal and traditional. Roughly translated, the guidelines explain that the closer the food comes from its place of production, the fewer resources it takes to transport and preserve it (Estonian NID, 2012).

The Sierra Leone Food Guide for Healthy Eating does not represent a typical Sierra Leonean plate, but rather, foods according to their availability and accessibility (Montagnese *et al.*, 2017). Consequently, the entire Sierra Leone FBDG promotes the consumption of local and seasonal foods: '...production and consumption of diverse nutrient-dense foods that are locally available, including locally produced or sourced foods from plants and animals' and gives specific examples from each food group such as '...examples of locally available fruits include oranges, mangoes, guava, watermelon, pineapple, star fruit, plums, berries and a whole range of wild fruits; while dark green leafy vegetables include potato and cassava leaves, okra and carrots'.

To encourage local production and seasonality, guidelines sometimes make the link to taste to drive consumption. However, only a few guidelines, such as the Estonian FBDGs seen above, explicitly state how this links to a more sustainable and healthy planet.

Meat and dairy

Reducing global livestock production, and the associated consumption of meat and dairy, is key to mitigating climate change.

The production of animal-source protein is extremely inefficient (Pimentel and Pimentel, 2003) in terms of energy inputs, water and land. Furthermore, conversion of land to pasture has contributed to accelerated deforestation and land and soil degradation (FAO, 2013). Estimates show that greenhouse gas emissions of cattle and sheep are on average 19–48 times higher than those of protein-rich plants such as legumes, seeds and grains. Although livestock other than cattle and sheep produce relatively less emissions, plant products are still significantly lower (Ripple *et al.*, 2014). In addition, animal proteins, saturated fats, processed meats and red meat have been linked to multiple health risks including cardiovascular disease, obesity and increased cancer rates (Pan *et al.*, 2012). In spite of all the scientific evidence, global meat consumption and associated greenhouse gas emissions are at all-time highs.

Consumers are often not aware of the issues, and thus FBDGs can act as an important platform to raise public awareness. Therefore, one of the main recommendations of sustainable FBDGs must be to reduce animal product consumption and shift towards plant-based diets (Allodi *et al.*, 2015). Consequently, guidelines have begun to explicitly promote, and link, the consumption of a plant-based diet to health and sustainability. These include Germany, Brazil and Qatar. Germany's guidelines recommend to 'Choose mainly plant-based foods. They have a health promoting effect and foster a sustainable diet.'

Chapter 7 of the Qatari guidelines 'Eat Healthy while Protecting the Environment' explicitly states the link between processed foods and meat consumption and the environment (The German Nutrition Society, 2013). Key recommendations are provided on how to reduce this impact: 'Choose fresh, home-made foods over highly processed foods and fast foods'. The EUPHA (2017) states that consumers should be encouraged to consume plant-based rather than meat-based proteins, reduce the portions of meat and eat meat less often.

The Mediterranean diet, low in saturated fat and red meat, has been promoted as both healthy and sustainable (Dernini *et al.*, 2017), and has been used as the model for sustainable diets around the world (Burlingame and Dernini, 2011).

In their latest dietary guidelines, the Chinese government recommends a slightly lower meat intake than it did in its 2007 guidance, even as the meat consumption in China is far higher than either the old or the new dietary guidelines recommend. Intake is estimated to be more than 300 g per person per day, with sharp increases projected in the coming decades (FAO, 2013). Adherence to the new (and similarly the old) dietary guidelines would therefore entail a substantial average decrease in meat consumption (Perignon *et al.*, 2016). If such reductions were to actually occur, it could result in significant reductions in adversely associated environmental impacts (Garnett and Wilkes, 2014; The Guardian, 2016).

The Swedish guidelines are among the very few recommending moderate consumption of dairy products due to environmental impacts (Swedish National Food Agency (Livsmedelsverket), 2015). They go as far as to explain the rationale for this, 'Dairy products come from cows, which release methane gas. This is bad for the environment, so it's a good idea not to consume too much cheese or other dairy products.'

Even in countries rejecting explicit references to sustainable diet characteristics in their recommendations, some small steps have been taken. In the UK, the name of the 'protein' food group has been updated to include beans and pulses along with fish, eggs, meat and other proteins, subtly highlighting the importance of plant-based proteins in a sustainable, healthy diet. The US DGAC stated that 'a dietary pattern that is higher in plant-based foods, and lower in animal-based foods is more health promoting and is associated with lesser environmental impact than is the current average U.S. diet'; although no definitive recommendation to reduce these products is included in the guidelines (Millen *et al.*, 2016).

It is important to note that livestock systems in some countries are managed sustainably, and that the imperative for a global reduction in livestock for meat and dairy is not meant as a requirement for all countries (HLPE, 2016).

Seafood

The UK guidelines state, 'Eat two portions of sustainably sourced fish per week, one of which is oily' (Public Health England, 2016). Although no further detail or rationale is provided, readers are guided to a link: 'Also www.msc.org/ for more guidance on sustainably sourced fish', where more information regarding sustainable fish practices and advice is provided. Qatar also specifies consumption of sustainable seafood and provides an online guide including, 'choose light tuna or salmon instead of white (albacore) tuna...are the healthiest and most environmentally friendly' (Qatar Ministry of Public Health, 2015).

Sweden's guidelines also provide tips for sustainable seafood:

> Seafood is largely a wild resource that is at risk of being depleted. There are also fishing methods and fish farming methods that can harm the environment. So not eating too much fish is good for the environment. Choosing sustainable fish makes it possible for us to continue eating fish in the future. Look out for ecolabels such as MSC, ASC and Krav, or use the WWF's fish guide.
>
> (Swedish National Food Agency (Livsmedelsverket), 2015)

Water

Although most FBDGs mention water, messages are typically only regarding water intake (e.g. substituting water for sugary drinks) and in relation to hydration during physical activity. For example, water fills the centre of the German food pyramid to highlight its importance.

However, an explanation of the 'water footprint', that is, the volume of freshwater required to produce the specific type of food or diet, is now included in some guidelines. The data for constructing the environmental pyramid of the double food and environmental pyramid model, developed by the BCFN Foundation, were based largely on the water footprint of foods and are used to guide FBDG development (Ruini *et al.*, 2015). For example, Brazil explains the environmental sustainability issues in the context of livestock production: 'Reduced consumption and thus production of animal foods will reduce... intensive use of water' (Ministry of Health of Brazil, 2015). Brazil also makes the links between water and highly processed foods.

Qatar, where water security is an issue, has FBDGs that state, 'In general, plant-based foods . . .use less water in their production than animal foods, such as beef', and 'conserve water in food preparation' (Qatar Ministry of Public Health, 2015).

Waste

Food waste was the primary motivation for one of the earliest FBDGs, issued by the US Food Administration in 1917, as shown in Fig. 16.1. All six recommendations contribute to the goal of not wasting food.

Other guidelines also address the issue of waste, in the context of food *per se*, natural resource waste in food production and preparation, and in the packaging for processed foods. One of China's core recommendations directly addresses the waste issue, stating (as the English translation): 'Eliminate waste and develop a new ethos of diet civilization' (Chinese Nutrition Society, 2016).

Behavioural/exercise

Many guidelines discuss the importance of exercise for health. However, the German FBDGs directly link exercise not only to health, but to the environment. 'For example, you can walk or take the bicycle from time to time. This protects the environment and promotes your health.'

Direct Link to Sustainability

As discussed, the majority of international FBDGs do not explicitly discuss sustainability or link recommendations to the environment or planetary health, in spite of several decades of calls to do so. Nevertheless, as the environmental impacts of diets, both production and consumption, are now impossible to ignore, FBDGs are being developed and implemented with sustainability in the forefront. Although many intergovernmental processes have specified that environmental sustainability should be included in FBDGs, many national governments find that the pressure from private sector vested interests makes this process difficult. Thus, many quasi-official FBDGs are stepping up with recommendations for sustainable diets (FAO, 2018).

The German government supports a myriad of alternative and additional guidelines aimed at a range of sub-populations and addressing a range of issues, for example, to inform purchasing decisions. The Sustainable Shopping Basket (RNE, 2013) promotes the messages that 'sustainable consumption is already possible today' and 'sustainable consumption means buying more thoughtfully and buying less'. Also available as an App, it presents information about a range of products and their impact on the environment. The main recommendations of the food segment are eating less meat and fish, eating five servings of fruit and vegetables a day, eating seasonal and regional products, and buying organic and Fairtrade products and drinks in recyclable packaging. It provides a seasonal calendar for fruit and vegetables and explains the different labels and certification schemes that German consumers might find on the packaging of products they buy. It also goes into further explanation regarding trade-offs between sustainably produced products that have then been transported long distances. It also discusses animal welfare, recommending the purchase of organic and Fairtrade products and advises choosing 'meat from animals raised under species-appropriate conditions' and not to buy eggs from hens that have been kept in battery cages.

Although the French dietary guidelines do not discuss environmental concerns, the French Agency for the Environment and Energy (ADEME), which works for the implementation of public policy in the areas of the environment, energy and sustainable development has produced a set of recommendations aimed at individuals and 'eco-citizens' to promote sustainable shopping habits. A section of their website called 'Mes Achats' (my purchases) provides four main messages: (i) to promote seasonal products, (ii) to 'adopt diets that combine health, environment and fun' (e.g. replace a meat dish by one based on grains or legumes once a week), (iii) 'buy products with environmental labels', and (iv) limit food waste.

Conclusions

Evidence is mounting that sustainable diets can be realized, which maintain nutritional adequacy and affordability, along with reduced environmental impacts. However, it is thought that no country currently meets basic dietary needs for its citizens at a globally sustainable level of resource use (Perignon *et al.*, 2016; O'Neill *et al.*, 2018).

The increasing awareness of environmental sustainability and overshooting planetary boundaries related to the ways food is produced and consumed, brings into clear focus the need for sustainable diets. One simple step in that direction is for countries to develop or revise their FBDG to include recommendations related to the impact of diets on ecosystems and natural resources. Early efforts to introduce elements of environmental sustainability into dietary guidelines had been successful in times of crisis, e.g. war efforts as illustrated in Fig. 16.1. The planetary crises now being faced related to climate, carbon, phosphorus, water, fisheries, livestock, biodiversity, land use, waste and more, make the case ever more strongly for countries to take on a range of policy and programme actions, one of the easiest of which should be to produce sustainable FBDGs. To do so contributes to meeting obligations contained in the Sustainable Development Goals, the Decade of Action for Nutrition, and many other international instruments agreed, ratified and endorsed by almost all countries in the world.

References

Albala, K. (2002) *Eating Right in the Renaissance*, Vol. 2. University of California Press, Berkeley, California, USA.

Allodi, M., Chikobava, D., Lappalainen, J. and Tarhonen, N. (2015) *Towards Sustainable Diets: Decreasing Meat Consumption*. Doctoral dissertation. University of Helsinki, Helsinki, Finland.

Arimond, M., Wiesmann, D., Becquey, E., Carriquiry, A., Daniels, M.C., *et al.* (2010) Simple food group diversity indicators predict micronutrient adequacy of women's diets in 5 diverse, resource-poor settings. *The Journal of Nutrition* 140(11), 2059S–2069S.

Boada, L.D., Henríquez-Hernández, L.A. and Luzardo, O.P. (2016) The impact of red and processed meat consumption on cancer and other health outcomes: epidemiological evidences. *Food and Chemical Toxicology* 92, 236–244.

Brown, C. and Miller, S. (2008) The impacts of local markets: a review of research on farmers markets and community supported agriculture (CSA). *American Journal of Agricultural Economics* 90(5), 1298–1302.

Brundtland, G.H. (1987) Report of the World Commission on Environment and Development: Our Common Future. Available at http://www.un-documents.net/our-common-future.pdf (accessed 28 June 2018).

Burlingame, B. (2014) Grand challenges in nutrition and environmental sustainability, Specialty Grand Challenge. *Frontiers in Nutrition*. doi: 10.3389/fnut.2014.00003.

Burlingame, B. and Dernini, S. (2011) Sustainable diets: the Mediterranean diet as an example. *Public Health Nutrition* 14(12A), 2285–2287.

Burlingame, B. and Dernini, S. (Eds) (2012) Sustainable Diets and Biodiversity: Directions and solutions for policy, research and action. *Proceedings of the International Scientific Symposium Biodiversity and Sustainable Diets United against Hunger*, 3–5 November 2010, FAO Headquarters, Rome http://www.fao.org/docrep/016/i3004e/i3004e.pdf (accessed 28 August 2018).

CBD (2006) COP 8 Decision VIII/23 Agricultural biodiversity. Cross-cutting initiative on biodiversity for food and nutrition. Convention on Biological Diversity. Available at https://www.cbd.int/decision/cop/?id=11037 (accessed 12 Feb 2018).

Chan, D.S., Lau, R., Aune, D., Vieira, R., Greenwood, D.C., Kampman, E. and Norat, T. (2011) Red and processed meat and colorectal cancer incidence: meta-analysis of prospective studies. *PLoS One* 6(6), e20456.

Chinese Nutrition Society (2016) Dietary guidelines for Chinese residents. Available in English at http://www.fao.org/nutrition/education/food-dietary-guidelines/regions/countries/china/en/ (accessed 3 March 2018).

da Costa Louzada, M.L., Baraldi, L.G., Steele, E.M., Martins, A.P.B., Canella, D.S., *et al.* (2015) Consumption of ultra-processed foods and obesity in Brazilian adolescents and adults. *Preventive Medicine* 81, 9–15.

Dernini, S. and Berry, E.M. (2015) Mediterranean diet: From a healthy diet to a sustainable dietary pattern. *Frontiers in Nutrition* 2, 15. DOI: https://doi.org/10.3389/fnut.2015.00015

Dernini, S., Berry, E.M., Serra-Majem, L., La Vecchia, C., Capone, R., *et al.* (2017) Med Diet 4.0: the Mediterranean diet with four sustainable benefits. *Public Health Nutrition* 20(7), 1322–1330.

Erve, I., Tulen, C.B.M., Jansen, J., Laar, A.V., Minnema, R., *et al.* (2017) Overview of Elements within National Food-Based Dietary Guidelines. Available at http://www.journalrepository.org/media/journals/EJNFS_30/2017/Jun/Wolvers712017EJNFS32645_.pdf. (accessed 24 February 2018).

Estonian NID (2012) How to eat properly. National Institute for Development. Available at http://toitumine.ee/ru/kak-pravilno-pitatsya (accessed 11 March 2018).

EUPHA (2017) Healthy and sustainable diets for European countries. Available at https://eupha.org/repository/advocacy/EUPHA_report_on_healthy_and_sustainable_diets_20-05-2017.pdf. European Public Health Association (accessed 1 February 2018).

FAO (2010) Greenhouse gas emissions from the dairy sector. A life cycle assessment. Available at http://www.fao.org/docrep/012/k7930e/k7930e00.pdf (accessed 3 March 2018).

FAO (2013) Tackling climate change through livestock: A global assessment of emissions and mitigation opportunities. Available at http://www.fao.org/3/a-i3437e.pdf (accessed 3 March 2018).

FAO (2018) Dietary guidelines and sustainability. Available at http://www.fao.org/nutrition/education/food-dietary-guidelines/background/sustainable-dietary-guidelines/en/ (accessed 8 January 2018).

Fischer, C.G. and Garnett, T. (2016) Plates, pyramids and planets: developments in national healthy and sustainable dietary guidelines: a state of play assessment. Available at http://www.fao.org/3/a-i5640e.pdf (accessed 28 June 2018).

Foote, J.A., Murphy, S.P., Wilkens, L.R., Basiotis, P.P. and Carlson, A. (2004) Dietary variety increases the probability of nutrient adequacy among adults. *The Journal of Nutrition* 134(7), 1779–1785.

Garnett, T. and Wilkes, A. (2014) Appetite for change: social, economic and environmental transformations in China's food system. Available at https://www.fcrn.org.uk/fcrn/publications/appetite-for-change (accessed 12 January 2018).

German Nutrition Society (2013) Ten guidelines for wholesome eating and drinking from the German Nutrition Society. Available at http://www.fao.org/nutrition/education/food-based-dietary-guidelines/regions/countries/germany/en/ (accessed 11 March 2018).

Gussow, J.D. and Clancy, K.L. (1986) Dietary guidelines for sustainability. *Journal of Nutrition Education* 18(1), 1–5.

Gustavsson, J., Cederberg, C., Sonesson, U., Van Otterdijk, R. and Meybeck, A. (2011) Global Food Losses and Food Waste. Food and Agriculture Organization of the United Nations, Rome, Italy, pp. 1–38.

HLPE (2016) Sustainable agricultural development for food security and nutrition: what roles for livestock? A Report by the High Level Panel of Experts on Food Security and Nutrition of the Committee on World Food Security. Available at http://www.fao.org/cfs/cfs-hlpe/reports/report-10-elaboration-process/en/ (accessed 13 March 2018).

IFMed (2015) Mediterranean Diet Pyramid, v. 4.0. Available at https://dietamediterranea.com/piramidedm/piramide_INGLES.pdf (accessed 3 March 2018).

Jackson, J.B., Kirby, M.X., Berger, W.H., Bjorndal, K.A., Botsford, L.W., *et al.* (2001). Historical overfishing and the recent collapse of coastal ecosystems. *Science* 293(5530), 629–637.

Jansen, M.C., Bueno-de-Mesquita, H.B., Feskens, E.J., Streppel, M.T., Kok, F.J. and Kromhout, D. (2004) Quantity and variety of fruit and vegetable consumption and cancer risk. *Nutrition and Cancer* 48(2), 142–148.

Kant, A.K., Schatzkin, A., Harris, T.B., Ziegler, R.G. and Block, G. (1993) Dietary diversity and subsequent mortality in the First National Health and Nutrition Examination Survey Epidemiologic Follow-up Study. *The American Journal of Clinical Nutrition* 57(3), 434–440.

La Vecchia, C., Muñoz, S. E., Braga, C., Fernandez, E. and Decarli, A. (1997) Diet diversity and gastric cancer. *International Journal of Cancer* 72(2), 255–257.

Melina, V., Craig, W. and Levin, S. (2016) Position of the Academy of Nutrition and Dietetics: vegetarian diets. *Journal of the Academy of Nutrition and Dietetics* 116(12), 1970–1980.

Merrigan, K., Griffin, T., Wilde, P., Robien, K., Goldberg, J. and Dietz, W. (2015) Designing a sustainable diet. *Science*, 350(6257), 165–166.

Michels, K.B. and Wolk, A. (2002) A prospective study of variety of healthy foods and mortality in women. *International Journal of Epidemiology* 31(4), 847–854.

Millen, B.E., Abrams, S., Adams-Campbell, L., Anderson, C.A., Brenna, J.T., *et al.* (2016) The 2015 Dietary Guidelines Advisory Committee Scientific Report: Development and Major Conclusions. *Advances in Nutrition* 7(3), 438–444.

Ministry of Health of Brazil (2015) Dietary guidelines for the Brazilian population 2014. Available at http://bvsms.saude.gov.br/bvs/publicacoes/dietary_guidelines_brazilian_population (accessed 11 March 2018).

Montagnese, C., Santarpia, L., Iavarone, F., Strangio, F., Caldara, A.R., *et al.* (2017) North and South American countries food-based dietary guidelines: A comparison. *Nutrition* 42, 51–63.

Monteiro, C.A., Levy, R.B., Claro, R.M., de Castro, I.R.R. and Cannon, G. (2010) Increasing consumption of ultra-processed foods and likely impact on human health: evidence from Brazil. *Public Health Nutrition* 14(1), 5–13.

NHMRC (2013) Australian dietary guidelines. Available at https://www.eatforhealth.gov.au/sites/default/files/content/n55_australian_dietary_guidelines.pdf (accessed 5 March 2018).

O'Neill, D.W., Fanning, A.L., Lamb, W.F. and Steinberger, J.K. (2018) A good life for all within planetary boundaries. *Nature Sustainability* 1, 88–95. DOI: 10.1038/s41893-018-0021-4.

Pan, A., Sun, Q., Bernstein, A.M., Schulze, M.B., Manson, J.E., *et al.* (2012) Red meat consumption and mortality: results from 2 prospective cohort studies. *Archives of Internal Medicine* 172(7), 555–563.

Perignon, M., Masset, G., Ferrari, G., Barré, T., Vieux, F., *et al.* (2016) How low can dietary greenhouse gas emissions be reduced without impairing nutritional adequacy, affordability and acceptability of the diet? A modelling study to guide sustainable food choices. *Public Health Nutrition* 19(14), 2662–2674. DOI: 10.1017/S1368980016000653

Pimentel, D. and Pimentel, M. (2003) Sustainability of meat-based and plant-based diets and the environment. The *American Journal of Clinical Nutrition* 78, 660–663.

Public Health England (2016) Eatwell guide. Available at https://www.gov.uk/government/publications/the-eatwell-guide (accessed 15 February 2018).

Qatar Ministry of Public Health (2015) Qatar Dietary Guidelines. Available at http://eservices.moph.gov.qa/qdgportal/home.jsp?lang=en (accessed 3 March 2018).

Richards, E. (1904) *First Lessons in Food and Diet*. Whitcomb and Barrows, Boston, Massachusetts, USA. Available at https://archive.org/stream/firstlessonsinf00richgoog#page/n5/mode/2up (accessed 12 Feb 2018).

Ripple, W., Smith, P., Haber, H.I., Montzka, S., McAlpine, C., Boucher, D. (2014) Ruminants, climate change and climate policy. Available at http://www.health.gov/dietaryguidelines/dga2015/comments/uploads/CID230_Ripple__2014_NatureClimateChange-Ruminants.pdf (accessed 12 April 2015).

RNE (2013) The sustainable shopping basket: a guide to better shopping. Available at https://www.bundesregierung.de/Content/EN/StatischeSeiten/Schwerpunkte/Nachhaltigkeit/Anlagen/shopping-basket.pdf?__blob=publicationFile&v=1 (accessed 11 March 2018).

Ruini, L.F., Ciati, R., Pratesi, C.A., Marino, M., Principato, L. and Vannuzzi, E. (2015) Working Toward Healthy and Sustainable Diets: The "Double Pyramid Model" Developed by the Barilla Center for Food and Nutrition to Raise Awareness about the Environmental and Nutritional Impact of Foods. *Frontiers in Nutrition* 2, 9. DOI: 10.3389/fnut.2015.00009

Scarborough, P., Appleby, P.N., Mizdrak, A., Briggs, A.D.M., Travis, R.C., *et al.* (2014) Dietary greenhouse gas emissions of meat-eaters, fish-eaters, vegetarians and vegans in the UK. *Climatic Change* 125(2), 179–192. DOI: 10.1007/s10584-014-1169-1

Sheehan, M.C., Burke, T.A., Navas-Acien, A., Breysse, P.N., McGready, J. and Fox, M.A. (2014) Global methylmercury exposure from seafood consumption and risk of developmental neurotoxicity: a systematic review. *Bulletin of the World Health Organization* 92, 254–269F.

Steyn, N., Nel, J., Nantel, G., Kennedy, G. and Labadarios, D. (2006) Food variety and dietary diversity scores in children: are they good indicators of dietary adequacy? *Public Health Nutrition* 9(5), 644–650.

Swedish National Food Agency (Livsmedelsverket) (2015) Find your way to eat greener, not too much and to be active! Available at https://www.livsmedelsverket.se/en (accessed 24 March 2018).

The Guardian (2016) China's plan to cut meat consumption by 50% cheered by climate campaigners. Available at https://www.theguardian.com/world/2016/jun/20/chinas-meat-consumption-climate-change (accessed 15 February 2018).

Tilman, D. and Clark, M. (2014) Global diets link environmental sustainability and human health. *Nature* 515(7528), 518.

Toledo, Á. and Burlingame, B. (2006) Biodiversity and nutrition: a common path toward global food security and sustainable development. *Journal of Food Composition and Analysis* 19(6–7), 477–483.

USDA (2015a) Dietary guidelines for Americans 2015–2020, 8th Edition. Available at https://health.gov/dietaryguidelines/2015/guidelines/ (accessed 23 March 2018).

USDA (2015b) Scientific report of the 2015 Dietary Guidelines Advisory Committee. Available at https://ods.od.nih.gov/pubs/2015_DGAC_Scientific_Report.pdf (accessed 28 June 2018).

USDA (2016) 2015 Dietary Guidelines: Giving You the Tools You Need to Make Healthy Choices. Media release. Available at https://www.usda.gov/media/blog/2015/10/6/2015-dietary-guidelines-giving-you-tools-you-need-make-healthy-choices (accessed 28 June 2018).

US Food Administration (1917) Food – don't waste it. Available at https://www.loc.gov/item/2002699356/ (accessed 3 July 2018).

Vandevijvere, S., De Vriese, S., Huybrechts, I., Moreau, M. and Van Oyen, H. (2010) Overall and within-food group diversity are associated with dietary quality in Belgium. *Public Health Nutrition* 13(12), 1965–1973.

WHO (1998) Preparation and use of food-based dietary guidelines. Technical Report Series 880. World Health Organization, Geneva. Available at http://apps.who.int/iris/bitstream/handle/10665/42051/WHO_TRS_880.pdf;jsessionid=0A3927E28E7EEC623FCCBA92E189C53B?sequence=1 (accessed 28 June 2018).

Zelber-Sagi, S., Ivancovsky-Wajcman, D., Fliss Isakov, N., Webb, M., Orenstein, D., *et al.* (2018) High red and processed meat consumption is associated with non-alcoholic fatty liver disease and insulin resistance. *Journal of Hepatology* 68(6), 1239–1246. DOI: 10.1016/j.jhep.2018.01.015

17 Costs and Benefits of Sustainable Diets: Impacts for the Environment, Society and Public Health Nutrition

Adam Drewnowski

Abstract

The four domains of food sustainability are nutrition, economics, society and the environment. Sustainable diets need to be nutritionally adequate, safe, affordable, acceptable and appealing, while sparing of both human and natural resources. Those multiple demands are contradictory and can be hard to satisfy at the same time. First, the most nutrient-rich diets are not necessarily the most affordable or environmentally friendly. It is empty calories of minimal nutritional value that are cheap. Second, the most nutrient-rich diets require more land, water and energy use; empty calories are more sparing of the environment. Third, some foods that are nutrient rich, affordable and environmentally friendly may not be socially or culturally acceptable. As a result, assessing the likely impact of sustainable diets on economic equity, food security and population health is a continuing challenge. Cost–benefit analyses rely on multiple inputs. Diet quality is measured through a variety of indices, both food- and nutrient-based. Affordability is measured in terms of calories and nutrients per penny. Cultural acceptance can be based on purchases and consumption frequencies across population groups. Environmental impact is measured in terms of land, water, and energy use, notably greenhouse gas emissions. However, relevant input data are scarce, especially at the local and regional level. Mainstream public health nutrition needs to pay more attention to food production and cost, sensory and cultural acceptance of foods, and the environmental impact of the recommended diets. The way forward is through multi-sector engagement and through sustainable food-based dietary guidelines.

Introduction

Sustainable diets, as defined by the Food and Agriculture Organization (2010), are those that are nutritionally adequate, economically affordable, culturally acceptable and environmentally friendly. That definition embodies the food systems approach, bringing nutrition and health closer to the allied social, economic, behavioural and environmental sciences (The Giessen Declaration, 2005). Consistent with food systems thinking, nutrition scientists are beginning to pay more attention to ways in which food is produced, processed, distributed, purchased and wasted (Auestad and Fulgoni, 2015). Population diets represent the direct link between personal and planetary health (Tilman and Clarke, 2014). Increasing demand for more sustainable diets can help shape the modern food supply and reduce the risk of all forms of malnutrition worldwide.

There is a fundamental dilemma. What is healthiest for people may not be optimal for the planet and vice versa. There is a pressing need for cost–benefit analyses to accompany current and future dietary guidelines (USDA, 2015). Some trade-offs between personal, public and planetary health will need to be made. Insisting that the environmental footprint should be the

 Sustainable Diets: Linking Nutrition and Food Systems (eds B. Burlingame and S. Dernini)

sole criterion for a healthy diet can be counterproductive. Of all plant crops, it is sugar that appears to have the lowest carbon footprint (MacDiarmid *et al.*, 2012; MacDiarmid, 2013; Masset *et al.*, 2014). More nutrient-rich foods have a greater impact on the environment.

There are multiple challenges to overcome. The infrastructure and operations of food systems range from the agricultural production to food security and health (FAO, 2017; International Food Policy Research Institute, 2017). The food supply chain, dominated by the global food industry, covers food production, processing, distribution and retail. Local and regional government food policies affect farming activities, commodity programmes, food processing and safety, and labour and environmental laws. Helping to shape consumer food demand are government regulations of marketing practices, price supports, tariffs, taxes, food assistance and nutrition guidance. Arguably, philanthropic foundations, donors, non-governmental organizations, civil society or academia are not an integral part of the global food system. However vocal they may be, none of them are directly involved in producing, processing, distributing or selling food to the public. Their indirect role has been to influence food system dynamics to shape food and nutrition policies, hoping to create more consumer demand for healthier foods.

The Sustainable Diet Challenge

Designing diets that are nutrient rich, low cost, culturally acceptable and environmentally friendly is a continuing challenge (Vieux *et al.*, 2012; Perignon *et al.*, 2016, 2017). The Food and Agriculture Organization definition of sustainable diets contains some inherent contradictions. First, the more nutrient-rich foods cost more per calorie (Darmon and Drewnowski, 2015), and have a proportionately higher impact on the environment (McDiarmid *et al.*, 2012; McDiarmid, 2013; Masset *et al.*, 2014). Compared to staple grain crops, animal-source foods, including meat and dairy, are more nutrient rich but also cost more (Darmon and Drewnowski, 2015) Based on analyses of retail food prices in multiple countries (Drewnowski, 2010a; Jones *et al.*, 2014; Mendoza *et al.*, 2017), empty calories were less expensive than were the recommended healthier foods. Reducing the price of fresh, healthy foods, the suggested panacea, is easier said than done.

Second, plant-based diets are not necessarily more nutritious or more environmentally friendly than are mixed options (McDiarmid *et al.*, 2012; McDiarmid, 2013; Masset *et al.*, 2014a, 2014b). What tends to be forgotten is that corn, soy, rice, wheat and sugar cane are all plants, as are leafy green vegetables, beans and legumes, nuts and seeds. The lowest greenhouse gas emissions per calorie were uniformly observed for grains, vegetable oils and sweets. Based on some reports, replacing meat with equicaloric amounts of vegetables was actually associated with higher carbon cost. However, accurate modelling of the environmental and health impacts of replacing current diets with vegan, vegetarian or Mediterranean diets has been hampered by the lack of high-quality economic or environmental data, especially at local or regional levels (Perignon *et al.*, 2016, 2017).

Third, the social drivers of food choice should not be underestimated. Consumers tend to select foods based on taste, cost and convenience. Nutritional value, variety and environmental impact tend to be valued less. Low-cost foods that meet nutrient-density or environmental standards can still be rejected by consumers if they deviate from population norms and are socially unacceptable (Maillot *et al.*, 2010). If healthy foods are rejected, then their potential impact on public health nutrition and the environment is largely beside the point. Trade-offs between the foods' nutritional value, cost, and sensory appeal will need to be made.

Measures of Diet Quality

Metrics of diet quality can be both food- and nutrient-based. The best-known healthy eating index (HEI) was developed by the USDA to track compliance with the Dietary Guidelines for Americans. The HEI scores contain both adequacy components (food groups to increase) and moderation components (food groups or nutrients to reduce). The latest iteration, HEI 2015 (National Cancer Institute, 2017), is a 13-component, 100-point score. The key food groups now include total and whole fruits, total and dark green vegetables and beans, low-fat dairy, whole grains, and seafood and plant protein foods. The HEI also specifies a healthy ratio of monosaturated and polysaturated to saturated fatty acids, and places limits on sodium and empty calories. The empty calories were redefined as added sugars and saturated fats, each accounting for 10 points of the score. The HEI score uses a nutrient-density approach to set standards: all dietary components were calculated per 1000 kcal.

Not surprisingly, more nutrient-rich diets and higher HEI scores have been associated with higher diet costs. In the United States, higher quality but more costly diets were selected by groups of higher education and income. The geographic distribution of how HEI 2010 scores varied across Seattle and King County neighbourhoods was recently modelled using residential property values, a novel index of accumulated wealth (Drewnowski *et al.*, 2016). The geographic distributions of obesity and type II diabetes across Seattle neighbourhoods showed the same economic gradient.

Dietary nutrient density is another measure of diet quality. Nutrient profiling (NP) models were initially designed to distinguish between individual foods that were energy dense and those that were nutrient rich (Drewnowski, 2005; Drewnowski and Fulgoni, 2008). The energy density and the nutrient density of foods are inversely linked. Whereas energy density is expressed in kcal/100g, nutrient density is typically expressed as nutrients per 100 kcal.

The energy density of foods depends almost entirely on their water content. The most energy-dense foods are foods that are dry, grains, candy and chocolate, and fats and oils. By contrast, fluid milk, juices, soft drinks, and fresh vegetables and fruit are mostly water (Drewnowski, 2010a). The dryness of foods has had recent implications for public health nutrition. Mexico has imposed taxes on non-essential snacks with energy density >275 kcal/100 g, whereas Chile has imposed warning labels on foods with energy density in excess of 350 kcal/g.

NP methods were recently reviewed by the World Health Organization Regional Office for Europe (WHO, 2015) and the Pan American Health Organization (PAHO, 2016). NP models rank or assign foods into categories based on their nutrient content relative to calories, derived from nutrient composition data (Drewnowski, 2005; Drewnowski and Fulgoni, 2008). Scores can be based on disqualifying nutrients (saturated fat, added sugar, sodium), on qualifying nutrients (protein, fibre, vitamins and minerals) or on some combination of both.

One example, the Nutrient Rich Foods (NRF9.3) index (Drewnowski, 2005; Drewnowski and Fulgoni, 2008), was based on nine qualifying nutrients: protein, fibre, vitamins A, C, and E, calcium, iron, potassium and magnesium. The three disqualifying nutrients were saturated fat, added sugar, and sodium. The NRF algorithm was the sum of percentage daily values (DVs) for the nine qualifying nutrients, minus the sum of %DVs for three disqualifying nutrients, each calculated per 100 kcal and capped at 100% DV (Drewnowski, 2010b). Nutrient profiling can be applied to single foods and beverages, meals, menus and the total diet.

Economic Cost and Cultural Acceptance

Sustainable diets need to be nutrient rich, affordable and socially acceptable. Food or diet costs can be measured in either calories or nutrients per penny (Drewnowski, 2010b). Per calorie, energy-dense grains and fats cost less, whereas lower energy density vegetables and fruit cost more. However, processed energy-dense snacks are not necessarily the lowest-cost options. Lower-cost diets in Mexico derived more calories from tortillas, tamales, beans, sugar and lard and were more likely to be consumed by the rural poor (Mendoza *et al.*, 2017). Higher income groups consumed more fast foods, more processed snacks, and more vegetables and fruit (Mendoza *et al.*, 2017).

Culturally appropriate foods may be the key to the adoption of healthier diets. Here, studies show that it costs less to satisfy nutrition requirements than it does to satisfy social norms. The famous Stigler diet, an early diet optimization problem, was designed to provide sufficient energy and meet the recommended daily allowances for protein, calcium, iron, vitamin A, thiamine, riboflavin, niacin and ascorbic acid at minimum cost. The resulting 'diet' consisted of wheat flour, evaporated milk, cabbage and dried navy beans (and calf liver in later iterations) that only cost US$0.11 a day in 1939. A similar exercise conducted in France in 2010 identified porridge, potatoes, low-fat milk, carrots, chicken livers and oil as key components of a healthful diet, for a total cost of €1.50 a day (Maillot *et al.*, 2010). However, as the modelled optimized diets began to resemble the current eating habits in France, they also became more expensive. The French conclusion was that nutrition criteria alone were meaningless; in order to succeed, healthy affordable diets also had to respect social norms (Maillot *et al.*, 2010).

However, respecting social desires and norms did not appear to be a universal concern. A parallel study from New Zealand (Wilson *et al.*, 2013) showed that nine foods: whole-meal flour, pasta, dried peas, eggs, sugar, milk powder, carrots, vegetable oil and kiwifruit, all costing just NZ$3.19 per day, satisfied all nutrient requirements for men. Compared to existing patterns, the optimized diets were higher in dietary fibre, potassium, iron, zinc, thiamine and vitamin E but were lower in total sugars, saturated fat and sodium. Increasing dietary variety and using more mainstream foods did increase daily cost up to NZ$6.75 per day, as occurred in the French study. However, the conclusion was diametrically different from the one drawn in France; here, the suggestion was to promote the attractiveness of 'poor' foods: flour, dried peas, milk powder, carrots and vegetable oil, using discounts or vouchers, or by engaging celebrity chefs.

The search for alternative proteins places the social aspects of nutrition into sharp focus, by illustrating what trade-offs among nutrition, economics, society and the environment will need to be made. Food preferences and eating habits are intricately linked with social and cultural identity (Ruby *et al.*, 2016). Meat, poultry and fish, as well as dairy, are acceptable sources of high-quality protein and other nutrients; however, their production is associated with an environmental cost. Plant proteins from pulses and soy are already well-established in human diets; by contrast, proteins from insects or from brown and green algae may have a more limited appeal. Selection of dietary protein from beef, pork or dairy may be further influenced by geography, religion or culture. While the search for sustainable diets continues, social and cultural constraints on consumption need to be addressed as well.

Social and Labour Issues

The continued emphasis on fresh, local and minimally processed foods should be viewed through a prism of middle-class and Eurocentric values. First, the harvest of fresh vegetables and fruit in the US (and not only in the US) depends almost entirely on an undocumented migrant labour force that is never mentioned in the dietary guidelines. Second, freshly prepared nutritious foods, a common enough demand, require a skilled and willing labour force, minimum wage, benefits and health insurance, not to mention sensible immigration policies.

The search for sustainable diets would benefit from a better understanding of the drivers of food choice, solid nutrient composition databases and extensive knowledge of local dietary intakes. The role of the global food industry needs to be addressed as well. For example, the advent of packaged, prepared and fast foods has everything to do with the entry of women into the global labour force. Negative attitudes toward 'ultra-processed' foods are very likely linked to the continuing availability of cheap domestic help.

Making the transition to a more sustainable global food supply should be a multi-sector effort. The global food industry has multiple roles and multiple responsibilities in making sure that the global food supply is nutritionally adequate, safe, affordable and appealing. In particular, the role of processed and fortified foods in promoting sustainable food and nutrition security needs to be addressed more fully. In particular, the current preference for fresh and home-cooked foods, as opposed to processed, seems inconsistent with notions of gender equality featured so prominently in the United Nations Sustainable Development Goals. The rules of engagement between international organizations, non-governmental organizations and the food industry would benefit from further development.

Environmental Impact Metrics

The production, processing, transportation, retail and storage of foods are each associated with greenhouse gas emissions (GHGEs) (Drewnowski *et al.*, 2015). The carbon footprint of foods (mostly methane gas) is often expressed in grams of CO_2 equivalents per unit weight. Other measures of environmental impact include the use of land and water resources, as well as industrial pollution. Many of these issues are complicated by global warming and climate change.

The present convention is to express the environmental cost per unit weight of food. However, foods can vary greatly in their energy and nutrient content, so that the base of calculation can make a big difference. Vegetables may have a low carbon footprint per unit weight, but many vegetables are mostly water, which provides no calories and no nutrients. GHGEs associated with the production of vegetables may be low when expressed per 100 g but become disproportionately high when expressed per 100 kcal (Drewnowski *et al.*, 2015). Given differences in energy density across food groups, the environmental cost of foods should be expressed per calorie, per nutrient, or per unit of high-quality protein, as opposed to per unit weight (Drewnowski *et al.*, 2015).

Conclusion

The key to sustainable food supply is multi-sector engagement. Global micronutrient deficiencies of greatest public health concern include vitamin A, iodine, vitamin D, vitamin B12, calcium, iron and zinc (FAO, 2017; International Food Policy Research Institute, 2017). Those deficiencies, endemic in countries that are still consuming traditional plant and grain-based diets, are effectively remedied by the consumption of meat and dairy. They can also be addressed by the introduction of processed fortified foods (FAO, 2017; International Food Policy Research Institute, 2017) and by more active engagement on the part of the food industry.

Eliminating micronutrient deficiencies will require joint efforts by both public and private sectors. Biofortification is one option; fortifying low-cost commodities with vitamins and minerals to assure high-nutrient density at a low cost is another. Assuring genetic biodiversity by promoting traditional and local plants takes advantage of local smallholder farming, still producing the bulk of the global food supply. A continuing focus on freshwater management is also critical for the future of agricultural production.

These and other options would benefit from sustained informed debate on sustainable diets in the context of economics, society and the environment. We need more comprehensive discussion on engaging with the global food system actors and the many segments of the food industry.

Declaration of Interest

AD has received grants, honoraria and consulting fees from numerous food, beverage and ingredient companies and from other commercial and non-profit entities with an interest in diet quality and nutrient density of foods. The University of Washington receives research funding from public and private sectors.

References

Auestad, N. and Fulgoni, V.L. (2015) What current literature tells us about sustainable diets: Emerging research linking dietary patterns, environmental sustainability, and economics. *Advances in Nutrition* 6(1), 19–36.

Darmon, N. and Drewnowski, A. (2015) Contribution of food prices and diet cost to socioeconomic disparities in diet quality and health: a systematic review and analysis. *Nutrition Reviews* 73(10), 643–660.

Drewnowski, A. (2005) Concept of a nutritious food: toward a nutrient density score. *American Journal of Clinical Nutrition* 82(4), 721–732.

Drewnowski, A. (2010a) The cost of US foods as related to their nutritive value. *American Journal of Clinical Nutrition* 92(5), 1181–1188.

Drewnowski, A. (2010b) The Nutrient Rich Foods Index helps to identify healthy, affordable foods. *American Journal of Clinical Nutrition* 91(4), 1095S–1101S.

Drewnowski, A. and Fulgoni, V. (2008) Nutrient profiling of foods: creating a nutrient-rich food index. *Nutrition Reviews* 66(1), 23–39.

Drewnowski, A., Rehm, C.D., Martin, A., Verger, E.O., Voinnesson, M. and Imbert, P. (2015) Energy and nutrient density of foods in relation to their carbon footprint. *American Journal of Clinical Nutrition* 101(1), 184–191.

Drewnowski, A., Aggarwal, A., Cook, A., Stewart, O. and Moudon, A. (2016) Geographic disparities in Healthy Eating Index scores (HEI-2005 and 2010) by residential property values: Findings from Seattle Obesity Study (SOS). *Preventative Medicine* 83, 46–55.

FAO (2010) Food and Agriculture Organization of the United Nations. International Scientific Symposium Biodiversity and Sustainable Diets United Against Hunger, Rome. Available at http://www.fao.org/ag/humannutrition/28506-0efe4aed57af34e2dbb8dc578d465df8b.pdf.2010 (accessed 1 July 2018).

FAO (2017) Nutrition and food systems: A report by the High Level Panel of Experts on Food Security and Nutrition (2017). Available at http://www.fao.org/fileadmin/user_upload/hlpe/hlpe_documents/HLPE_Reports/HLPE-Report-12_EN.pdf (accessed 1 July 2018).

International Food Policy Research Institute (2017) Global nutrition report. Available at https://www.globalnutritionreport.org/the-report/ (accessed 1 July 2018).

Jones, N.R.V., Conklin, A.I., Suhrcke, M. and Monsivais, P. (2014) The growing price gap between more and less healthy foods: Analysis of a novel longitudinal UK dataset. *PLoS One* 9(10), e109343.

Macdiarmid, J.I. (2013) Is a healthy diet an environmentally sustainable diet? *Proceedings of the Nutrition Society* 72(1), 13–20.

Macdiarmid, J.I., Kyle, J., Horgan, G.W., Loe, J., Fyfe, C., *et al.* (2012) Sustainable diets for the future: Can we contribute to reducing greenhouse gas emissions by eating a healthy diet? *American Journal of Clinical Nutrition* 96(3), 632–639.

Maillot, M., Darmon, N. and Drewnowski, A. (2010) Are the lowest-cost healthful food plans culturally and socially acceptable? *Public Health Nutrition* 13(8), 1178–1185.

Masset, G., Soler, L.G., Vieux, F. and Darmon, N. (2014a) Identifying sustainable foods: the relationship between environmental impact, nutritional quality, and prices of foods representative of the French diet. *Journal of the Academy of Nutrition and Dietetics* 114(6), 862–869.

Masset, G., Vieux, F., Verger, E.O., Soler, L.G., Touazi, D. and Darmon, N. (2014b) Reducing energy intake and energy density for a sustainable diet: a study based on self-selected diets in French adults. *American Journal of Clinical Nutrition* 99(6), 1460–1469.

Mendoza, A., Perez, A.E., Aggarwal, A. and Drewnowski, A. (2017) Energy density of foods and diets in Mexico and their monetary cost by socioeconomic strata: analyses of ENSANUT data 2012. *Journal of Epidemiology and Community Health* 71(7), 713–721.

National Cancer Institute (2017) Comparing the HEI-2015, HEI-2010 and HEI-2005. Available at https://epi.grants.cancer.gov/hei/comparing.html (accessed 1 July 2018).

PAHO (2016) Pan American Health Organization Nutrient Profile model. Available at http://iris.paho.org/xmlui/handle/123456789/18621 (accessed 1 July 2018).

Perignon, M., Masset, G., Ferrari, G., Barré, T., Vieux, F., *et al.* (2016) How low can dietary greenhouse gas emissions be reduced without impairing nutritional adequacy, affordability and acceptability of the diet? A modelling study to guide sustainable food choices. *Public Health Nutrition* 19(14), 2662–2674.

Perignon, M., Vieux, F., Soler, L. G., Masset, G. and Darmon, N. (2017) Improving diet sustainability through evolution of food choices: review of epidemiological studies on the environmental impact of diets. *Nutrition Reviews* 75(1), 2–17.

Ruby, M.B., Alvarenga, M.S., Rozin, P., Kirby, T.A., Richer, E. and Rutsztein, G. (2016) Attitudes toward beef and vegetarians in Argentina, Brazil, France, and the USA. *Appetite* 96, 546–554.

The Giessen Declaration (2005) *Public Health Nutrition* 8(6A), 783–786.

Tilman, D. and Clark, M. (2014) Global diets link environmental sustainability and human health. *Nature* 515(7528), 518–522.

USDA (2015) Scientific Report of the 2015 Dietary Guidelines Advisory Committee. Available at https://health.gov/dietaryguidelines/2015-scientific-report/PDFs/Scientific-Report-of-the-2015-Dietary-Guidelines-Advisory-Committee.pdf (accessed 1 July 2018).

Vieux, F., Darmon, N., Touazi, D. and Soler, L.G. (2012) Greenhouse gas emissions of self-selected individual diets in France: Changing the diet structure or consuming less? *Ecological Economics* 75, 91–101.

WHO (2015) World Health Organization Regional Office for Europe Nutrient Profile Model. Available at http://www.euro.who.int/en/health-topics/disease-prevention/nutrition/publications/2015/who-regional-office-for-europe-nutrient-profile-model-2015 (accessed 1 July 2018).

Wilson, N., Nghiem, N., Ni Mhurchu, C., Eyeles, H., Baker, M.G. and Blakely, T. (2013) Foods and dietary patterns that are healthy, low-cost, and environmentally sustainable: A case study of optimization modeling for New Zealand. *PLoS ONE* 8(3), e59648. DOI: 10.1371/journal.pone.0059648

18 The One Planet Sustainable Food Systems (SFS) Programme as a Multi-stakeholder Platform for a Systemic Approach

Michael Mulet Solon, Patrick Mink, Sandro Dernini, Marina Bortoletti and James Lomax

Abstract

Food production and consumption has a failing performance in terms of food security, nutrition and health, but also equality, environmental protection and climate change mitigation, posing a serious sustainability challenge as the planet's population grows while consuming beyond planetary boundaries, compromising future generations' well-being. Responses to the wicked problem posed by securing humanity's food are more likely to succeed if built on two pillars, as championed by the One Planet Sustainable Food Systems (SFS) Programme. The first pillar refers to the need to adopt a food-systems approach, which enables identifying and addressing more holistic solutions. The second pillar proposes that multi-stakeholder, inclusive approaches are more likely to succeed, especially if they fulfil conditions for collective action, if they overcome polarization by embracing the inherent conflict in a locked-in system, and if they adopt a mindset that focuses on innovation. The SFS Programme has been built on both pillars, adopting five focus themes and organizing its work across four areas. Through its governance structure, the programme has launched a series of core initiatives that are participated in by coalitions of organizations from diverse sectors, and they were developed building on pre-existing projects, expertise and resources in order to leverage synergies and avoid effort duplication. They address key problems related to SFS and link several elements from production to consumption. The core initiatives are inclusive, enabling faster learning through constant communication and overall coordination, becoming mutually reinforcing activities to accelerate the shift to SFS, in support of the implementation of the Agenda 2030.

Introduction

With over 800 million people suffering hunger, 1.9 billion overweight, of which 600 million are obese, 30% of food going to waste, and 71% of deforestation caused by commercial agriculture (FAO, 2016a), current global food production and consumption has a non-satisfactory performance in terms of food security, nutrition and health, but also equality, environmental protection and climate change mitigation. The approach that has dominated in the past focused primarily on solving food insecurity through interventions based on agricultural productivity (e.g. the Green Revolution). While this strategy made some progress, the goal was far from achieved and, furthermore, it entailed serious negative outcomes on the natural resource base and on social welfare, becoming responsible for undermining food security itself.

Responses to this 'wicked' challenge are more likely to succeed if they are built on two pillars. The first pillar refers to the critical need to adopt a food-systems perspective. The second pillar proposes that action by multi-stakeholder networks, in comparison to practitioners in individual hierarchically organized institutions, also offer advantages for resolving a wicked problem such as our unsustainable food systems.

After outlining the rationale behind these two pillars, this chapter introduces the Sustainable Food Systems (SFS) Programme of the United Nation's 10-Year Framework of Programmes on Sustainable Consumption and Production Patterns, established in 2015 with the goal of accelerating the shift towards more sustainable food systems. The One Planet SFS Programme vision is that 'All food systems are sustainable, delivering food security and nutrition for present and future generations' (SFSP, 2016). It depicts how the SFS Programme's wireframe is rooted in the two mentioned pillars.

The chapter continues with an exposition of the programme's key features, governance set up, work areas and the organization of its activities, and then presents two examples of its core initiatives: the first intervening at the normative level, developing voluntary guidelines for assessing the sustainability of diets at country level to enable policy makers; the second addressing governance frameworks, developing an analytical framework that aims to inform and institutionalize a food-systems approach in food policy planning at any scale.

Challenges

Today, the main challenge for the food and agriculture sector is to simultaneously provide enough food – both in quantity and quality – to meet everyone's nutritional needs, while conserving the natural resources to produce food for present and future generations.

Despite the fact that the world is producing enough food to feed its entire population, in 2016 almost 815 million people are undernourished, up from 777 million in 2015; 155 million under-5 year olds are estimated to be stunted (FAO *et al.*, 2017); and about two billion are malnourished, lacking the essential micronutrients they need to lead healthy lives (IFPRI, 2016). At the same time, the number of overweight people has reached more than 1.9 billion adults globally – representing about 30% of the total adult population, of which 600 million are obese (FAO *et al.*, 2017). Around 30% of the food produced worldwide – about 1.3 billion tons – is lost or wasted every year (FAO, 2011). In addition, food market dynamics can generate issues such as price volatility, capable of causing major impacts on well-being like the 110 million people driven into poverty in the 2008 food price crisis (FAO, 2009).

Current pressures on the planet's natural resources will further increase with population and economic growth, unless consumption and production patterns become more efficient and less polluting and are brought to operate within planetary boundaries. In many parts of the world, water resources are under increasing stress, and irrigated agriculture is by far the largest water user globally, accounting for 70% of water withdrawals globally (HLPE, 2015). Global food systems are extremely vulnerable to climate change and biodiversity loss, but still they are responsible for between a quarter and a third of global greenhouse gas emissions (Vermeulen *et al.*, 2011), 73% of deforestation in tropical and subtropical regions (FAO, 2016a), and about 90% of wild capture fisheries are overfished or depleted (FAO, 2016b).

Why we Need a Food Systems-based Approach

The complexity of these challenges demands an approach that looks at the food system as a whole, as championed by the One Planet SFS Programme. In this section we lay out the rationale for shifting to such a food-systems framework on the premise that there has been a propensity by the scientific and policy making communities to address food-related challenges in isolation as well as a tendency to overlook the power relations that play a major role in casting food systems (IPES Food, 2015).

The concepts that in past decades have dominated the debate revolved, mainly, around solving the food insecurity challenge with production-level approaches. The Green Revolution is the iconic intervention of this supply-side focus. However, even though laudable progress was achieved on food production, food insecurity persisted with hundreds of millions afflicted by hunger, malnutrition and overweight/obesity. These shortcomings caused a shift in the terms of the debate, which evolved to incorporate the social dimension notions of availability, access and utilization of food. Debate was further shaped by an increasing awareness of the impact on the environment (UNEP, 2016). With the incorporation of these dimensions, a growing consensus affirmed that not only did the supply-side approach not accomplish its goal of feeding the world, but that it simultaneously brought along serious negative impacts to the social and environmental dimensions, which became responsible, in turn, for *undermining food security itself* (Ericksen, 2008).

This two-way feedback loop between: (i) the natural and social resource bases; (ii) the food-systems activities that rely on the former; and (iii) the outcomes (impacts) of these activities on the natural and social dimensions is the notion that sits at the core of the food-systems concept. In this sense, a food system 'gathers all the elements (environment, people, inputs, processes, infrastructures, institutions, etc.) and activities that relate to the production, processing, distribution, preparation and consumption of food and the outputs of these activities, including socioeconomic and environmental outcomes' (HLPE, 2014). A SFS is a food system that ensures food security and nutrition for all in such a way that the economic, social and environmental bases to generate food security and nutrition of future generations are not compromised (HLPE, 2014). In other words, the complex set of food chain activities has a complex set of implications for social and environmental welfare, while the latter also influence food chain activities, with interactions that can be multi-scale (e.g. in time) and/or multi-level (e.g. in terms of jurisdictional level) (Ericksen, 2008). Figure 18.1 attempts to illustrate this complex set of interdependencies that characterize food systems (UNEP, 2016; CNS-FAO, 2016).

Rooted in systems theory, which appeared in the 1950s with the aim of reducing the growing isolation caused by specialization of disciplines (Jordan, 1998), a new holistic analytical framework was crafted, which has several advantages over the more compartmentalized frameworks that have persisted in the past: (i) it provides a lens with which to look at food-systems issues as 'component parts' of broader systemic problems (e.g. addressing food waste regulation to influence food security and nutrition);

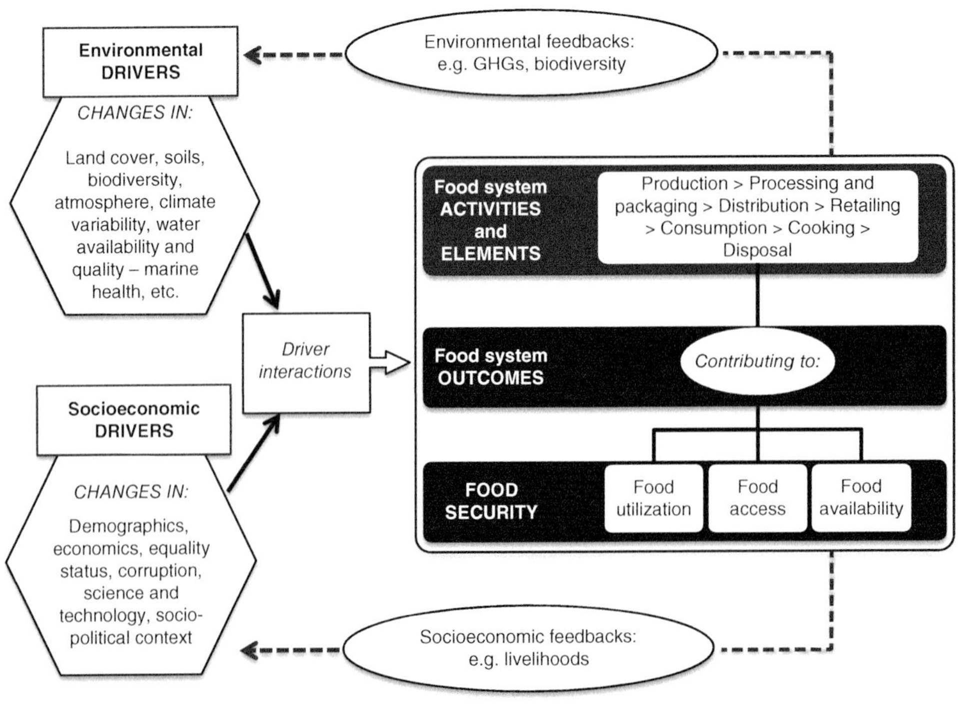

Fig. 18.1. The interactions between environmental and socioeconomic drivers, food-system activities and outcomes.

(ii) it offers a lens to deal with food system's complexity; (iii) it provides a lens that incorporates a range of perspectives, disciplines or functions of logic; (iv) the multi-scale and multi-level treatment that a systemic approach to food-systems analysis requires facilitates a more adequate evaluation of trade-offs; (v) it provides a lens with which to identify synergies and leverage points for implementing context-specific solutions (UNEP, 2016; Ericksen, 2008). In other words, a systems perspective allows reconsidering the causes of a system's flaws, such as the lock-ins that afflict food systems (IPES FOOD, 2016).

The mentioned advantages are essential to counter persisting views that use a narrowing lens in the framing of the food security challenge. Issues like climate change or inequality are still being treated as phenomena that exist alongside the priority goal of food security, as if they were not all inherently connected (IPES FOOD, 2015). This narrow view eclipses the fact that 'the only food system to be secure is that which is sustainable, and the route to food security is by addressing sustainability' (Lang and Barling, 2012). In this regard, the international community has recently taken an important step to widen its lens and frame the interconnectedness of global challenges. Agenda 2030 on Sustainable Development and its list of 17 interrelated Sustainable Development Goals (SDGs) advocate the need for an integrated approach, encouraging inter-ministerial and multi-stakeholder collaboration to address the planet's sustainability challenges. The food-systems approach is a perfect example of how the SDGs can be tackled in a holistic way, since food systems cut across the majority of SDGs (in particular SDGs 2 and 12) but also goals related to poverty eradication, health, economic development, biodiversity and ecosystems, climate change, and so on.

The Need for a Multi-stakeholder Approach for Collective Impact to Transform Food Systems

Multi-stakeholder networks like the SFS Programme offer advantages over traditional hierarchic organizational approaches for addressing wicked problems by providing actors with a 'co-owned' space that can enable a shift in power relationships in which every organization is not only accountable for its own performance but for the whole system's 'health'. The new environment acts as a space of experimentation for innovation where new rules are developed based on the specific needs for tackling the wicked problem. In other words, the locked-in system of entrenched power starts a process of disentanglement through the multi-stakeholder network (Waddell *et al.*, 2013).

Similarly, collective impact theory, popularized by Kramer and Kania in 2011, is also a proposition for tackling wicked problems. Their theory criticizes the generalized problem-solving approach based on individual organizations crafting predetermined solutions. They argue that collaborative approaches are better equipped to achieve collective impact through emergent solutions if five conditions are met: (i) a common agenda; (ii) shared measurement; (iii) mutually reinforcing activities; (iv) continuous communication; and (v) backbone support organization. Fulfilling these five conditions, a network of stakeholders can lead to progressive alignment of perspectives, continuous learning feedback loops, and a mechanism for early adoption of changes (Kania and Kramer, 2013). As will be described in the next sub-section, the SFS Programme fulfils these five conditions.

However, as sustained by critics of collective impact theory, this is not a silver bullet. Nevertheless, criticisms such as the statement that collective impact-type initiatives tend to 'remain silent in regard to policy advocacy' are possible to transcend (Hoey *et al.*, 2017). In the SFS Programme experience this was proven by the issuing of the Pretoria Resolution, an outcome of the 1st Global Conference of the SFS Programme that contained policy recommendations and calls to action (SFSP, 2017). Another criticism to multi-stakeholder approaches, which the SFS Programme strives to overcome, states that the inherent diversity of positions held by the different actors renders a network incapable of dealing explicitly with power dynamics and status quo, which prevents effective collaboration. Pereira and Drimmie (2016) suggest that a key response to this obstacle should be to embrace the conflict, and trust experimentation and emergence. Bringing people to the table in acknowledgement of existing untenable positions and power differentials might be the best way to have a transparent, decision-making process. Overcoming polarization and promoting inclusiveness are paramount conditions to enhance multi-stakeholder collaboration for the transformation of food systems, as discussed by the Swiss National FAO Committee (CNS-FAO, 2016).

The One Planet Sustainable Food Systems Programme's Multi-Stakeholder Approach to Promote Food Systems Thinking

The SFS Programme was purposefully established as a collaborative multi-stakeholder partnership that promotes a systemic approach to accelerate the shift towards more sustainable food systems. Both aspects can increase the likeliness of success in tackling the wicked problem of unsustainable food systems.

Stakeholders from different sectors and at multiple levels – national, regional and global – from around the globe, across society and throughout food systems are actively involved in a governance structure, which comprises three tiers:

- the co-leads, who provide secretarial functions as well as overall guidance and coordination;
- a multi-stakeholder advisory committee (MAC), providing guidance and advice to the co-leads;
- partners, which are part of the active 'community of practice', bringing in their own activities or connecting with other members to build new synergies and collaborations.

The 4 co-leads and the 23-member MAC actively participate in the elaboration, implementation and monitoring of the SFS Programme's work plan, recommend and promote actions, proactively engage stakeholders, suggest and evaluate new or existing activities, seek and foster synergies, deliver (technical) advice, and fundraise for the programme, acting as a sort of 'backbone support structure', taking the terminology from the collective impact theory. Concretely, the MAC convenes at least three times per year – with a minimum one annual face-to-face meeting. The co-leads prepare for these meetings and carry out follow-up work through monthly coordination calls. In addition, the MAC has recently set up four task forces to promote the implementation of the SFS Programme work plan with regard to its four work areas.

These four broad work areas through which the One Planet SFS Programme aims to achieve its goal are:

- Work area 1: Raising awareness on the need to adapt sustainable consumption and production patterns in food systems.
- Work area 2: Building enabling environments for SFS.
- Work area 3: Increasing the access to and fostering the application of actionable knowledge, information and tools to mainstream sustainable consumption and production (SCP) in food systems.
- Work area 4: Strengthening collaboration among food system stakeholders to increase the sector's SCP performance.

Cutting across these work areas as illustrated in Fig. 18.2, the SFS Programme's MAC identified five 'focus themes' that allow member organizations to coalesce around for activity coordination. Taken together, these focus themes cover all aspects of the food-system concept. The focus themes of the SFS Programme are as follows:

- sustainable diets;
- sustainability along all food value chains;
- reduction of food losses and waste;
- local, national, regional multi-stakeholder platforms; and
- resilient, inclusive, diverse food production systems.

The programme set out to create so-called 'core initiatives' as the principal projects of the network – interventions that are developed and implemented jointly by a group of two or more programme members and which report to the programme's MAC and co-leads. Core initiatives are diverse in nature, and typically work across different levels and scales. They address core problems related to sustainable food systems and link several elements of food systems from production to consumption, with strategies in line with the SFS Programme's work areas. They build on existing projects, resources and expertise of organizations that decided to open up to each other to foster new synergies, and develop and/or

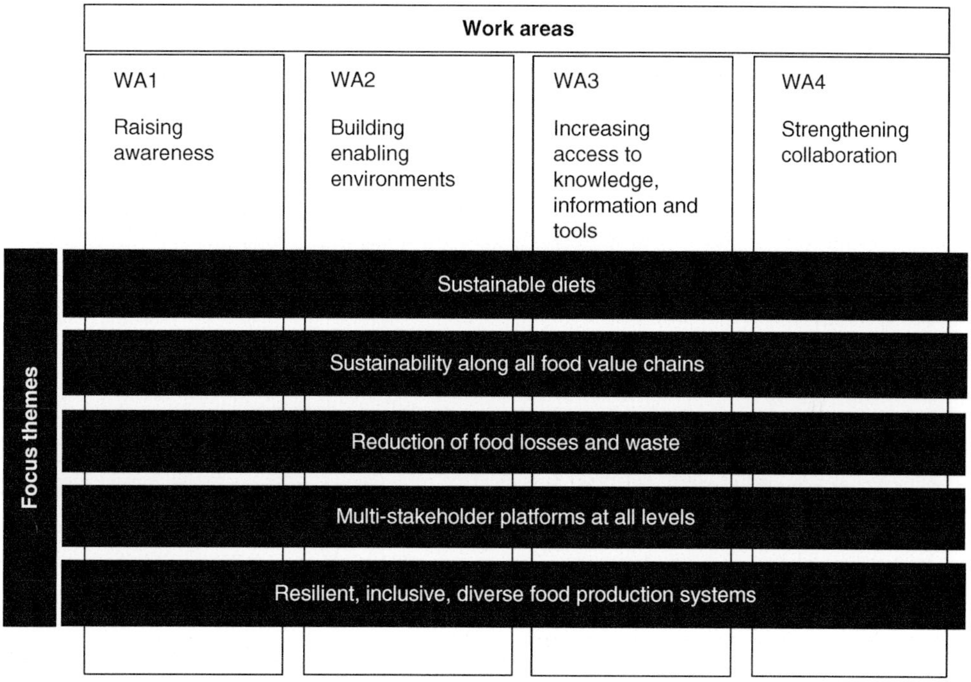

Fig. 18.2. Core initiatives as building blocks for a systems approach.

disseminate innovative solutions. In this sense, core initiatives are projects that have become wider 'coalitions' of partners, enabling a 'widening of their lens' and of participants' 'accountability'. Initiatives have become more inclusive, enabling faster learning through constant communication and overall coordination, becoming 'mutually reinforcing activities'. The 'shared measurement' that occurs through an agreed upon indicator framework is the fifth and last collective impact theory condition met by the SFS Programme.

As of November 2017, eight core initiatives were under implementation, classified under the five focus themes of the programme:

- Sustainable diets:
  1. Sustainable diets in the context of SFS.
  2. Sustainable and healthy gastronomy as a key driver for SFS.
- Sustainability along all food value chains:
  3. Sustainability along all value chains: identifying and promoting local initiatives linking small-scale producers and consumers.
  4. Complementing existing value chain sustainability assessments: Measuring, communicating and valuing biodiversity in food systems.
- Reduction of food losses and waste:
  5. Delivering SDG Target 12.3 on Food Loss and Waste Reduction.
- Local, national, regional multi-stakeholder platforms:
  6. Setting the Table for our Children – exploring the path to more SFS through multi-stakeholder action.
- Resilient, inclusive, diverse food production systems:
  7. Sustainable food systems – what is in it for farmers?
  8. The Organic Food System Programme: Organic food systems as models and living laboratories for transformation processes towards SFS.

Two core initiatives are presented in more detail in Box 18.1.

Conclusion

The world faces an array of complex and interconnected economic, social and environmental challenges linked to food and food security, and

Box 18.1. Core Initiative examples.

1. Sustainable diets in the context of sustainable food systems – a core initiative of the One Planet SFS Programme (Co-led by FAO, UN Environment, with the collaboration of UN Standing Committee on Nutrition, Hebrew University, CIHEAM, ENEA, WWF)

The objective of this core initiative is to strengthen the sustainability of current dietary patterns contributing to a broader sustainable food demand to drive more sustainable food systems (SFS), linking food consumption to food production. It will advance the existing knowledge-sharing tools and mechanisms for improving the sustainability of current dietary patterns and in this way to contribute to the shift towards more SFS. The core initiative will facilitate the sharing of such knowledge by taking into account the multi-dimensional nature of food systems and diets and will promote multi- and trans-disciplinary research and multi-stakeholder engagement. It will serve to make progress to better understand the relations between sustainable diets and SFS, in both developed and developing countries. It will take into account the four dimensions of sustainable diets: health and nutrition, environment including biodiversity, economy and socio-cultural factors. By taking into consideration the multi-dimensional nature of diets and food systems, the project will rely on multi- and trans-disciplinary approach within a multi-stakeholder environment. The core initiative will take into account current knowledge collection and will be developed on a solid scientific and policy knowledge base, with particular regards to the rural-urban interface, smallholders, indigenous people, gender and social inclusion issues. Its purpose is to promote effective and reliable scientific communication, in order to advance the existing knowledge-sharing tools and mechanisms for improving the sustainability of current diets while improving SFS. The core initiative will confront different perspectives and methodologies to assess the sustainability of diets in both developed and developing countries and will provide relevant case studies. Main trends and drivers will be identified and highlighted. It will allow the dissemination and promotion of results through a wide range of seminars, workshops and conferences. It will build capacity and increase awareness on the impact of sustainable diets in the context of the implementation of the UN international agenda of the Decade of Action on Nutrition by developing effective communication and advocacy activities.

2. The One Planet SFS Programme Core Initiative 'Setting the Table for Our Children' – a core initiative of the One Planet SFS Programme (Biovision, Hivos and UN Environment)

The core initiative 'Setting the Table for Our Children' promotes a food-system approach and engages stakeholders to improve policy and governance of food systems. Together they aim to develop more integrated strategies, transformative roadmaps, as well as enabling conditions towards more SFS at local and national levels, also enhancing the coordination between those two levels.

As collaboration across the food system is rare with silo thinking predominant within the public sector and stakeholders, the initiative is currently developing a coherent and universally applicable analytical framework for SFS that will inform and institutionalize a food-systems approach in food policy planning at any scale. The framework will recommend on key policy levers, methodologies, tools and collaborative activities across the food system that should be taken by public and private actors determined by the countries themselves.

The coalition is currently present in Kenya, Uganda, Zambia, Senegal, Indonesia and Bolivia. As an example, in Zambia, national government officers have been receiving training on T21-integrated policy planning and monitoring tool for sustainable food systems. In Chongwe District, Lusaka Province, the initiative has been supporting a multi-stakeholder platform, called the Food Change Lab, which convenes varied actors (citizens, farmers, journalists, entrepreneurs, civil society and civil servants) to collectively re-assess Zambia's food system and develop policy recommendations towards sustainable food systems.

which pose a threat to future generations. The complexity of these challenges demands an approach that looks at the food system as a whole, as championed by the One Planet SFS Programme. The food-system concept allows dealing with complexity, across multiple levels and scales, displaying the interconnectedness between the components of food systems, as required by the Agenda 2030 SDGs.

Multi-stakeholder partnerships, such as the SFS Programme, establish a new environment with new rules of the game that can address the power dynamics in a locked-in system. The SFS Programme fulfils the five conditions defined in collective impact theory to enable the emergence of solutions to a wicked problem. The programme's members have all reached consensus on the definition of the problem and on a common goal, work areas and focus themes (common agenda); common indicators have been put in place (shared measurement); a work plan consisting of a portfolio of jointly implemented 'core initiatives' has been developed (mutually reinforcing activities); a coordination body steers the programme (backbone support organization), requiring regular structured communication and knowledge sharing (continuous communication). But the SFS Programme acknowledges that fulfilling these five conditions may be 'necessary but not sufficient' to achieve impact. Multi-stakeholder approaches have inherent issues linked to power. Nevertheless, embracing conflict and adopting a mindset that focuses on innovation and inclusiveness is seemingly the best formula to enable transparent decision-making to disentangle the locked-in food system.

Therefore, the SFS Programme is a means to 'overcome polarization' and 'foster inclusiveness'. The SFS Programme's governance structure and its core initiatives are key tools to advance the coalescing of multi-stakeholders around a common agenda in order to achieve progress in the mandate of the SFS Programme, accelerating the shift to SFS, in support of the implementation of the Agenda 2030.

References

CNS-FAO (2016) Working towards Sustainable Agriculture and Food Systems, a discussion paper. Available at https://www.blw.admin.ch/blw/en/home/international/institutionen/multistakeholder-partnerschaften/cns-fao.html (accessed 17 November 2017).

Ericksen, P.J. (2008) Conceptualizing food systems for global environmental change research. *Global Environmental Change* 18, 234–245. DOI: 10.1016/j.gloenvcha.2007.09.002

FAO (2009) *The State of Food Insecurity in the World*. Food and Agriculture Organization of the United Nations, Rome, Italy.

FAO (2011) Global food losses and food waste: extent, causes and prevention. Available at http://www.fao.org/docrep/014/mb060e/mb060e00.htm (accessed 20 October 2017).

FAO (2016a) *State of the World's Forests 2016. Forests and Agriculture: Land-Use Challenges and Opportunities*. Food and Agriculture Organization of the United Nations, Rome, Italy.

FAO (2016b) *The State of World Fisheries and Aquaculture 2016. Contributing to Food Security and Nutrition for All*. Food and Agriculture Organization of the United Nations, Rome, Italy.

FAO, IFAD, UNICEF, WFP and WHO (2017) *The State of Food Security and Nutrition in the World 2017. Building Resilience for Peace and Food Security*. Food and Agriculture Organization of the United Nations, Rome, Italy.

HLPE (2014) Food losses and waste in the context of sustainable food systems. Available at http://www.fao.org/3/a-i3901e.pdf (accessed 20 October 2017).

HLPE (2015) Water for food security and nutrition. Available at http://www.fao.org/fileadmin/user_upload/hlpe/hlpe_documents/HLPE_Reports/HLPE-Report-9_EN.pdf (accessed 20 October 2017).

Hoey, L., Colasanti, K., Pirog, R. and Fink Shapiro, L. (2017) Implementing collective impact for food systems change: reflections and adaptations from Michigan. *Journal of Agriculture, Food Systems, and Community Development*. 7(2), 101–115. DOI: 10.5304/jafscd.2017.072.014

IFPRI (2016) *Global Nutrition Report 2016: From Promise to Impact: Ending Malnutrition by 2030*. International Food Policy Research Institute, Washington, DC, USA.

IPES FOOD (2015) The new science of sustainable food systems. Overcoming barriers to food system reform. Available at http://www.ipes-food.org/images/Reports/IPES_report01_1505_web_br_pages.pdf (accessed 17 November 2017).

IPES FOOD (2016) From uniformity to diversity: a paradigm shift from industrial agriculture to diversified agroecological systems. Available at http://www.ipes-food.org/images/Reports/UniformityToDiversity_FullReport.pdf (accessed 17 November 2017).

Jordan, J.S. (1998) *Systems Theories and A Priori Aspects of Perception*. Elsevier Science, Amsterdam, The Netherlands, pp. 47–74.

Kania, J. and Kramer, M. (2013) Embracing emergence: How Collective Impact addresses complexity. *Stanford Social Innovation Review* 1–7.

Lang, T. and Barling, D. (2012) Food security and food sustainability: reformulating the debate. *The Geographical Journal* 178(4), 313–326.

Pereira, L. and Drimie, S. (2016) Governance arrangements for the future food system: addressing complexity in South Africa. *Environment: Science and Policy for Sustainable Development* 58(4), 18–31. DOI: 10.1080/00139157.2016.1186438

SFSP (2016) Annex to the Rio+20 conference outcome document A/CONF.216/5. Available at http://www.oneplanetnetwork.org/sites/default/files/10yfp_sfsp_programme_document.pdf (accessed 26 June 2018).

SFSP (2017) Conference 'Pretoria' Resolution. In: 1st Global Conference of the 10YFP. Sustainable Food Systems (SFS) Programme: Sustainable Food Systems for All – Catalyzing Change through Multi-Stakeholder Action [online]. Available at http://www.scpclearinghouse.org/sites/default/files/1st_global_10yfp_sfs_programme_conference_-_final_resolution.pdf (accessed 8 December 2017).

UNEP (2016) Food systems and natural resources. A report of the Working Group on Food Systems of the International Resource Panel. Available at www.resourcepanel.org/file/395/download?token=JqcqyisH (accessed 20 October 2017).

Vermeulen, D.J., Campbell, B.M. and Ingram, J.S.I. (2012) Climate change and food systems. Available at http://www.annualreviews.org/doi/abs/10.1146/annurev-environ-020411-130608 (accessed 20 October 2017).

Waddell, S., McLachlan, M. and Dentoni, D. (2013) Learning and transformative networks to address wicked problems: a golden invitation. *International Food and Agribusiness Management Review* 16(A), 23–32.

19 The Med Diet 4.0 Framework: a Multidimensional Driver for Revitalizing the Mediterranean Diet as a Sustainable Diet Model

Sandro Dernini, Denis Lairon, Elliot M. Berry, Gianluca Brunori, Roberto Capone, Lorenzo M. Donini, Massimo Iannetta, Dalia Mattioni, Suzanne Piscopo, Lluis Serra-Majem, Andrea Sonnino and Milena Stefanova

Abstract

The Mediterranean diet (MD), despite the fact that it is acknowledged as one of the healthiest diets in the world, is paradoxically becoming less the diet of choice in most Mediterranean countries. This process of erosion of the MD is alarming as it has undesirable impacts not only on health, but also on social, cultural, economic and environmental domains in the Mediterranean area. The Med Diet 4.0 has been developed as a multidimensional framework to revitalize the MD. It characterizes the MD as a sustainable diet model, through four interdependent sustainable benefits, with country-specific variations: (i) well-documented nutrition and health advantages, preventing chronic and degenerative diseases and reducing public health costs; (ii) low environmental impacts and richness in biodiversity, reducing pressure on natural resources and climate change; (iii) positive local economic returns, reducing rural poverty; and (iv) high social and cultural food values, increasing appreciation, mutual respect and social inclusion. All these elements interact and feed into each other synergistically, contributing to holistic well-being of individuals and communities. The Med Diet 4.0 has the broader scope to catalyze a renewed multi-stakeholder interest in the MD as a sustainable driver connecting food consumption to production towards more Mediterranean sustainable food systems. It will allow a new awareness among Mediterranean people of the multiple sustainable values and benefits of the MD, thereby facilitating its revitalization. The Med Diet 4.0 reshapes a contemporary knowledge of the MD and its appreciation in terms of a more holistic vision of sustainability linked to nutritional well-being and food security. The complexity of interdependent challenges, within the radical transformation of the contemporary Mediterranean and global scenario, requires new forms of transdisciplinary and intercultural dialogues, strategies and research, at different levels, for the revitalization of the MD. Within such complexity, the Med Diet 4.0 provides a synthesis to better understand and enhance the MD as a sustainable diet model in the context of the improvement of the sustainability of Mediterranean food systems, reconnecting diets, food consumption, food production, food security and sustainability in the Mediterranean region. It provides useful insights to tackle the challenging policy issue of balancing human and planetary health, within an interconnected, globalized food system.

Introduction

Interest in sustainable diets has increased markedly during the last decade as an emerging public health nutrition challenge, as well as a critical issue for sustainable food systems, within the international debate on sustainability, food security and nutrition (Berry *et al.*, 2015; HLPE, 2017). This debate has also emerged in the Mediterranean region, where food and nutrition security is still one of the utmost concerns (FAO/CIHEAM, 2017; CIHEAM/FAO, 2015). In this global multifaceted discussion, the Mediterranean diet (MD) has recently been investigated as a model of sustainable diet (FAO/CIHEAM, 2012; FAO, 2015). The MD concept has undergone a progressive evolution over the past 50 years, from that of a healthy dietary pattern to the model of a sustainable diet (Gussow, 1995; Burlingame and Dernini, 2011; Dernini and Berry, 2015; Dernini *et al.*, 2017).

Food insecurity and nutrition are still problems in many Mediterranean countries, especially southern and eastern ones, while obesity and overweight are also becoming a challenge in the entire Mediterranean region in response to unsustainable dietary shifts (CIHEAM/FAO, 2015). Traditional ways of consuming and producing food have changed considerably, mainly due to economic, social, cultural, demographic and technological trends, increasing urbanization and globalization and shifting lifestyles. As a result, the Mediterranean region is passing through a 'nutritional transition' in which problems of under-nutrition coexist with overweight, obesity and food-related chronic diseases (Belahsen, 2014). The increasing erosion of the MD heritage is alarming because it has undesirable impacts not only on health, but also on other social, cultural, economic and environmental dimensions related to food in the Mediterranean. '*Food*' is a source of identity for the Mediterranean people, a time-place for dialogue and exchange, most important for its individual cultural, social and economic values in each country of the Mediterranean region (CIHEAM/FAO, 2015). The MD, understood as a lifestyle in continual evolution through time, is a complex system of shared knowledge related to food and people, as a result of this particular environmentally and historically diverse geographic region. The dietary patterns embraced by the Mediterranean people over the centuries (Berry *et al.*, 2011), expressed within the MD notion, are the result of a number of factors: food production availability, seasonality, the use of small-scale technologies, the wide variety of local cultivars grown, the freshness of the food consumed, their homemade preparation, the conviviality of meals, and a physically active lifestyle (Serra-Majem and Medina, 2014).

The Development Process of the Med Diet 4.0 Framework

Following a long standing participative process on a MD redefinition, started in the early 2000s in Calabria, Barcelona, Athens and Rome (Dernini *et al.*, 2012), in the 2009 3rd CIISCAM International Conference on 'The Mediterranean diet today: a model of sustainable diet' was held in Parma (Italy), with the dual purpose of reaching a consensus on a new updated pyramid of the MD as well as to present the MD as a sustainable diet model, with country-specific and culturally appropriate versions. The traditional MD pyramid (Willet *et al.*, 1995) was revised to incorporate positive lifestyle changes, with serving sizes based on frugality, local habits, conviviality, culinary activities and physical activity. It also included new characteristic sustainability elements, such as biodiversity, seasonality, traditional, local and eco-friendly food products. This revised MD pyramid, as 'a lifestyle model for today', was further finalized at the 8th Mediterranean Diet International Conference, organized in 2010 in Barcelona (Bach-Faig *et al.*, 2011).

As follow up at the 2010 FAO/Bioversity symposium 'Biodiversity and sustainable diets: United against hunger', held in Rome at the Food and Agriculture Organization (FAO), an entire session of the programme was devoted to the MD as an example of a sustainable diet (Burlingame and Dernini, 2012). Subsequently, since 2011, the MD was identified by FAO and CIHEAM-Bari as a joint case study for characterization and assessment of the sustainability of food consumption patterns and diets in the Mediterranean region. Through a series of international workshops, reports and scientific publications, a methodological multidimensional approach started to be developed towards the assessment of the sustainability of the MD.

In 2015, at the Expo Milan international conference 'Does the Mediterranean diet still exist?', the International Foundation of Mediterranean Diet-IFMeD and the Forum on Mediterranean Food Cultures, in collaboration with CIHEAM-Bari presented the Med Diet 4.0 as a multidimensional framework for the characterization of the MD as a sustainable diet model. It highlighted four interdependent sustainable benefits of the MD, with country-specific variations: (i) well-documented nutrition and health advantages, preventing chronic and degenerative diseases and reducing public health costs; (ii) low environmental impacts and richness in biodiversity, reducing pressure on natural resources and climate change; (iii) positive local economic returns, reducing rural poverty; and (iv) high social and cultural food values, increasing appreciation, mutual respect and social inclusion (Fig. 19.1).

These interdependent benefits operate at different levels, varying by country contexts. They all interact and feed into each other in a *synergistic way*, reversing current unsustainable dietary shifts and contributing to achieve a wider nutritional well-being, food security and sustainable food systems in the Mediterranean region (Fig. 19.2).

The Med Diet 4.0 highlights the concept that 'health is a state of complete physical, mental and social *well-being* and not merely the absence of disease or infirmity', as stated in the constitution of the World Health Organization (WHO, 1948).

In 2015, a Med Diet 4.0 pyramid installation, as a travelling education and communication campaign project, was developed by IFMeD and the Forum on Mediterranean Food Cultures, with the support of CIHEAM-Bari, to promote the multiple MD sustainable benefits towards its revitalization. Considering that young generations are the majority of the populations in the Southern and Eastern Mediterranean countries, the model aimed to encourage young people to make more healthy food choices by rediscovering the MD as a contemporary active, diversified healthy and eco-friendly lifestyle, in balance with the well-being of the individual, the community and the planet.

In 2016, at the 1st World Mediterranean Diet Conference on 'Revitalizing the Mediterranean diet: from a healthy dietary pattern to a healthy Mediterranean sustainable lifestyle', organized by IFMeD with CIHEAM-Bari and European Federation of Nutrition Societies, the 2016 Call for Action for the Revitalization of the Mediterranean Diet was endorsed by more than 35 national and international scientific societies and institutions,[1] stressing the need to revitalize the MD, by rethinking it as: (i) a significant part of Mediterranean food systems, from consumption to production, and no longer just a diet, but an expression of the diversity of Mediterranean food cultures and culinary systems; (ii) a pivotal element for sustainable food systems in the countries of the Mediterranean region within the 2030 agenda for sustainable development; and (iii) a way of living of the Mediterranean people, a complex web of interdependent cultural aspects, from nutrition to the economy, involving law, history, politics and religion, while also being strongly linked to local territories and environments.

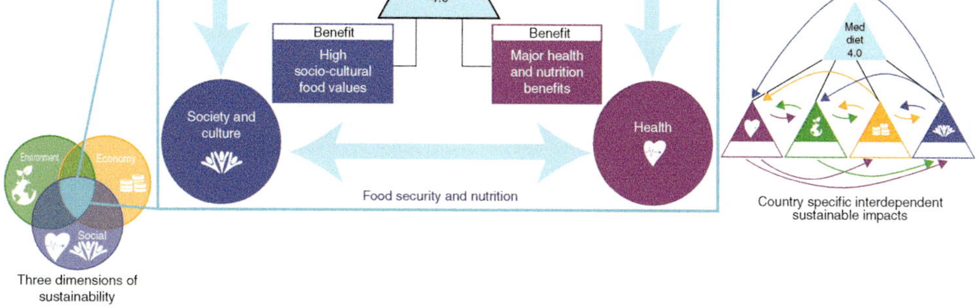

Fig. 19.1. The Med Diet 4.0 multidimensional framework. Source: adapted from Dernini *et al.* (2017).

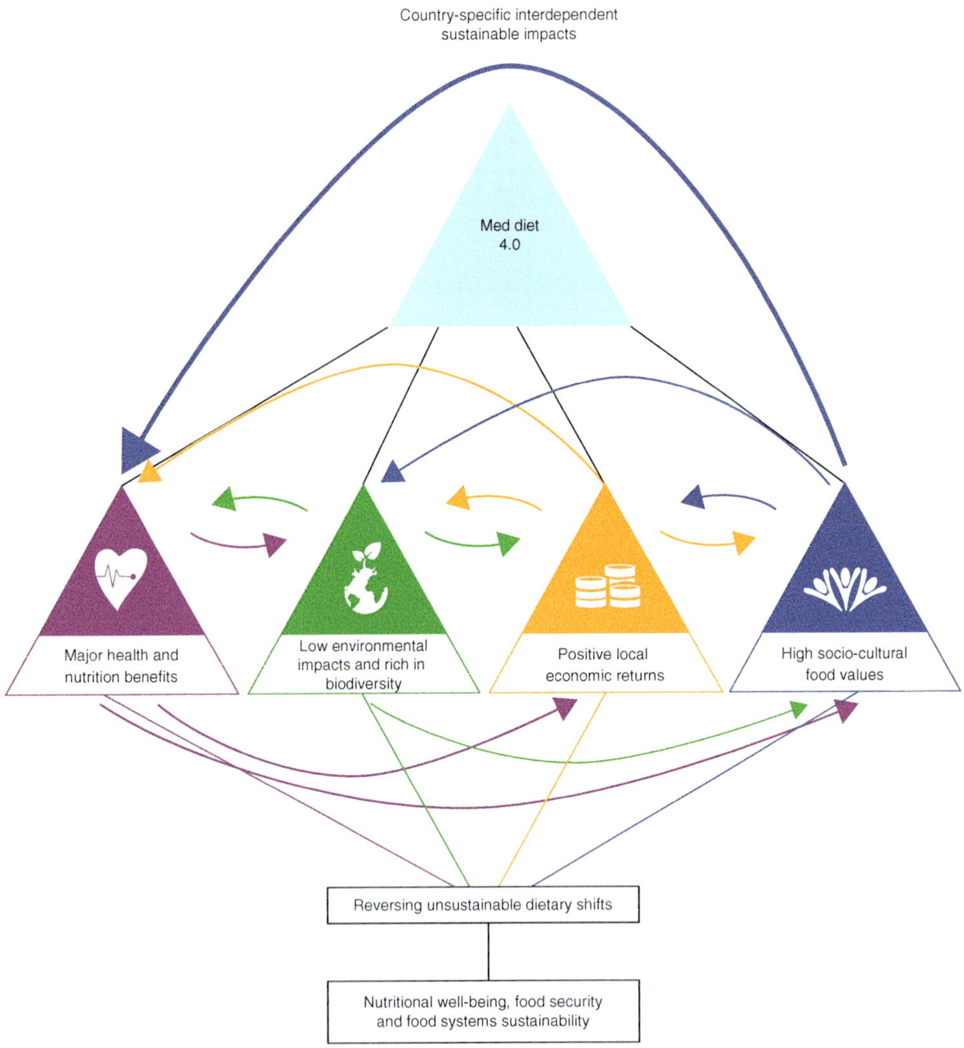

Fig. 19.2. The Med Diet 4.0 framework reversing dietary shifts.

The Mediterranean Diet as a Well-documented Healthy Diet

Many scientific papers have highlighted the nutritional quality and health benefits of the MD to prevent chronic and degenerative diseases. Subjects who adhere closely to a MD pattern have been shown to fulfil requirements for most fibres, minerals and vitamins much better than persons on a typical Western diet (Castro-Quezada *et al.*, 2014). This supports the concept that this plant-based dietary pattern is effective in improving nutritional status to promote health by integrating the positive metabolic inputs driven by the variety of typical Mediterranean foods (Serra-Majem *et al.*, 2009; Trichopoulou *et al.*, 2009).

In the past decades, the number of relevant prospective epidemiological studies, and clinical and community trials increased exponentially, raising the level and the quality of the evidence on the health benefits of the MD (Sofi *et al.*, 2010). Surveys have consistently shown that adherence to a MD pattern is associated with a reduced body weight and lower prevalence of overweight and obesity (Gotsis *et al.*, 2015; Lairon, 2015). In this context, several surveys have shown that such adherence is associated with a lower incidence of the metabolic syndrome (Babio *et al.*, 2009; Kastorini *et al.*, 2011; Kesse-Guyot , 2013) and of type 2 diabetes (Salas-Salvadó *et al.*, 2011). Since the pioneer Seven Countries Study, a traditional Mediterranean dietary pattern has been linked to a markedly reduced incidence of coronary heart disease mortality (Keys, 1970). An impressive series of further surveys confirmed this key protective effect on cardiovascular risk and mortality (Buckland *et al.*, 2008). The relevance of these associations has also been confirmed by controlled intervention trials showing a significantly reduced risk of cardiovascular risks and events in at-risk subjects who followed a Mediterranean-type diet (DeLorgeril *et al.*, 1994; Estruch *et al.*, 2013; Rosato *et al.*, 2017). The data from a series of case-control studies have also indicated that a high intake of foods typical of the traditional MD pattern were associated with a reduced risk for developing various types of cancers (La Vecchia, 2009) and considered favourable for combatting the major non-communicable diseases due to a high level of fibre and nutrient-rich plant foods and products (Serra-Majem *et al.*, 2009). A protective effect of a MD pattern has also been shown against cognitive decline in various contexts (Aridi *et al.*, 2017). Altogether, these specific protective effects of a MD pattern against major current pathologies result in a demonstrably reduced overall mortality (Trichopoulou *et al.*, 2009).

The Mediterranean Diet as a Sustainable Environmental Model for Redesigning the Supply Side of Mediterranean Food Systems

Investigating the environmental performances of diets cannot be performed without reference to the corresponding food systems, which drive and are driven by them (Meybeck, 2015). The environmental benefits of the MD are mainly recognized in relation to its plant-based dietary pattern. Indeed, shifting dietary patterns towards MD dietary patterns may result in environmentally beneficial changes on the supply side of food systems, through reducing livestock production and shifting land use and trade patterns (Aleksandrowicz, 2016; Tilman and Clark, 2014). On the other hand, food systems connected with the MD have additional environmental benefits beyond plant-based dietary patterns, which constitute a critical sociocultural heritage to preserve and valorize within a sustainable development model (Agnoletti, 2014). When considering the MD as a dietary model for sustainable food systems, the question relevant to food supply is: what should be changed in current food production, distribution and trade systems to reconfigure Mediterranean food supply in a more sustainable way? In this respect, the MD can drive modifications in existing Mediterranean food systems by reconfiguring them to change the interfaces between: (i) agriculture and the environment; (ii) actors of the food *value chain* that connect production and consumption; (iii) urban and rural areas; and (iv) food supply and food consumption (i.e. the food environment) (Sonnino and Stefanova, 2018). It is necessary to create a favourable enabling environment, which comprises cultural and behavioural aspects, tacit and explicit norms and standards for knowledge creation, use and distribution, private and public policies, institutions and governance mechanisms. In Table 19.1, a number of features of food systems connected with the MD are associated with different environmental benefits. These characteristics can be interpreted from a contemporary point of view and system theoretic perspective (Therond *et al.*, 2017; Ramage and Shipp, 2009) in order to construct an ideal target model, which can guide food system transitions by simultaneously reconfiguring interfaces and relationships to achieve desired environmental benefits. It is also important to select and interpret the relevant characteristics on the basis of contemporary evidence. For example, the results of a recent cohort survey of 22,900 subjects (Seconda *et al.*, 2017) show that adhering to an organic MD pattern, made of high-quality and environment-friendly foods, may provide optimal health and environmental benefits. A fundamental importance in this framework is given to the values underpinning the MD. They are necessary to allow transitions at the level of enabling environments, which, through their institutional settings, can support the creation of corresponding cultural, societal or market-oriented values. In this way, integrating several characteristics into a single target model (the MD diet) and expanding system boundaries could facilitate the design of system-level innovations,

Table 19.1. The environmental aspects of the Mediterranean diet.

Mediterranean diet characteristics	Values	Environmental benefits	Impacted interfaces	Contemporary interpretation
Mosaic-type landscape	High value attached to land and biological resources	High biodiversity in agro-ecosystems and associated ecosystem services	Agriculture–environment	Biodiversity-based, farming systems (i.e. agro-ecology); diet diversity; environment and health protection
Agricultural biodiversity	High value attached to land and biological resources	Locally adapted plant and animal breeds associated with lower resource consumption (e.g. water and toxic chemicals)	Agriculture–environment; value chain	High-quality traditional products; diet diversity
Frugality, recycling of food residues	High value attached to food and to other bio-products	Food waste is not an issue; closing of nutrient cycles at farm level, management of food-energy nexus	Rural–urban areas; value chain	Circular bio-economies at level of territorial production systems; supply-chain efficiency
Local markets, seasonal and fresh products, traditional storage and conservation methods	Value of proximity relations	Adapted to scarcity of energy resources needed for transportation and storage/conservation	Rural–urban areas; food environment	Alternative food networks and short supply chains
Plant-based dietary pattern	High value attached to scarce natural and biological resources	Livestock-based production is not exerting environmental pressures	Agriculture–environment; food environment	Shift in (global) land use and trade patterns; supply-chain efficiency

while overcoming path-dependencies and existing lock-ins (Meynard *et al.*, 2016).

It is therefore of fundamental importance to: (i) shift the focus of analysis from *filliere*/sector based approaches in disentangling the value chain concept (FAO, 2014), toward integrated landscape approaches targeting the whole economy of a territory (Therond *et al.*, 2017); and (ii) restore the values attached to land, local biodiversity resources, food and other bio-products. This requires revisiting current practices in environmental assessment of diets, which tend to focus only on plant-based dietary patterns associated with the MD locking in this way the analysis into trade-off identification and maximization. Instead, considering the MD as a reference target model, which integrates more desirable characteristics and enlarging system boundaries towards a more unified territorial-level approach, could allow for the identification of options that seek to maximize synergies at the level of food value chains, while shifting the trade-off resolution at the level of enabling environments by beginning to question different values which different stakeholders attach to food and diets (Freidberg, 2015).

The Mediterranean Diet Model as an Engine of Local Rural Economic Growth: the Role of Alternative Food Networks

In recognition of its sociocultural importance, UNESCO declared, in 2010, the MD as an Intangible Cultural Heritage of Humanity, thus acknowledging the MD as being a 'direct expression of the Mediterranean lifestyle, a way of living everyday life' (UNESCO, 2010). As opposed to common food-based descriptions of dietary patterns and intake, the description of the MD includes many sociocultural aspects concerning 'the way of selecting, cooking and eating' such as moderation, conviviality, the importance of cooking, and of choosing seasonal, biodiverse, traditional and local products (Bach-Faig, 2011).

However, people today adhere less and less to the traditional MD. Globalization has led to the development of a mass food culture and a transformation of taste and food choices that are partly formed *outside* the family. In contrast to previous decades, family units have 'shrunk' with the decline of the traditional extended family and consequently the knowledge that was once kept and passed on through female family members is now lost (Hachem *et al.*, 2016).

As people take on different identities related to their different geographical domiciles – citizen of the world, of one's country or city – they express them through food, thus giving rise to different food practices that coexist side-by-side, often without contradiction, and known as 'food polytheism' (Montanari, 2004). What this means is that although the practice of MD has weakened it has not disappeared and many of its elements persist. 'Not all is lost', as the cultural grounds on which the MD rests are powerful and deep (González-Turmo, 2012). In the current Mediterranean food environment, people are 'carriers' of different food practices due to the simultaneous existence of many different elements (Table 19.2).

The MD, intended as a model, provides material, symbolic and social resources to change the current Mediterranean food environment (Table 19.3) within a shared narrative.

In the last decades, a number of alternative ways of 'using' food have emerged at the grassroots level to challenge conventional food systems on the grounds of solidarity, health and environmental benefits (Dowler *et al.*, 2009; Brunori

Table 19.2. Elements of current Mediterranean food environment.

	Material	Competence	Meaning
Buying food	Steady decline in traditional markets, but a rise in GAS (solidarity purchase groups) and farmers' markets. Role of cars, refrigerators, packaging	Relative decline in knowledge of different types of fresh foods	Type of food: fast food growing – youth finds it more 'cool'
Cooking	Evolution of cooking tools, conservation, packaging, etc.	Relative decline in cooking skills	Greater symbolic meaning given to ready-made foods (sign of innovation, progress)
Organizing meals	Growing role of processing and pre-cooking technologies	Loose meal patterns as loss of routines and gap of knowledge on nutritional implications	Conviviality still important in spite of lack of time

Table 19.3. Resources of the Mediterranean diet that still remain and can form the basis for its revitalization.

	Material	Competence	Meaning
Buying food	Shifting to a MD food basket	Revitalize knowledge of different varieties of fresh and seasonal MD foods	MD as a narrative for healthy and sustainable living
Cooking	Evolution of cooking tools, conservation, packaging, etc.	Rebuilding food knowledge	Rediscovering food practices as leisure and culture
Organizing meals	Developing 'soft' food technologies	Mobilizing residual local knowledge	Fostering occasions for conviviality

et al., 2012; Goodman *et al.*, 2012). These range from farmers' markets and on-farm direct sales, to community supported agriculture and local public procurement. Such movements have also emerged in the Mediterranean region as a means to valorize and strengthen Mediterranean food practices. Thanks to their role in linking urban consumers to local producers, they have been an important means of generating revenue at the local level (Fonte, 2013). By participating in alternative food networks people have a chance to strengthen those food practices more closely tied to their MD heritage roots – access to minimally processed foods, predominance of seasonal fresh products among those that are exchanged – while at the same time fuelling local rural economic growth.

Considering that typical agri-food products are the cornerstone of the MD, for an effective valorization of these products, it is important to combine tradition, innovation and sustainability. A successful pilot project along these lines was developed by CIHEAM-Bari who launched a voluntary sustainability certification for the enhancement of typical agri-food quality products of the Apulian region (Capone *et al.*, 2016).

Enhancing Mediterranean Diet Education

As has been highlighted previously, the value of the MD goes beyond its nutritional role in sustenance, health promotion and disease prevention. It has a broader sociocultural value that is imbued with the richness of the nations in which it thrives and survives. In analysing these cultural aspects of food, concepts such as symbolic value and social identity come to the fore. The meaning we give to food and, consequently, how the food we eat identifies who we are is a complex but powerful process (Berry and De Geest, 2012). When we are choosing what to eat, we are consciously or unconsciously transmitting intrapersonal or interpersonal messages and fulfilling particular needs. Applying Warde's (1997) values guiding consumption to the dietary domain, when people select food they are considering its exchange value (the economic aspect), its use value (to satisfy needs and wants) and its identity value (boosting one's personal or perceived image) (Table 19.4).

Moreover, the meaning of food we select to eat involves a complex interplay of images, memories and emotions that are in turn influenced by the different settings – comprising, individuals, groups of people and contexts – where consumption choices and actions occur.

Thus, at any meal it is not only the food itself, but also who produced it for us, where and how it was cultivated or grown, who prepared it for us and the equipment and implements used, where we are eating it and with whom that infuse the food with multiple meanings.

All this has implications for the planning of MD education, where mainstreaming the adoption of the diet goes beyond emphasizing its healthful properties (Piscopo, 2009). Awareness of the symbolic value and social identity value of food is important for designing MD messages that are meaningful, motivational and have the potential to be assimilated and adopted. Due to the role of children as potential 'trend-setters' or 'taste-makers' (Fieldhouse, 1995), MD education should be given priority during all the years of schooling, starting from kindergarten. However, it should not be limited to children and

Table 19.4. Values guiding consumption.

Value	Meaning
Exchange value	Concerned with monetary price; how much one is willing to pay for an item with particular characteristics
Use value	Concerned with the final act of consumption; the satisfaction of needs and wants which change over time and can be symbolic rather than practical
Identity value	Concerned with orientations of actions, the desire: (i) to impress members of a different status group; (ii) to impress members of their own status group; and (iii) to impress themselves

Source: adapted from Warde (1997).

teenagers. Adults as parents, grandparents, employers and employees can all be targeted with particular messages relevant to their responsibility and power to bring about changes in dietary habits. The values typically associated with the MD, including frugality, simplicity, freshness, authenticity, nourishment, conviviality and tradition among others, can be utilized as emotive and practical marketing and valorizing tools. Table 19.5 offers a selection of messages and actions that could be emphasized with different audiences, within country-specific contexts, when doing MD education.

By weaving recognition of the MD and fostering its multiple values in educational strategies to promote the uptake of the MD, a stronger and more long-term impact on behaviour change for holistic well-being may be achieved. Programmes and initiatives that emphasize the sociocultural value of the MD will help reverse its erosion and strengthen its foothold in Mediterranean food systems (Coderoni *et al.*, 2017; Phull, 2015).

Conclusions

The Med Diet 4.0, as a multidimensional framework, valorizes the MD for its multiple sustainable benefits, reshaping knowledge of the MD and its appreciation in terms of a more contemporary and holistic vision of sustainability linked to nutritional well-being and food security.

The complexity of interdependent challenges within the radical transformation of the contemporary Mediterranean and global scenario requires new forms of transdisciplinary and intercultural dialogues, strategies and research at different levels for the revitalization of the MD. Within such complexity, the Med Diet 4.0 provides a synthesis to better understand and enhance the MD as a driver for the improvement of the sustainability of Mediterranean food systems, reconnecting the sociocultural space of food consumption to the environmental and economic place of food production.

Through its comprehensive multidimensional perspective, the Med Diet 4.0 provides an integrated framework for better understanding and appreciating the multiple and interdependent sustainable benefits of the MD, catalyzing new multi-stakeholder partnerships and innovative inter-sectorial efforts, and thus paving the way to reverse unsustainable dietary shifts in the Mediterranean region.

As highlighted in the recommendations of the 2017 HLPE report, the MD provides useful insights as a sustainable diet model to tackle the challenging policy issue of balancing human

Table 19.5. Mediterranean diet education for different groups.

Target population	Knowledge and understanding	Action
Children	The producers of our food Children's role as part of the food system Healthy growth and sustainable development	Visiting where food is produced Growing, producing, cooking, buying, choosing food themselves
Parents and grandparents	The value of cooking from scratch Awareness of their responsibility to be role models for the MD and transmitting MD traditions to the younger generations	Organizing to cook from scratch Learning food preparation skills using traditional and new sustainable food products
Food producers and chefs	Awareness of traditional, typical and quality MD foods and dishes Modification of MD foods and dishes to make them healthier and sustainable	Producing traditional MD foods and dishes which are tasty, easy to cook, healthy, sustainable, convenient, fun, come in a broad price range
Health professionals, educators, policymakers	Awareness of the various dimensions of the MD which make it healthy, sustainable, acceptable and desirable	Taking responsibility to offer practical, usable guidance and creating policies, incentives and structures which will favour adoption of the MD by the general population, with special emphasis on children

and planetary health, within an interconnected, globalized food system (HLPE, 2017).

Further development of the Med Diet 4.0 education and communication campaign, based on key simple messages (Serra-Majem *et al.*, 2017) and country-specific adaptations, can contribute to fully acknowledging the MD as a common ground among Mediterranean people, residing on all its shores, increasing mutual understanding, intercultural and inter-religious dialogue, towards more social cohesion, stability and peace in the Mediterranean region.

Note

[1] The 2016 Call for Action on the Revitalization of the Mediterranean Diet, Endorsed by International Foundation of Mediterranean Diet (IFMeD); International Centre for Advanced Mediterranean Agronomic Studies (CIHEAM); European Federation of Nutrition Societies(FENS)-African Federation of Nutrition Societies (FANUS); Italian Society of Human Nutrition; Spanish Society of Community Nutrition (SENC); Société Française de Nutrition (SFN); Society for Nutrition Education and Behaviour; Interuniversity International Study Centre on Mediterranean Food Cultures (CIISCAM); Forum on Mediterranean Food Cultures-Al Quds Public Health Society; Braun School of Public Health, Hebrew University; Hadassah Medical School; IUIBS/ULPGC Instituto Universitario de Investigaciones Biomédicas y Sanitarias of the University of Las Palmas de Gran Canaria; Hellenic Health Foundation; International Commission on Food Anthropology (ICAF)- UNESCO Chair on Food, Culture and Development, Open University of Catalonia; CNR-ENEA-CREA, Sapienza University of Rome; Dipartimento di Scienze Cliniche e di Comunita', Universita' di Milano; PREDIMED Network; University of Malta; Aix-Marseille Université; University of Valencia; Chouaib Doukkali University; Hasan Kalyoncu University Department of Nutrition and Dietetics; American University of Beirut; Centre Català de la Nutrició de l'Institut d'Estudis Catalans - CIBEROBN 'Centro de Investigación Biomédica en Red-Fisiopatología de la Obesidad y Nutrición'; FIN Nutrition Research Foundation; AEN 'Spanish Academy of Nutrition and Food Science'; IACON 'International Association of Community Nutrition'; Institut Català d'Oncologia; the IRCCS Istituto Neurologico Mediterraneo 'Neuromed'; Assessorato della Salute, Regione Siciliana.

References

Agnoletti, M. (2014) Rural landscape, nature conservation and culture: some notes on research trends and management approaches from a (southern) European perspective. *Landscape and Urban Planning* 126, 66–73.

Aleksandrowicz, L., Green, R., Joy, E.J., Smith, P. and Haines. A. (2016) The impacts of dietary change on greenhouse gas emissions, land use, water use, and health: A systematic review. *PLoS One* 11(11), e0165797.

Aridi, Y.S., Walker, J.L. and Wright, O.R.L. (2017) The Association between the Mediterranean Dietary Pattern and Cognitive Health: a Systematic Review. *Nutrients* 9(7), 674. DOI: 10.3390/nu9070674

Babio, N., Bulló, M., Basora, J., Martinez-González, M.A., Fernández-Ballart, J., *et al.* (2009) Adherence to the Mediterranean Diet and Risk of Metabolic Syndrome and its Components. *Nutrition, Metabolism and Cardiovascular Diseases* 19 (8), 563–570.

Bach-Faig, A., Berry, E., Lairon, D., Reguant, J., Trichopoulou, A., *et al.* (2011) Mediterranean diet pyramid today. Science and cultural updates. *Public Health Nutrition* 14(12A), 2274–2284.

Belahsen, R. (2014) Cultural diversity of sustainable diets. Nutrition transition and food sustainability. *Proceedings of the Nutrition Society* 73(3), 385–388.

Berry, E.M. and De Geest, S. (2012) Tell me what you eat and I will tell you your sociotype: coping with diabesity. *Rambam Maimonides Medical Journal* 3(2), e0010.

Berry, E.M., Arnoni, Y. and Aviram, M. (2011) The Middle Eastern and biblical origins of the Mediterranean diet. *Public Health Nutrition* 14 (12A), 2288–2295.

Berry, E.M., Dernini, S., Burlingame, B., Meybeck, A. and Conforti, P. (2015) Food security and sustainability: can one exist without the other? *Public Health Nutrition* 18(13), 2293–2302.

Brunori, G., Rossi, A. and Guidi, F. (2012) On the new social relations around and beyond food. Analysing consumers' role and action in Gruppi di Acquisto Solidali (Solidarity Purchasing Groups). *Sociologia Ruralis* 52(1), 1–30.

Buckland, G., Bach, A. and Serra-Majem, L. (2008) Obesity and the Mediterranean diet: a systematic review of observational and intervention studies. *Obesity Review* 9(6), 582–593.

Burlingame, B. and Dernini, S. (2011) Sustainable diets: the Mediterranean diet as an example. *Public Health Nutrition*, 14(12A), 2285–2287.

Burlingame B. and Dernini, S. (Eds) (2012) Sustainable Diets and Biodiversity: Directions and solutions for policy, research and action. *Proceedings of the International Scientific Symposium Biodiversity and Sustainable Diets United against Hunger*, 3–5 November 2010, FAO Headquarters, Rome http://www.fao.org/docrep/016/i3004e/i3004e.pdf (accessed 28 August 2018).

Capone, R., El Bilali, H. and Bottalico, F. (2016) Assessing the sustainability of typical agro-food products: insights from Apulia region, Italy. *New Medit* 15(1), 28–36.

Castro-Quezada, I., Román-Viñas, B. and Serra-Majem, L. (2014) The Mediterranean diet and nutritional adequacy: a review. *Nutrients* 6(1), 231–248.

CIHEAM/FAO (2015) Mediterranean food consumption patterns: Diet, environment, society, economy and health. A white paper Priority 5 of Feeding Knowledge Programme Expo Milan 2015. CIHEAM-Bari/FAO, Rome, Italy.

Coderoni, S., Perito, M.A. and Cardillo, C. (2017) Consumer behaviour in Italy. Who spends more to buy a Mediterranean Diet? *New Medit* 16 (2), 38–46.

DeLorgeril, M., Renaud, S., Salen, P., Monjaud, I., Mamelle, N., *et al.* (1994) Mediterranean alpha-linolenic acid-rich diet in secondary prevention of coronary heart disease. *The Lancet* 343(8911), 1454–1459.

Dernini, S. and Berry, E.M. (2015) Mediterranean diet: from a healthy diet to a sustainable dietary pattern. *Frontiers in Nutrition* 2(15), 1–7.

Dernini, S., Berry, E.M., Bach-Faig, A., Belahsen, R., Donini, M.L., *et al.* (2012) A dietary model constructed by scientists: The Mediterranean diet. In: *Mediterra*, CIHEAM–SciencesPo Les Presses, Paris, France, pp. 71–88.

Dernini, S., Berry, E.M., Serra-Majem, L., La Vecchia, C., Capone, R., *et al.* (2017) Med Diet 4.0: the Mediterranean diet with four sustainable benefits. *Public Health Nutrition* 20(7), 1322–1330.

Dowler, E., Kneafsey, M., Cox, R. and Holloway, L. (2009). Doing food differently: reconnecting biological and social relationships through care for food. *The Sociological Review* 57(S2), 200–221.

Estruch, R., Ros, E., Salas-Salvadó, J., Covas, M.I., Corella, D., *et al.* (2013) Primary prevention of cardiovascular disease with a Mediterranean diet. *New England Journal of Medicine* 368(14), 1279–1290.

FAO/CIHEAM (2012) *Towards the Development of Guidelines for Improving the Sustainability of Diets and Food Consumption patterns: The Mediterranean Diet as a Pilot Study*. A FAO/CIHEAM discussion paper. Food and Agriculture Organization of the United Nations, Rome, Italy.

FAO/CIHEAM (2017) *Development of Voluntary Guidelines for the Sustainability of the Mediterranean Diet in the Mediterranean Region.* Food and Agriculture Organization of the United Nations, Rome, Italy.

FAO (2015) *Assessing Sustainable Diets within the Sustainability of Food Systems. Mediterranean Diet, Organic Food: New Challenges*. Food and Agriculture Organization of the United Nations, Rome, Italy.

FAO (2014) *Developing Sustainable Food Value Chains: Guiding Principles*. Food and Agricultural Organization of the United Nations, Rome, Italy.

Fieldhouse, P. (1995) *Food and Nutrition: Customs and Culture*. Chapman & Hall, London, UK.

Fonte, M. (2013) Food consumption as social practice: Solidarity Purchase Groups in Rome, Italy. *Journal of Rural Studies* 32, 230–239.

Freidberg, S. (2015) From behind the curtain: talking about values in LCA. *The International Journal of Life Cycle Assessment*, 23(7), 1410–1414.

González-Turmo, I. (2012) The Mediterranean diet: consumption, cuisine and food habits. In: *Mediterra* CIHEAM/Presses de Sciences Po (PFNSP), Paris, France, pp. 115–132.

Goodman, D., DuPuis, M. and Goodman, M. (2012) *Alternative Food Networks.* Routledge, New York, USA.

Gotsis, E., Anagnostis, P., Mariolis, A., Vlachou, A., Katsiki, N., *et al.* (2015) Health benefits of the Mediterranean diet: an update of research over the last 5 years. *Angiology* 66(4), 304–318.

Gussow, J.D. (1995) Mediterranean diets: are they environmentally responsible? *American Journal of Clinical Nutrition* 61, 1383S–1389S.

Hachem, F., Capone, R., Yannakoulia, M., Dernini, S., Hwalla, N., *et al.* (2016) The Mediterranean diet: A sustainable consumption pattern. In: *Mediterra*. FAO/CIHEAM/ Presses de Sciences Po (PFNSP), Paris, France, pp. 243–261.

HLPE (2017) *Nutrition and Food Systems*. Report of the High Level Panel of Experts on Food Security and Nutrition of the Committee on World Food Security, Rome, Italy.

Issa, C., Darmon, N., Salameh, P., Maillot, M., Batal, M., *et al.* (2011) A Mediterranean diet pattern with low consumption of liquid sweets and refined cereals is negatively associated with adiposity in adults from rural Lebanon. *International Journal of Obesity* 35(2), 251–258.

Kastorini, C.M., Milionis, H.J., Esposito, K., Giugliano, D., Goudevenos, J.A., *et al.* (2011) The effect of Mediterranean diet on metabolic syndrome and its components: a meta-analysis of 50 studies and 534, 906 individuals. *Journal of the American College of Cardiology* 57(11), 1299–1313.

Kesse-Guyot, E., Ahluwalia, N., Lassale, C., Hercberg, S., Fezeu, L., *et al.* (2013) Adherence to Mediterranean diet reduces the risk of metabolic syndrome: a 6-year prospective study. *Nutrition, Metabolism and Cardiovascular Diseases* 23(7), 677–683.

Keys, A.B. (1970) Coronary heart disease in seven countries. *Circulation* 41(1), 1–200.

Lairon, D. (2015) The Mediterranean diet and adiposity. In: Preedy, V. and Watson, R.R. (eds) *The Mediterranean Diet: an Evidence-Based Approach*. Elsevier Inc., Philadelphia, Pennsylvania, USA, pp. 303–312.

La Vecchia, C. (2009) Association between Mediterranean dietary patterns and cancer risk. *Nutrition Reviews* 67(s1).

Meybeck, A. (2015) Understanding sustainable diets: from diets to food systems, from personal to global. In: *Assessing Sustainable Diets within the Sustainability of Food Systems. Mediterranean Diet, Organic Food: New Challenges.* Food and Agriculture Organization of the United Nations, Rome, Italy, pp. 207–214.

Meynard, J.M., Jeuffroy, M.H., Le Bail, M., Lefèvre, A., Magrini, M.B., *et al.* (2016) Designing coupled innovations for the sustainability transition of agrifood systems. *Agricultural Systems* 157, 330–339.

Montanari, M. (2004) *Il Cibo come Cultura*. Editori Laterza, Rome/Bari, Italy.

Phull, S. (2015) The Mediterranean diet: socio-cultural relevance for contemporary health promotion. *The Open Public Health Journal* 8, 35–40.

Piscopo, S. (2009) The Mediterranean diet as a nutrition education, health promotion and disease prevention tool. *Public Health Nutrition* 12(9A), 1648–1655.

Ramage, M. and Shipp, K. (2009) *Systems Thinkers*. Springer, London, UK.

Rosato, V., Temple, N.J., La Vecchia, C., Castellan, G., Tavani, A., *et al.* (2017) Mediterranean diet and cardiovascular disease: A systematic review and meta-analysis of observational studies. *European Journal of Nutrition* Nov 25. DOI: 10.1007/s00394-017-1582-0

Salas-Salvadó, J., Martinez-Gonzalez, M.A., Bulló, M. and Ros, E. (2011) The role of diet in the prevention of type 2 diabetes. *Nutrition, Metabolism and Cardiovascular Diseases* 21, B32–B48.

Seconda, L., Baudry, J., Allès, B., Hamza, O., Boizot-Szantai, C., *et al.* (2017) Assessment of the sustainability of the Mediterranean diet combined with organic food consumption: an individual behaviour approach. *Nutrients* 9, 61. DOI: 10.3390/nu9010061

Serra-Majem, L. and Medina, F.X. (2014) The Mediterranean diet as an intangible and sustainable food culture. In: Preedy, V.R. and Watson, D.R. (eds) *The Mediterranean Diet: An Evidence-Based Approach*. Academic Press-Elsevier, London, UK, pp. 37–46.

Serra-Majem, L., Bes-Rastrollo, M., Román-Vinas, B., Pfrimer, K., Sánchez-Villegas, A., *et al.* (2009) Dietary patterns and nutritional adequacy in a Mediterranean country. *British Journal of Nutrition* 101(2), S21–S28.

Serra-Majem, L., Aranceta Bartrina, J., Ortiz-Andrellucchi, A., Ruano-Rodriguez, C., Gonzalez-Padilla, E., *et al.* (2017) Decalogue for sustainable food and nutrition in the community: Gran Canaria declaration 2016. *Journal of Environmental and Health Sciences* 3(2), 1–5.

Sofi, F., Abbate, R., Gensini, G.F. and Casini, A. (2010) Accruing evidence on benefits of adherence to the Mediterranean diet on health: an updated systematic review and meta-analysis. *American Journal of Clinical Nutrition* 92(5), 1189–1196.

Sonnino, A. and Stefanova, M. (2018) Rural transformation, innovation and sustainable agro-food systems. *Energia, Ambiente e Innovazione, €conomie*, vol. 1. January–March (2018). DOI 10.12910/EAI2018-023

Therond, O., Duru, M., Roger-Estrade, J. and Richard, G. (2017) A new analytical framework of farming system and agriculture model diversities: a review. *Agronomy for Sustainable Development* 37, 21. DOI: 10.1007/s13593-017-0429-7

Tillman, D. and Clark, M. (2014) Global diets link environmental sustainability and human health. *Nature* 515(7528), 518–522.

Trichopoulou, A., Bamia, C. and Trichopoulos, D. (2009) Anatomy of health effects of Mediterranean diet: Greek EPIC Prospective Cohort Study. *British Medical Journal* 338, b2337.

UNESCO (2010) *Inscription of Mediterranean Diet on the Representative List of the Intangible Heritage of Humanity.* Intergovernmental Committee for the Safeguarding of the Intangible Cultural Heritage. UNESCO, Paris, France.

Warde, A. (1997) *Consumption, Food and Taste*. Sage, London, UK.

WHO (1948) Constitution of the World Health Organization. World Health Organization, Geneva.

Willett, W.C., Sacks, F., Trichopoulou, A., Drescher, G., Ferro-Luzzi, A., *et al.* (1995) Mediterranean diet pyramid: a cultural model for healthy eating. *American Journal of Clinical Nutrition* 61(6), 1402S–1406S.

20 Traditional Foods at the Epicentre of Sustainable Food Systems

Antonia Trichopoulou

Abstract

Patterns of food production and consumption have changed in ways that profoundly affect ecosystems and human diets. The accelerated speed of loss of food biodiversity and degradation of most ecosystems forces us to examine the role of traditional foods in sustainable food systems, since the notion of a food system generally focuses on food. Local traditional foods are an important component of a sustainable diet in many areas of the world, and consequently of a sustainable food system. The general concept of traditional foods includes the preservation of traditional farming knowledge, local crop and animal varieties, and native forms of sociocultural organization. Importantly, for the production of traditional foods, local products are generally used, thus their cultivation contributes to the employment of local people. Traditional foods, apart from being vehicles of our culture, may also possess health qualities, since tradition rarely honours foods that are not palatable and healthy. In 2010, in a Food and Agriculture Organization report on biodiversity in sustainable diets, it was stated that: 'Countries, communities and cultures that maintain their own traditional food systems are better able to conserve local food specialties, with a corresponding diversity of crop varieties and animal breeds. They are also more likely to show a lower prevalence of diet-related chronic diseases. The Mediterranean diet offers a clear example'. The health effects of the Mediterranean diet and indeed its identity can be partly attributed to the traditional foods that this diet integrates.

Introduction

Patterns of food production and consumption have changed in ways that profoundly affect ecosystems and human diets. For centuries, traditional farmers have developed diverse and locally adapted agricultural systems, managing them with ingenious practices that often result in both community food security and the conservation of agrobiodiversity (Altieri, 2004). In fact, the accelerated speed of loss of food biodiversity and degradation of most ecosystems forces us to examine the role of traditional foods to sustainable food systems.

Recently, the High-Level Panel of Experts on Food Security and Nutrition (HLPE) proposed a comprehensive, descriptive definition:

> A food system gathers all the elements (environment, people, inputs, processes, infrastructures, institutions, etc.) and activities that relate to the production, processing, distribution, preparation and consumption of food, and the outputs of these activities, including socio-economic and environmental outcomes.
> (HLPE, 2014)

Thus, traditional foods are an integral part of sustainable food systems and their role should be clarified.

Diets and food systems are closely linked. However, the notion of food systems is generally focused on food; in this chapter we revisit the sustainable diets that are considered those diets that have 'low environmental impact and are respectful of biodiversity while optimizing natural and human resources' (Burlingame, 2012).

Sustainable Diets

In the early 1980s, the notion of 'sustainable diets' was described to recommend diets that are healthier for the environment as well as for consumers (Gussow and Clancy, 1986).

The concept of a 'sustainable diet', borrowed from 'sustainable agriculture', promoted activities that minimized the waste of natural resources and addresses food production for local and seasonal consumption (Burlingame and Dernini, 2011). Local traditional foods are an important component for a sustainable diet in many areas of the world, and consequently for a sustainable food system, which includes the preservation of traditional farming knowledge, local crop and animal varieties, and native forms of sociocultural organization. Maintaining and, when necessary, returning to local crops and traditional food systems is essential for the conservation of biodiversity.

The introduction into the food market of poor imitations of traditional foods is misleading the consumers. A prerequisite of minimizing this misguiding is the registration and standardization of traditional foods. Registration permits the proper definition of the food, whereas standardization assures that manufactured traditional foods maintain the organoleptic, physicochemical and microbiological properties that characterize it.

Traditional Foods

Definition

According to the European Union EuroFIR project, 'traditional foods' refer to practices or specifications established prior to the Second World War.

Traditional food is a food of a specific feature or features, which distinguish it clearly from other similar products of the same category in terms of the use of 'traditional ingredients' (raw materials or primary products) or 'traditional composition' or 'traditional type of production and/or processing method'. The selected time limit 'prior to the Second World War' implies 'prior to the era of mass food production' and it delineates the period when most population groups still applied simple, time-honoured approaches. This is before the large-scale introduction of technological innovations that substantially altered the food production processes (Trichopoulou, 2006).

There is a need to protect the cultural, nutritional and industrial elements of traditional foods, since the term 'traditional' is frequently being used as a commercial tool for marketing purposes. The EuroFIR definition of European traditional foods, developed for scientific purposes, could form the basis for a fruitful collaboration with the responsible public authorities, in order to protect the identity of traditional foods, which is inseparable from its cultural dimension (Trichopoulou *et al.*, 2007). Following World War II, Europe entered the era of mass food production and importation, and consequently the dietary patterns of European populations were abruptly disrupted.

This was reflected in a steady decline of culinary habits that populations had been culturally and gastronomically attached to for centuries. There had been other nutritional revolutions in the past even before the Second World War (e.g. the introduction of potatoes, coffee, cocoa, and so on, into Europe). However, following the Second World War there were – in a short time period – dramatic changes in food production processes and availability of raw materials, when middle- and low-income countries achieved stunning increases in agricultural production principally by using new varieties of crops such as wheat, rice and maize, and increased use of pesticides and oil-based fertilizers combined with mechanization. This was a part of the post-war 'second agricultural revolution' that increased yields in industrialized nations from 1945 to 1970 (Hardin, 2008).

In general, the Northern and Central European countries experienced the above referred alterations much earlier than the Southern European countries. However, the food industrial revolution, which was gradually expanding in Europe after the Second World War, finally reached the Southern European countries, causing a serious disruption in the traditional dietary pattern of the Mediterranean populations.

Traditional foods and health

Traditional foods, apart from being vehicles of our culture, may also possess health qualities, since tradition rarely honours foods that are not palatable and healthy.

The health implications of traditional foods are not addressed nor implied by the EuroFIR definition. In fact, traditional foods do not necessarily associate with health benefits. In Mediterranean countries, traditional foods are generally considered as healthy foods. The longevity associated with the Mediterranean diet could be partly attributed to Mediterranean traditional foods, which this diet incorporates.

The micronutrient content of certain traditional Greek foods was investigated in relation to international recommendations. Many of these foods showed a rich micronutrient profile, indicating that in order to meet micronutrient requirements, at least in the Mediterranean countries, a simple approach would be to adhere to traditional dietary patterns and reinstate traditional foods into the daily diet (Vasilopoulou and Trichopoulou, 2009). It has also been shown that many Greek traditional foods contain high amounts of a variety of antioxidants (Trichopoulou *et al.*, 2005).

In vivo experiments, with ancient wheat varieties have shown convincing beneficial effects on various parameters linked to cardio-metabolic diseases such as lipid and glycaemic profiles, as well as inflammatory and oxidative status. However, given the limited number of human trials, it is not possible to definitively conclude that ancient wheat varieties are superior to all modern counterparts in reducing chronic disease risk (Dinu *et al.*, 2017). Significant improvements of irritable bowel syndrome symptoms and the inflammatory profile were reported after the ingestion of ancient wheat products (Sofi *et al.* 2014). Traditional Doogh and yoghurt show a higher abundance of total bacteria and lactobacilli and a higher bacterial diversity, respectively. Considering diversity and higher probiotic bacteria content in traditional Doogh, consumers' health could be promoted with these products (Sayevand *et al.*, 2017). It has been demonstrated that a Greek weekly menu, largely composed of traditional foods, provides the macronutrients and a wide range of micronutrients that meet recommended daily allowances, developed by the Scientific Committee for Food of the European Commission (Vasilopoulou *et al.*, 2013).

Traditional food environments and sustainable diets

In 2010, in a Food and Agriculture Organization report on biodiversity in sustainable diets, it was stated:

> Countries, communities and cultures that maintain their own traditional food systems are better able to conserve local food specialties, with a corresponding diversity of crop varieties and animal breeds. They are also more likely to show a lower prevalence of diet-related chronic diseases. The Mediterranean diet offers a clear example.
>
> (FAO, 2010)

The importance of the Mediterranean diet for the rest of the world lies less on its specific foods and nutrients, and more in the methods used to capture its essence and the culture and philosophy of sustainability that lies at its core. These same methods can be used to characterize sustainable diets in other ecosystems and food systems. The way the Mediterranean diet has been conceptualized and studied in relation to health and other parameters can be considered to be one of the models for the study of sustainable diets and a reference for addressing some of the challenges faced in many of the middle and low-income regions around the globe.

The health effects of the Mediterranean diet, and indeed its identity, can be partly attributed to the traditional foods, which this diet integrates. For the production of traditional foods, local products are generally used. Cultivation of local products contributes to a sustainable environment and employment of local people. The Mediterranean diet, a system rooted in the respect for the territory, ensures the conservation and development of traditional activities and crafts linked to fishing and farming, thereby guaranteeing the balance between the territory and the people (Dernini *et al.*, 2017). Indeed, the analysis of the water footprint (WF) based upon a Mediterranean dietary pattern, indicated a reduction of WF associated with adherence to Mediterranean diet (Vanham *et al.*, 2016).

The dietary patterns of the Mediterranean people are the result of various factors: food production and availability, seasonality, the use of traditional and often small-scale technologies, the wide variety of local cultivars, the freshness of the foods, their homemade preparation, the frugality and the conviviality of meals, besides a physically active lifestyle (Trichopoulou and Lagiou, 1997).

Conclusion

Globalization in food production and consumption is well recognized and there is no denial that globalization is an ongoing process, unavoidable and, in some respects, even desirable. It entails, however, some risks and disadvantages that can best be demonstrated by looking at the value and the fate of sustainable diets and traditional foods across the globe.

Nutrition globalization will continue, but it is important to preserve diversity, cultural and collective identity. More generally, however, there is a need to highlight biodiversity, food production and food consumption as interconnected elements, with the purpose of promoting a broader assessment of the link between local food products, nutrition, food safety and sustainability with traditional foods at the epicentre (Trichopoulou, 2012).

References

Altieri, M.A. (2004) Linking ecologists and traditional farmers in the search for sustainable agriculture. *Frontiers in Ecology and the Environment* 2(1), 35–42. DOI: 10.1890/1540-9295(2004)002[0035:LEATFI]2.0.CO;2

Burlingame, B. and Dernini, S. (2011) Sustainable diets: the Mediterranean diet as an example. *Public Health Nutrition* 14(12A):2285e7. DOI: 10.1017/S1368980011002527

Burlingame, B. (2012) Preface. In: Burlingame, B. and Dernini, S. (eds) *Sustainable Diets and Biodiversity. Directions and Solutions for Policy, Research and Action*. Food and Agriculture Organization of the United Nations, Rome, Italy, p. 7.

Dernini, S., Berry, E.M., Serra-Majem, L., La Vecchia, C., Capone, R., *et al.* (2017) Med Diet 4.0: the Mediterranean diet with four sustainable benefits. *Public Health Nutrition* 20, 1322–1330.

Dinu, M., Whittaker, A., Pagliai, G., Benedettelli, S. and Sofi, F. (2017) Ancient wheat species and human health: Biochemical and clinical implications. *Journal of Nutritional Biochemistry* 52, 1–9. DOI: 10.1016/j.jnutbio.2017.09.001

FAO (2010) Technical Workshop Biodiversity In Sustainable Diets Rome, 31 May – 1 June 2010. Available at http://www.fao.org/ag/humannutrition/24994-064a7cf9328fbe211363424ba7796919a.pdf (accessed 24 June 2018).

Gussow, J.D. and Clancy, K.L. (1986) Dietary guidelines for sustainability. *Journal of Nutrition Education* 18, 1–5.

Hardin, L.S. (2008) Meetings that Changed the World: Bellagio 1969: The Green Revolution. *Nature* 455(7212), 470–471. DOI: 10.1038/455470a

HLPE (2014) *Food Losses and Waste in the Context of Sustainable Food Systems*. A Report by the High Level Panel of Experts on Food Security and Nutrition of the Committee on World Food Security. Food and Agriculture Organization of the United Nations, Rome, Italy.

Sayevand, H.R., Bakhtiary, F., Pointner, A., Remely, M., Hippe, B., *et al.* (2017) Bacterial diversity in traditional doogh in comparison to industrial doogh. *Current Microbiology* 75(4), 386–393. DOI: 10.1007/s00284-017-1392-x

Sofi, F., Whittaker, A., Gori, A.M., Cesari, F., Surrenti, E., *et al.* (2014) Effect of *Triticum turgidum* subsp. turanicum wheat on irritable bowel syndrome: a double-blinded randomised dietary intervention trial. *British Journal of Nutrition* 111(11), 1992–1999. DOI: 10.1017/S000711451400018X

Trichopoulou, A. (2012) Diversity v. globalization: traditional foods at the epicentre. *Public Health Nutrition* 15(6), 951–954.

Trichopoulou, A. and Lagiou, P. (1997) Healthy traditional Mediterranean diet: an expression of culture, history, and lifestyle. *Nutrition Reviews* 55, 383–389.

Trichopoulou, A., Vasilopoulou, E. and Georga, K. (2005) Macro- and micronutrients in a traditional Greek menu. *Forum of Nutrition* 57, 135–146.

Trichopoulou, A., Vasilopoulou, E., Georga, K., Soukara, S. and Dilis, V. (2006) Traditional foods: Why and how to sustain them. *Food Science and Technology* 17, 498–504.

Trichopoulou, A., Soukara, S. and Vasilopoulou, E. (2007) Traditional foods: a science and society perspective. *Trends in Food Science Technology* 18, 420–427.

Vanham, D., del Pozo, S., Pekcan, A.G., Keinan-Boker, L., Trichopoulou, A. and Gawlik, B.M. (2016) Water consumption related to different diets in Mediterranean cities. *Science of the Total Environment* 573, 96–105. DOI: 10.1016/j.scitotenv.2016.08.111

Vasilopoulou, E. and Trichopoulou, A. (2009) The micronutrient content of traditional Greek foods. *Mediterranean Journal of Nutrition and Metabolism* 2, 97–102. DOI: 10.1007/s12349-009-0045-4

Vasilopoulou, E., Dilis, V. and Trichopoulou, A. (2013) Nutrition claims: a potentially important tool for the endorsement of Greek Mediterranean traditional foods. *Mediterranean Journal of Nutrition and Metabolism* 6, 105–111. DOI: 10.1007/s12349-013-0123-5

21 Globally Important Agricultural Heritage Systems (GIAHS): a Legacy for Food and Nutrition Security

Parviz Koohafkan

Abstract

In many countries specific agricultural systems and landscapes have been created, shaped and maintained by generations of farmers and herders based on diverse species and their interactions and using locally adapted, distinctive and often ingenious combinations of management practices and techniques. Globally important agricultural heritage systems (GIAHS) represent a unique sub-set of these agricultural systems, which exemplify customary use of globally significant agricultural biodiversity and merit to be recognized as a heritage of mankind. Agricultural heritage systems throughout the world testify to the inventiveness and ingenuity of farmers in their use and management of the finite resources, biodiversity and interspecies dynamics, and the physical attributes of the landscape, codified in traditional but evolving knowledge, practices and technologies. However, GIAHS are rapidly shrinking victims of globalization, urbanization and unsustainable technological and economic changes. In order to safeguard and support the world's agricultural heritage systems, the author conceptualized and presented on behalf of Food and Agriculture Organization a partnership initiative on 'Conservation and Adaptive Management of Globally Important Agricultural Heritage Systems' that was adopted by the World Summit on Sustainable Development, in Johannesburg South Africa. The initiative seeks to promote the international recognition, conservation and adaptive management of these systems, including support for local and indigenous communities in developing enabling environment and appropriate policies for dynamic conservation of GIAHS. The concept of GIAHS has already laid the foundation for the recognition of traditional food systems as food heritage and its contribution to sustainable diets. Recognizing traditional food systems as national or global heritage, not only gives utmost pride to the custodians of the traditional food systems (i.e. the small-scale family farmers, traditional food processors and distributors), but it would also encourage their collaboration and participation in programmes to improve efficiency and productivity within the food systems.

Introduction

The story of world agriculture is closely interwoven with that of the evolution of human civilization and of its diverse cultures and communities across the globe. In many countries, agricultural and rural life to this day is considerably influenced by the society's ancient cultural traditions and local community institutions and values, which are mostly conditioned by natural endowments, wealth and breadth of accumulated knowledge and experience in the management and use of natural resources.

In many countries specific agricultural systems and landscapes have been created, shaped and maintained by generations of farmers and herders based on diverse species and their interactions and using locally adapted, distinctive and often ingenious combinations of management practices and techniques. Building on generations of accumulated knowledge and experience, these ingenious agri-'cultural' systems reflect the evolution of humanity and its profound harmony with nature. They have resulted not only in outstanding aesthetic beauty, maintenance of globally significant agricultural biodiversity, resilient ecosystems and valuable cultural inheritance but, above all, in the sustained provision of multiple goods and services, food and livelihood security and quality of life.

Such agricultural and agro-silvo-pastoral and fisheries systems can be found, in particular, in highly populated regions or in areas where the population has, for various reasons, had to establish complex and innovative land-use/management practices, for example, due to geographic isolation, fragile ecosystems, political marginalization, limited natural resources and/or extreme climatic conditions. These systems reflect often rich and sometimes unique agricultural biodiversity, within and between species but also at ecosystem and landscape level. Having been founded on ancient agricultural civilizations, certain of these systems are linked to important centres of origin and diversity of domesticated plant and animal species, the conservation of which is of great global value.

Their ecosystem resilience and robustness has been developed and adapted to cope with change (natural events and social, technological and political context) so as to ensure food and livelihood security and alleviate risk. The dynamic human management strategies and processes that allow the maintenance of biodiversity and essential ecosystem services are characterized by continuous technological and cultural innovation, transfer between generations and exchange with other communities and ecosystems. The wealth and breadth of accumulated knowledge and experience in the management and use of resources is a globally significant resource that needs to be preserved and allowed to evolve.

Globally Important Agricultural Heritage Systems: Context and Definition

The Globally Important Agricultural Heritage Systems (GIAHS) Partnership Initiative[1] was conceptualized by the author and launched in Johannesburg South Africa, during the World Summit on Sustainable Development in 2002 (FAO, 2008). It was adopted as a programme in 2014 by the Food and Agriculture Organization (FAO) governing bodies and has developed a solid institutional mechanism including a multi-donor programme covering up to this date 19 countries and 39 GIAHS designated sites.[2]

GIAHS are defined as: 'Remarkable land-use systems and landscapes which are rich in globally significant biological diversity evolving from the co-adaptation of a community with its environment and its needs and aspirations for sustainable development' (Koohafkan and Altieri, 2011).

GIAHS introduces the need for efforts to promote public understanding and recognition of the agricultural legacy, in which the multiple goods and services provided by the smallholders, family farming and indigenous communities are distinct in many ways. It provides food sovereignty, health and nutrition to many poor, helpless and isolated people, and contributes to maintenance of globally significant biodiversity and genetic resources for food and agriculture, ecosystem services through functional diversity, products and services diversity, collective and individual knowledge systems and cultural diversity. As such, it effectively underpins the concept of sustainable diets in all its elements (FAO, 2012).

GIAHS Selection Criteria and Essential Considerations

Similar to any designated world heritage site, the description and selection of a GIAHS site also follows some practical and basic criteria. The selection criteria was formulated through international consultations and based on 100 case studies gathered from individuals, institutions and through online search. In order to be referred as a GIAHS, the site must be of global importance based on its fundamental values and inherent characteristics. Global (or national) importance is a composite criterion, under which the overall value is established for a traditional/historic agricultural system, represented by a particular site, as a heritage of human kind (or a country). It synthesizes its overall global (or national/local) 'public good' value described under the five inter-related criteria. By combining the five criteria and their complex relationships, and positive connectivity and linkages between the systems' elements the valuation of the system is meaningful through an integrated and holistic approach. These criteria are similar to UNESCO's 'universal value' in the World Heritage Convention but somehow are more subjective and difficult to judge in the case of agricultural evolving systems and communities.

A summary of the global importance of the individual and combined characteristics of the system/site, with its intrinsic resilience and capacity to strike a social–environmental balance, by its historic and contemporary relevance for human development and by whether the site is a unique or outstanding example of the agricultural system, represents and testifies an agricultural heritage system in comparison with similar systems and sites.

The outstanding (or unique) features of the system should be summarized in terms of their relevance to global concerns addressing sustainable development and ecosystems management and their cultural and agricultural heritage value. The five basic criteria for the selection of the GIAHS represents the totality of tangible and intangible values/benefits, functionalities, goods and services provided by the system are as follows:

1. Food and livelihood security: the proposed agriculture system should contribute to the food, nutrition and livelihood security of local communities (often indigenous), representing the majority of their livelihood provisions. This includes provisioning and exchange among local communities to create a relatively stable and resilient food and livelihood system. This criterion is the most important basis for the selection of GIAHS and the monitoring of the sustainability of the system, as it brings about the number of people depending on traditional agriculture (Clawson, 1985). In spite of rapid globalization, urbanization and industrialization of agriculture, numerous traditional agriculture systems are providing the basic livelihood and food security of billions of farmers around the world particularly in developing countries.

2. Biodiversity and ecosystem function: agricultural biodiversity and genetic resources (species, varieties and breeds), as well as other biodiversity such as wild relatives, pollinators and wildlife associated with the agricultural system and landscape (Brookfield and Padoch, 1994). The system/site should be endowed with globally (or nationally) significant biodiversity and genetic resources for food and agriculture (e.g. endemic, rare, endangered species of crops and animals).

3. Knowledge systems and adapted technologies: maintain invaluable knowledge, ingenious technology and management systems of natural resources, including biota, land, water; and social organizations and institutions including customary institutions for agro-ecological management, normative arrangements for resource access and benefit sharing.

4. Cultures, value systems and social organizations: cosmo-vision, value systems and agricultural practices associated with environment and agricultural calendar; festivities and rituals as knowledge transfer. Local institutions play a critical role in balancing environmental and socioeconomic objectives, in creating resilience and in the reproduction of all elements and processes critical to the functioning of the agricultural system. Some may ensure conservation of and promote equity in the use and access to natural resources; some transmit traditional knowledge systems and critical values that promote custodianship of biodiversity, land and water; some facilitate planning, cooperation and innovation/adaptation. Such institutions may take the form of ceremonial and religious beliefs and practices, including taboos, ceremonies and festivities; of customary law and conflict resolution, including on resource tenure; of kinship, marriage and inheritance systems; of forms of leadership, decision making and cooperation; of oral and written traditions; of games and other forms of education and instruction; of division of roles and distribution of labour, including gender roles and specialized functions (intangibles).

5. Remarkable landscapes: landscape diversity is a basic characteristic of resilient agroecosystems. The landscape features resulting from human management, that provide particularly ingenious or practical solutions to environmental or social constraints, such as land-use mosaics, irrigation/water management systems, terraces, particular adaptive architecture, which might provide for resource conservation/efficiency or provide habitats for valued biodiversity, recreational values collective or non-commercial valuable uses (aesthetic, artistic, educational, spiritual, and/or scientific values of ecosystems) are among remarkability of the landscapes.

The Multi-faceted Nature of GIAHS

A socio-ecological landscape for food and livelihood security

GIAHS throughout the world testify to the inventiveness and ingenuity of people in their use and management of biodiversity, interspecies dynamics and, more importantly, utilizing the physical attributes of the landscape where they live, codified in traditional but evolving knowledge, practices and technologies. Ingenious agro-ecosystems reflect human evolutionary transitions intimately linking sociocultural systems with biophysical systems. They use traditional knowledge systems, 'trial-and-error' and experiential learning, insights and innovations. Their ingenuity has resulted in well-balanced agro-ecological systems in marginal, extreme or very specific ecologies, which could not otherwise have sustainably supported human life and agrobiodiversity. These systems are organized and managed through highly adapted social and cultural and customary practices and institutions. These agricultural 'landscapes' typically evolve in parallel with their associated 'lifescapes'. They are characterized by continuous technological and cultural innovations, as well as adjustment of management practices and uses of resources and ecosystems, through their transfer between generations, exchanges with other communities and ecosystems and in response to natural events and to changing social, technological and political context.

A reservoir of agro-biodiversity and associated biodiversity

The biodiversity that underpins agricultural systems spans a continuum from simple human use of wild species (whether directly for sustenance or indirectly for increasing yields from desired species) to the creation and intensive management of genetically modified organisms. Within this spectrum, 'agricultural biodiversity' represents that group of organisms which has been domesticated, maintained and adapted in a process of co-evolution with human management systems. Thus, land races and wild species of animals and plants are the essential source of genetic variability for responding to biotic and abiotic stress through genetic adaptation.

Agricultural practices in many parts of the world have led to landscape-scale ecosystem variation, and provide mosaics of micro-habitats, that support associated plant and animal communities, whose viability depends largely on continued and innovative management. In many regions of the world, especially where natural conditions of climate, soil, accessibility and human presence militate against intensification, there still persist agro-ecosystems and landscapes that are maintained by traditional practices developed by generations of farmers and herders. Based on a high diversity of species and their interactions, the use of locally adapted, distinctive and often ingenious combinations of management practices and techniques (e.g. agricultural systems) testify to millennia of co-evolution of human societies with their natural environments. These systems often reflect rich and globally unique agricultural biodiversity, within and between species but also at ecosystem and landscape level.

Having been founded on ancient agricultural civilizations, certain of these systems are linked to important centres of origin and diversity of domesticated plant and animal species, the *in situ* conservation of which is of great importance and global value. A growing body of scientific evidence demonstrates that indigenous and traditional agricultural systems feature a high degree of diversity of plant and animal genetic resources for food and agriculture (Craats, 2005). GIAHS often reflect rich and globally unique agricultural biodiversity within and between species, but also at ecosystem and landscape level. For instance, tropical agro-ecosystems composed of agricultural and fallow fields, multi-storey farming practices, complex home gardens and agroforestry plots commonly contain well over 100 plant species per field. Their biodiversity products are used for construction material, firewood, tools, medicines, livestock feed and, more importantly, for sustainable diets. The richness of biodiversity in any form and given condition, however, can only be effectively maintained, adapted and conserved with the human management systems that have created it, including indigenous knowledge systems and technologies, specific forms of social organization, customary or formal law and other cultural practices. Having been founded on ancient agricultural civilizations, GIAHS are linked to important centres of origin and diversity of domesticated plant and animal species, the *in situ* conservation of which is of economic importance and global value.

A body of traditional knowledge

In many regions of the world, especially where natural conditions of climate, soil, accessibility and human presence militate against intensification, there still persist ecosystems and landscapes that are maintained by traditional knowledge and practices developed by generations of farmers, forest dwellers, fisher folks and herders. As such, these systems have evolved and co-evolved with the human communities, handed down from one generation to another, refined and continuously fine-tuned, primarily as a response to the specific natural environmental change to satisfy the needs of people and to gain their livelihood. GIAHS are unique systems with a set of practices, knowledge, institutions, technologies, skills, traditions, beliefs and values proper to belonging farming communities. The traditional and indigenous knowledge systems employed in GIAHS are the foundation and basis of managing these agro-ecosystems, including processes and functions, maintaining ecosystem and landscape integrity. GIAHS throughout the world testify to the inventiveness and ingenuity of people in their use and management of the finite resources, biodiversity and interspecies dynamics, and the physical attributes of the landscape, codified in traditional but evolving knowledge, practices and technologies.

Cultural diversity

GIAHS have other values beyond production of foods, fibres, maintenance and conservation of plant and animal genetic resources for food and agriculture, and other provisioning services. These living and evolving systems and communities have kept their distinct identities intact on the strength of unifying values such as nature, family, community, history, and a sense of belonging to their natural habitats. What sets apart the agricultural heritage systems from the UNESCO world heritage sites is a unique feature of outstanding universal value, that GIAHS are not static or frozen in time or space. They represent a living, dynamic, socioeconomic, cultural and institutional mosaic of how man has adapted over the centuries to the demands of dramatic advances in human civilization, while preserving and conserving to this day a rich heritage of customs, livelihood patterns and landscapes. Their cultural diversity is also a factor which reinforces the heritage characteristics of GIAHS. These systems are bonded by a common thread of distinct identities, language use, ethnicity, aesthetics, and a respect for nature and ecosystem. GIAHS are agricultural legacies, of not only important agro-ecosystems, landscapes or landmarks of historical value, but also living and evolving family farming communities, institutions and ecological and cultural heritage.

Remarkable landscape with aesthetic beauty

GIAHS have evolved over time to specific and highly adapted forms of social organization through which the ecosystems and landscapes management takes place, and cultural identity is preserved. These indigenous and traditional agricultural systems have resulted in outstanding landscapes with remarkable aesthetic beauty. Some of these GIAHS landscapes appear to satisfy the objectives of the UNESCO Convention concerning Protection of the World recognized as world heritage sites. The Ifugao Rice Terraces of the Philippines is one example of a GIAHS and a world heritage site. This system is an epitome of an agricultural legacy dated from more than 2000 years ago. The spectacular rice terraces' landscapes allows protection and conservation of significant and important agricultural biodiversity and associated biodiversity, features marvellous engineering systems and innovativeness, promotes tourism, as well as expressing the conquered and conserved harmony between humankind and the environment. The system is also dubbed as the 'Living Cultural Heritage'.

The Contemporary Relevance of GIAHS

As has been demonstrated throughout history, the memory of agriculture and associated civilizations is carried by people, landscapes, seeds, plants, animals and by farmers' knowledge and technologies, but also by oral traditions, languages, arts, rituals, culinary traditions and unique forms of social organization. Together, these elements are integral parts of the living agricultural systems and associated cultures. Numerous agricultural heritage systems around the world have proven their robustness and resiliency and have passed the test of time. They offer solutions for present and future generations and environmental sustainability. They contain a wealth of biological resources, knowledge systems and management practices that can help to ensure food security and quality of life for humanity and to cope with challenges of today and tomorrow. Building on generations of accumulated knowledge and experience, these ingenious agricultural systems reflect the evolution of humanity and its profound harmony with nature. They have resulted not only in outstanding aesthetic beauty, maintenance of globally significant agricultural biodiversity, resilient ecosystems and valuable cultural inheritance but, above all, in the sustained provision of multiple goods and services, food and livelihood security and quality of life for the most poor and remote communities.

Such agricultural, forestry, fishery and agro-silvo-pastoral systems can be found, in particular, in highly populated regions or in areas where the population has, for various reasons, had to establish complex and innovative land-use and management practices, for example, due to geographic isolation, fragile ecosystems, political marginalization, limited natural resources, and/or extreme climatic conditions (FAO, 2012).

Indeed, the myriad of our agricultural heritage systems, and particularly the GIAHS, represent a unique sub-set of traditional agricultural systems, family farming and indigenous farming practices that exemplify customary use of globally significant agricultural biodiversity (Article 8J of the Convention on Biological Diversity) and merit to be recognized as a heritage of the mankind.

Not only does our agricultural heritage carry the accumulated wisdom and memory of the past, but they are also building blocks and essential foundation for our future in a rapidly changing world. In addition, GIAHS designation goes beyond the mere identification of interesting agricultural systems and their transformation into attractive snapshots of aesthetic landscapes. As has been demonstrated by UNESCO's World Heritage Convention, people take pride in the value given to their heritage and are extremely proud when the system they have nurtured is singled out as a world heritage site or as a GIAHS.

The new challenges arising from globalization are making it increasingly important to redefine the relationship between culture and development or, to be more precise, between cultural diversity, biological diversity and development. The 'agri'-cultural diversity as a source of innovation, creativity and exchange is humanity's guarantee for a mutually enriching and sustainable future. As such, it ranks alongside biodiversity as a key means of securing the sustainability of every form or expression of development, tangible and intangible. Together, agricultural diversity and biological diversity hold the key to ensuring resilience in both social and ecological systems. On the other hand, the opportunities of a globalized world with market access could add economic value and generate income for the local communities to enable them to access national and international niche markets, labelling opportunities and sustainable tourism and alternatively the world citizen's access to exotic and biodiverse food and sustainable diets derived from neglected crops and vegetables, wild species and traditional medicine.

Threats and Driving Forces to the GIAHS

The industrial agriculture and the focus on increasing agricultural production through price subsidies, intensive farming, specialization and the rapid technological change and internationally marketed commodities and associated neglect of externalities, has led to a generalized neglect of integrated agricultural systems that has often adapted to the extreme ecologies. The lack of promotion of diversified and environmentally friendly farming and integrated management practices, and the neglect of research and development and rural services for indigenous and ingenious systems threatens the foundation of agricultural 'culture' and associated biodiversity. Moreover, the urbanization and rapidity and extent of today's technological and economic changes threaten many of these agricultural heritage systems, including the biodiversity on which they are based and their societies. These threats are: erosion of rural values and adoption of unsustainable practices, overexploitation of resources and declining productivity, as well as imports of exotic domesticated species, leading to severe genetic erosion and loss of local knowledge systems. This poses the risk of loss of unique and globally significant agricultural biodiversity and associated knowledge, land degradation, poverty and threats to livelihoods and food security of many unique farming systems (FAO, 2012). In some areas, there are spill over effects from marginalization and increasing poverty in productive landscapes, onto wild biodiversity. The social and environmental integrity and resilience of such livelihood systems, and their associated biodiversity, depends on the adaptive capacity of concerned communities but also, on the enabling environment provided by policies and development strategies.

The driving forces of the adoption of unsustainable practices, overexploitation of resources, genetic erosion, loss of local knowledge, and associated risks of impoverishment, non-viable livelihood systems and socioeconomic instability, vary from one system to another. They essentially include population pressure and poverty, inappropriate policies and legal environment, especially insecure land tenure and external market forces, and lack of capacity to adapt land-use–livelihood systems to the rapidly changing environment while preserving the cultural and natural heritage. The root causes may include *inter alia*:

- Market incentives and economic policy environments that focus exclusively on short-term economic goals rather than long-term socioeconomic and environmental goods and services and sustainable agricultural and rural development.
- Reduced community involvement/empowerment in landscape/resource management decision making processes.
- Inadequate attention to local knowledge and experience, and inadequate valuation of GIAHS and their associated biodiversity by research and development services and policy and strategic frameworks.
- Inadequate support for the conservation and sustainable use of significant agricultural biodiversity (within and between species and at ecosystem level).
- Lack of marketing expertise and incentives to ensure that adequate value is placed on local cultivars and races and local produce, and benefit-sharing mechanisms and so forth.
- De-legitimization of local, customary institutions for the management of natural resources, particularly the normative frameworks for access, use and benefit sharing of natural resources. Such trends occur in the context of land reform, individuation of common property systems and policies that promote national cultural homogeneity.

To halt the rapid degradation of GIAHS their dynamic nature must first be recognized. Their resilience depends on their capacity to adapt to new challenges without losing their biological and cultural wealth and productive capacity. This requires continuous agro-ecological and social innovation combined with careful transfer of accumulated knowledge and experience across the generations. Trying to conserve GIAHS by 'freezing them in time' would surely lead to their degradation and condemn their communities to poverty. The GIAHS dynamic conservation should centre on the human development and knowledge systems, including the socio-organizational, economic and cultural features that underpin the conservation and adaptation processes in GIAHS without compromising their resilience, sustainability and integrity (Fig. 21.1).

GIAHS and Sustainable Food Systems

Increasing and evolving patterns of human food consumption, together with high rates of urbanization, unregulated development expansion, unsustainable use of natural resources, spread of invasive species, erosion of agrobiodiversity and

Fig. 21.1. Saving unique agricultural heritage systems at risk.

of local varieties, and environmental and climate change are all threats to the world's food diversity and nutrition security, and thus, sustainable diets (ETC Group, 2009). Despite the increased public, political and scientific interest in conserving plant genetic resources, many countries tend to overlook the nutritional and quality food production in their quest for increased agricultural production to feed the growing population.

When the member countries of the United Nations adopted the sustainable development goals (SDGs) in 2015, they committed to a world free of poverty and hunger by 2030, and in which all life can thrive. This will require countries to develop sustainable food systems and new ways of managing natural resources, including genetic diversity, in order to build a viable future for humankind.

Achieving sustainable diets for all is critically important to the delivery of the SDGs. Healthy diets provide a foundation to support successful progress toward targets in health, agriculture, inequality, poverty and sustainable consumption. Poor nutrition is associated with low educational attainment, poor physical growth and low labour productivity.

The new models of food systems that humanity will need to include are the food that has its roots in peoples' culture and the forms of farming that are more ecological, biodiverse, local, sustainable and socially just. This means that they should be rooted in the ecological rationale of traditional agricultural heritage systems, representing long-established examples of successful community-based local agriculture. There should be closer connections between producers and consumers, therefore local production and consumption and increased link between rural and urban areas.

Since the early 1980s, hundreds of agroecologically based projects have been promoted throughout the developing world, demonstrating that by blending elements of both traditional knowledge and modern agricultural science, the productivity and sustainability of small farming systems can be optimized and thus can enhance the conservation of natural resources and local and national food security.

Small-scale, family farming and more traditional forms of agriculture and food systems could significantly address many problems of sustainable agriculture today and in the future. Smallholder and family farmers have adapted their systems and adopted new practices to economic and environmental changes at scale. They continue to supply most basic food commodities at local, national and global levels (FAO, 2014). Their small-scale farms offer an array of environmental, economic, social and cultural services, and remain a source of employment, nutritious food, cultural value and quality of life (Koohafkan and Altieri, 2010, 2017; Altieri and Koohafkan, 2013, Koohafkan, Altieri and Gimenez, 2012).

These systems have been managed with time-tested resilience, ingenious combinations of techniques and practices that have typically led to food sovereignty, sustained resources and incomes, and the conservation of natural resources and biodiversity. Indeed, agricultural systems with high levels of social and human assets are able to innovate and adapt to uncertainties. Family farmers make the majority of contribution to agricultural production worldwide and are thus acknowledged as the key actors for ensuring future food and nutrition security. At regional levels, the smallholders provide up to 80% of the food supply in Asia and sub-Saharan Africa. Women have a key role since they account for 43 % of the agricultural labour force of developing countries, rising up to almost 50 % in Eastern and South-eastern Asia and sub-Saharan Africa (FAO, 2011, 2014). Their model of farming is based on means of organizing agricultural production mainly relying on family labour, including both women and men. In family farming and traditional agriculture, the central approach is the integration of the production activities with the local landscape in which the resources are complementary, integrated and sustainably valued. The indigenous knowledge and traditional techniques that smallholder farmers apply indeed derive from the deep knowledge of the context they interact with and are adapted to. Such knowledge, apart from being the backbone of diverse and environmentally sustainable production systems, makes farmers more able to adapt to their changing environment and thus more resilient to the changes expected in light of climate change and other pressures. In spite of these strengths, though, smallholder farmers are threatened by unfair trade and market forces and competitive pressure from globalization.

However, in many countries, consumers are already willing to pay more for products that come from sustainable agricultural systems, organically produced food and traditional landscapes because of health and environmental concerns. Product certification is one of the most commonly used instruments to identify and add value to such products and can provide a price premium for producers. The market for certified organic products has been growing by 20% a year since the early 1990s, a lot faster than the rest of the food industry both in developed and developing nations. Estimates of future growth range from 10% to 50% annually depending on the country.

There is great potential to develop markets for underutilized or wild species, given the wide availability of crops, livestock and fish that have not been (fully) domesticated or commercially exploited. Such developments would support the conservation through use of a wider range of genetic resources while providing farmers with opportunities to diversify livelihood options and increase their incomes, which is particularly relevant in dealing with global changes.

The concept of GIAHS (see Fig. 21.2) has already laid the foundation for the recognition of traditional food systems as food heritage. Recognizing traditional food systems as national or global heritage not only gives utmost pride to the custodians of the traditional food systems (i.e. the

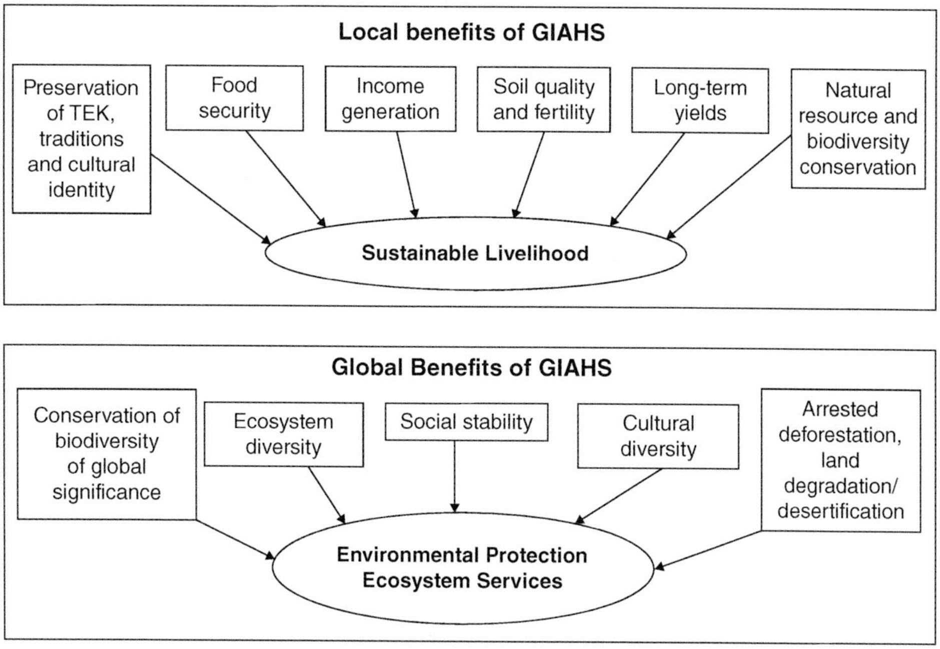

Fig. 21.2. Global and local goods and services of GIAHS.

small-scale farmers/family farmers, traditional food processors and distributors), but it would also encourage their collaboration and participation in programmes to improve efficiency and productivity within the food systems.

Conclusions and the Way Forward

The GIAHS initiative is the first that brings the link between agriculture and cultural heritage to the forefront of the nexus of poverty reduction, food and nutrition security and biodiversity conservation not only because of the great heritage value of the outstanding agricultural systems, but in view of their historical, current and potential future contribution to sustainable development.

GIAHS is also about the pillars of sustainability – environmental, economic and social. It is about meeting the needs of the present generation without compromising the ability of future generations to meet their own needs. It is about livelihood and food security of small holders, family farmers and indigenous people and their way of life but also about food security of the most vulnerable groups and the empowerment of such local communities. It is also about the conservation and sustainable utilization of our biodiversity, using our local knowledge to manage and enhance our capacity, and to live in harmony with nature.

The dynamic concept of agricultural heritage as the product of a long-term, scientifically demonstrable interactive process of co-existence between humans and nature embraces several strands running through the idea of heritage for future sustainability. It has helped to appreciate that nature/human interaction can produce extremely interesting results, visually, scientifically and in conservation. The concept of dynamic conservation is a management response and practical solution to the imperatives of conservation and necessities of change. As various aspects of development threaten to degrade and destroy our agricultural heritages and their inherent values, it is necessary to take up the challenge of dynamic conservation of these fragile, crucial and non-replaceable resources for the benefit of current and the future generations. It is now widely agreed that heritage with its value for identity, and as a repository of historical, cultural and social memory, preserved through its authenticity, integrity and 'sense of place', forms a crucial aspect of the sustainable development, sustainable food systems and sustainable diets.

Notes

[1] GIAHS was registered under the UN *Partnerships for Sustainable Development in 2004.* Available at https://sustainabledevelopment.un.org/partnership/?p=2309 (accessed 14 June 2018).
[2] www.giahs.org.

References

Altieri, M.A. and Koohafkan, P. (2012) Enduring Farms: Climate Change, Smallholders and Traditional Farming Communities. Third World Network. Available at http://www.fao.org/docs/eims/upload/288618/Enduring_Farms.pdf (accessed 28 June 2018).

Brookfield, H. and Padoch, C. (1994) Appreciating geodiversity: a look at the dynamism and diversity of indigenous farming practices. *Environment* 36(5), 6–45.

Clawson, D.L. (1985) Harvest security and intraspecific diversity in traditional tropical agriculture. *Economic Botany* 39, 56–67.

Craats, R. (2005) *Indigenous Peoples: Massai.* Weigl Publishers, Inc., New York, USA.

ETC Group (2009) Who will feed us? Questions about the food and climate crisis – 2009. Action Group on Erosion, Technology and Concentration, Val David, Canada.

FAO (2008) *Conservation and Adaptive Management of Globally Important Agricultural Heritage Systems (GIAHS)* GCP/GLO/212/GFF project document. The Food and Agriculture Organization of the United Nations, Rome, Italy.

FAO (2011) *The State of the World's Land and Water Resources for Food and Agriculture (SOLAW).* The Food and Agriculture Organization of the United Nations, Rome, Italy.

FAO (2012) *Conservation and adaptive management of Globally Important Agricultural Heritage Systems (GIAHS).* Project progress implementation report. The Food and Agriculture Organization of the United Nations, Rome, Italy.

FAO (2014) *The State of Food and Agriculture: Innovation in Family Farming*. The Food and Agriculture Organization of the United Nations, Rome, Italy.

Koohafkan, P. and Miguel A. Altieri, M.A. (eds) (2011) GHIAS-Globally Important Agricultural Heritage Systems. A Legacy for the Future. Food and Agriculture Organization of the United Nations Rome. http://www.fao.org/fileadmin/templates/giahs/PDF/GIAHS_Booklet_EN_WEB2011.pdf (accessed 9 September 2018).

Koohafkan, P., Altieri, M. A., & Gimenez, E. H. (2012). Green Agriculture: foundations for biodiverse, resilient and productive agricultural systems. International Journal of Agricultural Sustainability, 10(1), 61–75.

Koohafkan, P. and Altieri, M.A. (2017) Forgotten Agricultural Heritage Connecting Food Systems and Sustainable Development. Routledge, London.

22 Sustainability Along All Value Chains: Exploring Value Chain Interactions in Sustainable Food Systems

Allison Marie Loconto, Pilar Santacoloma, Roberto Azofeifa Rodríguez, Emilie Vandecandelaere and Florence Tartanac

Abstract

The value chain, as an analytical tool, has been used for more than 50 years as a way to better understand how agri-food products move and gain value from the farm gate to the table. Over the past 20 years, increasing attention has been paid to questions of sustainability within value chains and even more recently there has been a push to try to better understand how the way through which food is provisioned can deliver diets that are also sustainable. In this chapter, we explore the recent advances in value chain theories and we illustrate how taking a horizontal network, systemic and territorialized approach to food provisioning systems contributes to this literature. We argue that by looking both within and across value chains, we can better identify innovations in actor arrangements that are bringing new values (particularly sustainability) into food systems. By refocusing our analytical lens away from specific commodities and towards new forms of organization – such as short supply chains, circular economies, gastronomy and geographical indications – we can better capture how they might contribute to promoting sustainable consumption and production in local food systems.

Introduction

First developed in the 1980s, the concept of 'sustainable diets' was solidified in 2010 by the Food and Agriculture Organization of the United Nations (FAO) and Bioversity International as 'those diets with low environmental impacts which contribute to food and nutrition security and to healthy life for present and future generations' (Burlingame and Dernini, 2012). The achievement of a sustainable diet should be the outcome of ensuring sustainable production and consumption patterns, which have recently been included in the sustainable development goals (i.e. SDG 12). Global trends tell us that income growth and urbanization will drive changes in dietary patterns with substantial increase in demand for cereals, milk and meat products. The shift to higher consumption of animal products and food rich in fat and sugars will increase the risk of overweight and obesity (FAO, 2013, 2017a). The imbalanced demand for some food products over others can have distorting effects on the distribution and production networks that sustain food systems.

For these reasons, we focus in this chapter upon the supply chains and food provisioning systems that are fundamental to ensuring that consumption and production patterns can become sustainable and deliver sustainable diets. We use the concept of value chains to explore the variety of ways through which food systems might reorganize production and consumption patterns in order to achieve sustainable diets.

Value Chains and Sustainability

There are generally three schools of thought that have contributed to the emergence of the value chain as a structuring concept for implementing sustainability. The first developed in the 1960s by Louis Malassis and was based within the field of industrial organization. The *filière* approach, as it was referred to, mapped and calculated the socioeconomic characteristics of agro-enterprises and the monetary value of product flows from production to consumption (Raikes *et al.*, 2000). The second was based on Immanuel Wallerstein's world system's theory, which used the concept of the periphery that supplies the centre to analyze the tropical commodity systems (Hopkins and Wallerstein, 1986; Friedland, 2001) that persisted following the end of colonialism and became a dominant feature of globalization. Here the focus was on understanding sociological questions of power and exploitation in these systems. Finally, the term value chain was coined by Michael Porter (1985) as a management tool that could help firms to identify their competitive advantage within an industry structure. This approach was quickly taken up in corporate social responsibility programs and over the years has been repackaged as 'creating shared value' among supply chain actors (Porter and Kramer, 2011). These schools of thought provided a mix of theoretical and practical tools that have since been further developed and tuned to focus on specific elements, such as upgrading and governance (Gereffi and Korzeniewicz, 1994).

Questions of sustainability have entered this literature also in two ways. First, the concept of sustainable or green value chains is often used to refer to those value chains where environmental and social indicators are taken into consideration in determining the sustainability of the supply chain (Carter and Rogers, 2008). These chains can range in coverage from those that focus purely on closed-loop supply chains that reduce their environmental footprint by recycling the used products back through the chain (Srivastava, 2007) and thus creating circular economies (Andersen, 2007), to an idea of sustainable sourcing that focuses on the purchasing of certified raw materials (SAI, 2013). Certified raw materials traditionally rely upon systems of third-party certification where private standards enable value chain actors to make claims as to the value(s) of the products (Loconto, 2010). Indeed, this approach to sustainable sourcing has become dominant in global value chains for tropical commodities (Loconto and von Hagen, 2016; OECD and FAO, 2016).

Second, sustainability is considered systemically. Inspired by Kaplinsky and Morris (2002), FAO defines a sustainable food value chain as:

> the full range of farms and firms and their successive coordinated value-adding activities that produce particular raw agricultural materials and transform them into particular food products that are sold to final consumers and disposed of after use, in a manner that is profitable throughout, has broad-based benefits for society, and does not permanently deplete natural resources.
>
> (FAO, 2014)

This vision implies that the chain is not only a logistical structure as some of the more instrumentalist approaches propose, but rather a chain of relationships where different actors along the chain are adding value as the product moves from one actor to the next within a food system. This approach provides a roadmap from which to trace the actors who, through different nodes of negotiation, are involved in creating values throughout the chain (Ponte and Gibbon, 2005; Ouma, 2015).

Recent advances in the study of value chains work from this notion of multiple values and networks of relations in order to reconceptualize what is being exchanged, how it is being valued and the effects of changing the relationships between actors that can influence both of these elements (Loconto, 2017). One focus of this work has been to re-embed the value chain within the food system concept (Ericksen *et al.*, 2010) and more specifically within approaches to understanding alternative agri-food networks and local food systems (CIRAD-SAR, 1996; Goodman *et al.*, 2012). In the localized food system approaches, focus has shifted from trying to identify actors' positions in linear value chains towards understanding their positionality within territorially anchored, horizontal networks (Bowen and Mutersbaugh, 2014). The preferred approach within this school has been the study and promotion of geographical indications (GIs), which have been shown to deliver positive economic and social impacts on rural development, such as increasing production, employment, food system resilience and sociocultural sustainability (Barham and Sylvander, 2011; FAO, 2018b). GIs primarily differentiate and add value to products with specific characteristics, qualities or reputation resulting essentially from their geographical origin; they protect both consumers and producers from misuse of the territorial name, while they also contribute to the preservation of public goods (Vandecandelaere, 2016). Collective action is at the heart of GI processes whereby producers and the local community are able to organize themselves around a local identity and heritage. Local producers elaborate their GI product specifications, allowing the rules to be adapted to local conditions (natural and human resources) instead of being imposed by downstream segments of the value chain. Producer organizations that develop GIs have also demonstrated an important dynamism that supports environmental conservation at the landscape level of their territory ('terroir') and promotes local culture and gastronomy.

More recently, the concept of *circuit court* (Chiffoleau, 2012) or short food supply chains has emerged to try to capture how proximity (geographic or shared values) is often a common denominator in creating a strong nexus between consumers and producers that contributes to the sustainability of the food system (Renting *et al.*, 2003; Aubry and Kebir, 2013). Proximity is seen as a starting point for the collective construction of a new vision and identity around food production and consumption for urban communities (Parker, 2005). Proximity has also been shown to have positive effects on reinforcing site-specific cultural identity and the ability of local actors to be actively engaged in new forms of food citizenship (Renting *et al.*, 2012), such as community-supported agriculture (Hinrichs, 2000) or consumer-driven food initiatives (Fonte, 2013). The construction of geographical or social/institutional proximity in food systems implies building conscious relationships between producers, consumers and other intermediary actors who are increasingly fundamental in ensuring that sustainable production and consumption activities can meet (FAO, 2016, 2018a). These approaches move out of a linear focus on one product or commodity towards 'baskets of goods' that offer diverse food options for closely linked consumers. We thus draw upon these understandings of value chains as networks of interactions so to explore the relationships among actors who are working together across value chains.

Sustainability Within and Across Value Chains

Ensuring sustainability in value chain interactions that occur within agri-food systems are both a desirable outcome and a complex condition of action that requires constructive participation of all system actors. If we are to truly understand how value chains can encourage sustainable consumption and production patterns, we need to better understand what is already occurring in a variety of contexts and learn from those forms of collaboration and organization that work to deliver the sustainable outcomes that policymakers and food system actors seek. In this section, we use empirical examples from the authors' work to illustrate how actors in Africa and Latin America are defining what is sustainable and how they are implementing sustainability in their consumption-production networks. We argue that these types of interactions within and across value chains are key to ensuring the sustainability of food systems.

Peri-urban agricultural heritage systems of Mexico City: valuing tradition in short supply chains

The Chinampa system, an emblematic and resilient pre-Columbian system located in peri-urban Mexico City is being threatened by rampant urbanization pressure generating competition for labour, land and water resources. The Chinampa system is made up of an articulated set of floating, tiny, artificial islands surrounded by canals or ditches and rows of *ahuetojes* (*Salix bonpladiana*), which is a species of willow that performs several functions, including: living fences that provide windbreaks, hosting living species and preventing soil erosion (Gobierno de la Ciudad de México, 2016). The Chinampa system is active only in 19% of the total area (7300 ha) but provides a great diversity of horticultural, staple crops and ornamental products to the metropolis. Nearly 12,000 families are directly involved in agriculture activities in the Chinampa and this generates nearly 35% of their income. The permanence of the system synchronizes specific ways of organization, lifestyle, traditional forms of community and technical skills conforming a type of *Chinampa stewardship* (Gobierno de la Ciudad de México, 2016).

Despite being an outstanding intensive and efficient food system able to feed more than a million inhabitants in pre-Hispanic times, the Chinampa system is now fighting to preserve its agricultural and environmental services and functions. The most evident threat is water salinization due to the reduction in available water, which is a consequence of changes in water concessions for use in the metropolis. Moreover, the predominance of Mexico City's wholesale market (one of the biggest in the world) that sells undifferentiated products is affecting the profitability of farmers who produce using the Chinampa system. A recent opportunity for valuing the Chinampa system is its recognition as a globally important agricultural heritage system (GIAHS) by FAO-UNESCO (FAO, 2017a). With this recognition, a number of local non-governmental organizations have begun to create short supply chains so that the products produced in this GIAHS can be sold in traditional *tianguis* (open-air bazaars) in Mexico City. This strategy of directly linking an indigenous production system with an indigenous market outlet offers the possibility to preserve the sociocultural heritage and agricultural values of a sustainable food system that have been eroded over the years.

Since 2016, FAO has been collaborating with SEDEREC (Secretary of Rural Development and Equity for Communities) to strengthen the linkages between these two systems, to help local actors develop autonomous and sustainable systems. The FAO-SEDEREC strategy builds on two food system elements:

1. A farmer's market established in a central site of the city where producers can get fair prices and consumers can acquire fresh products with traceability of origin and cultural identity. Valued features of this market are transparency and information, cleanness, diversity and regularity of products. In this chain, the market coordinator's role is significant. They should manage information and communication technologies, build and manage market governance by communicating with both sides, and provide technical assistance to producers and processors to assure product quality.

2. Support producer market alliances with emphasis on participatory diagnosis, added-value and identification of main constraints and solutions. Support for strengthening collective action, entrepreneurial skills and provision of basic infrastructure – stalls or gathering centres – are deemed essential in the overall strategy (FAO, 2017b).

By combining an approach that focuses on sustainable production within a culturally and agroecologically important territory, farmers' markets that encourage direct exchanges between producers and consumers (and support services for intermediaries), this approach has been able to create reinforcing interdependencies among the actors. The need for a variety of products to supply markets and the reliance upon diverse ecosystem services to produce those products means that value chain specialization is not an option. Instead, building upon local knowledge to manage these flows is what will ensure the sustainability of this food system.

Gastronomy in Costa Rica: creating value chains that link chefs and producers

Since the early 1990s, Costa Rica has been at the forefront of the movement towards efficient and environmentally responsible production systems in both the regulation of and collaboration with the private sector and through specific public-sector policies, programs and projects (Azofeifa, 2015). Costa Rica has adopted a two-pronged approach to sustainable production and consumption in its agri-food systems. On the production side, Costa Rica has been moving towards efficient and environmentally responsible production systems that include good agricultural practices, organic production, low carbon agri-food systems, organic residues for energy and source of fertility, among others, to improve efficiency in farming systems. The results of these efforts are very clear as Costa Rica is consistently included within the top countries for sustainably certified farms and forests (Potts *et al.*, 2014; Lernoud *et al.*, 2017).

However, on the side of sustainable consumption, the efforts to influence consumption behaviour to promote sustainable diets through consumer awareness and information have been very weak. Bringing producers and consumers closer together in networks of proximity has been a far greater challenge than originally imagined. In Costa Rica, the dominance of unhealthy diets based on unsustainably produced food and highly processed products are a major reason for poor health, loss of biological and cultural diversity and environmental degradation in the country. A strong gastronomy sector that can create the interconnections between healthy and sustainable consumption habits and sustainable production systems has been lacking.

In 2012, a multi-stakeholder initiative launched a National Plan for Healthy and Sustainable Gastronomy in order to reverse this trend in Costa Rica. The goal is to initiate a new, healthy and sustainable national cuisine that can act as a driver for agricultural development and strengthen production opportunities for diversified family farming systems. The hope is that by doing so, this initiative can influence national action and the global debate around sustainable diets.

The National Plan for Healthy and Sustainable Gastronomy is positioned in the context of the efforts to close sustainable food production and consumption loops. Specifically, a national platform made up of public and private actors administers this plan by facilitating the exchange of ideas, talent and interests in the implementation of partner activities. The plan strengthens the role of consumers and consumer behaviour that can promote sustainable diets. Among other relevant aspects, the plan has set up a series of activities that will turn consumer demand for seasonal products into the driving force for organic and sustainable production that can increase agrobiodiversity, reduce food losses and waste, and develop local markets and value chains.

This approach strengthens production opportunities for family farming by encouraging direct collaboration between chefs and producers. Activities such as promoting the consumption of local fruits and vegetables and enhancing the local cuisine by incorporating more edible plants and diversified food into gourmet meals can inspire broader incorporation of these local products into consumers' diets. By focusing on local and indigenous varieties of food found in Costa Rica, the gastronomic approach uses consumer interest to generate solutions to the problem of decreasing food biodiversity. The creation of direct provisioning networks between urban restaurants and family farming communities diversifies local economies and strengthens their local resilience to sustainability shocks. Finally, the focus on high profile chefs and the emerging food culture influences consumers' eating habits and encourages sustainable diets.

While the plan has triggered important progress in this sector, a lack of information and awareness about health and environmental impacts of food has been identified as a major obstacle for achieving greater impact. Therefore, efforts are being taken to increase the publicity of stakeholders who are engaging in these networks by organizing cooking events where consumers can meet the chefs and the producers. The focus on developing both geographic and values proximity in networks through these types of exchanges can begin to build long-term relationships that can outlast any food fad that is often associated with the idea of sustainable gastronomy. This collective approach enables actors to share practical advice about how they can improve their diets, reduce food losses and waste and build preferences for sustainably produced products. In the end, it is via these new short supply chains and collaborations that trust is built between actors and that consumer lifestyle changes occur.

Songhai Center in Benin: turning a value chain into a circular economy

Established as a youth training center in 1985, the Songhai Center incorporates three key sectors of the economy into a single organizational form. It is organized in such a way as to create synergy and complementarity between sustainable production methods based on an integrated production system that includes vegetable, pulse, cereals and fruit crop production, livestock raising, aquaculture and biogas production. It includes an industrial cluster model where artisanal and modern food processing takes place (e.g. fruit juice, snacks, popcorn, baked goods, bread, fresh cuts and cured meats, soap, plastics recycling, plastic buckets). The Songhai Center also organizes the production and sale of sustainable inputs (seeds, manure, compost and effective micro-organisms (EM)), provides agro-tourism and internet services, and is involved in developing appropriate technologies for sustainable production.

The Songhai network in Benin is currently made up of the main demonstration site in Porto Novo and five satellite centers in regional urban centers that source, when necessary, from surrounding rural farms. No link functions without a relationship to one or more of the other links and the satellites are governed through a centralized, hierarchical, chain of command that permits horizontal linkages between network members. There is a central procurement and marketing service that organizes the procurement of raw materials for processing and the sales of processed products from the Porto Novo hub. However, each satellite is also responsible for local sales of their fresh produce and artisanal processed goods. In 2014, 54% of the value of finished products was sold within the network and 46% constituted product sales with a value of US$7 million, of which the off-farm sales of finished products accounted for US$2.5 million (Loconto and Vicovaro, 2015).

Within the Songhai model, the actors in the network have had a role in defining what organic means in the country through their use of consumer-facing labels, Songhai has taken over running some of the Ministry of Agriculture's youth training activities and Songhai has successfully created an organizational model that is being replicated in other countries. In fact, the greatest revenue in 2014 came from the corporate licensing fees they received from the Nigerian operations. This mobilization has occurred through the establishment of a multi-actor innovation platform that focuses the attention of the actors in the network on sustainable agriculture technologies. Innovation intermediaries are highly influential in this system as the interactions between producers and consumers take place in the regional satellite centers (Agossou *et al.*, 2016).

The Songhai model of production is maintained by consumer demand for the qualities of its products. These qualities are communicated by word of mouth, with posters and direct communications by the employees at the sales points, through direct experience with the agricultural techniques either by attending the center's training programme or through visits to the demonstration site; by consuming the food in the on-farm restaurants or by reading the on-packet labels. The consumer-facing labels of Songhai products make claims about the product 'qualities' including: organic, healthy, medicinal properties of certain crops, and nutritional properties. According to research conducted in 2015 (FAO, 2018a), all types of actors believe that the local food system is rather sustainable, with producers being the most optimistic. This suggests that a horizontal network model, with both central and distributed production, processing and sales, that is managed by a core intermediary, has been able to effectively maintain the communication of sustainable values from production to consumption.

Geographical indications that support sustainable production and consumption

Coffee is a major cash crop for Guinea and a source of income for thousands of small-scale farmers. Guinean coffee is not well-established in the international coffee market, because of low quality, and is mainly exported to African countries (Senegal, Morocco and Algeria) (UNCTAD 2015). Nevertheless, the Ziama-Macenta coffee has gained a good reputation in the market, because of the high-quality orientation of its major producer, the Woko cooperative, and because of the influence of its *terroir* on the organoleptic characteristics of the coffee. Technical assistance and public support through the African Intellectual Property Organization (OAPI) and French Development Agency project PAMPIG (Projet de mise en place des Indications Géographiques dans les Etats membres de l'OAPI) supported the registration of the GI 'Ziama-Macenta coffee' for green beans, which has contributed to the economic, social and environmental sustainability of the local food system.

The GI area is found in the forest perimeter of the Ziama Mountains, which represents a refuge for several rare, vulnerable and threatened species, and a habitat for endemic species of the large upper Guinean forest block. The environmental factors (microclimate with importance of rain and low temperature, dense forests and secondary forests, located between 500 m and 1000 m altitude, and geological substrate on mountain slopes) strongly contribute to the territorial link of the GI. This coffee comes from traditional and hybrid varieties of Robusta coffee, with a tangy and slightly bitter flavor that is close to Arabica.

'Café Ziama-Macenta' for green coffee was registered in 2013 by the Ministry of Industry in Guinea and by the OAPI in June 2014. The GI specification includes specific production practices linked to the agroforestry system of shade grown coffee (including organic fertilization and no use of chemical pesticides) that protect the environment around the Ziama Mountains.

The creation of the GI has also structured the value chain and strengthened cooperation among the actors within the local region. The Woko cooperative has been strengthened and a second cooperative called Diani has been established. The two cooperatives were working with 38 formalized producer groups in 2014 (compared to 17 before the GI registration), with an additional 1116 producers engaged in the GI strategy. Three groups of collectors and sellers have also been formalized to promote the sale of the GI products. All of these actors collaborated to establish the inter-professional association ADECAM, which manages the GI. Its objectives are: to facilitate coordination and not competition among producers; to increase the reputation of the GI on the coffee market; and to sensitize the local population about the importance of forest conservation. Economically, the GI's impact is important. The 2013/2014 campaign showed a price increase compared to the non-GI coffee (GRET, 2015). The first exported container (18 tons) in 2013 benefited from a premium of 13% compared to the Guinean coffee market price and 22% for the second container in 2015 (UNCTAD, 2015).

The internal control system implemented by ADECAM played a key role in increasing and guaranteeing the coffee quality, offering it a better place in the international market. The project and public recognition of the GI has enhanced collaborations and synergies with regards to export procedures, public funding (research and national projects) and development of a local coffee market for the GI. This local market has created job opportunities locally, with investment in local infrastructure and ecotourism development, allowing the promotion of other local products and handicrafts.

The importance of the quality linked to origin defined in the specification, the focus on local organization, the awareness raising among the citizens of the area through general assemblies and the market development have all worked together to contribute to local sustainable development.

Conclusions

As we explored in this chapter, the focus on a sustainable diet is often not at the forefront of value chain interventions – particularly if there is a focus on global value chains. Each empirical example explained how sustainability was defined in the context with some more focused on production (like in Mexico and Guinea) and others on consumption (in Costa Rica and Benin). However, some of the commonalities that we see are found by looking across value chains, rather than along them. As we illustrated in the beginning of the chapter, the literature has been focused mostly on value chains for tropical commodities and very little work has been completed on products that are needed for sustainable (and diversified) diets. For example, to be more sustainable could mean increasing the economic viability of local production while preserving traditional methods that are environmentally friendly. Alternatively, the focus may be on stimulating consumer interest in traditional or healthy food by offering direct contact with producers who are able to explain the importance of their sustainable practices. In all cases, there is a need to better strengthen the organizational arrangements that bring sustainable production and consumption practices together in specific territories as these arrangements provide the catalysts for action.

A distinctive feature of value chains that contribute to sustainable food systems, as illustrated in this chapter, is the social construction of an enriched range of attributes generally used to define food quality that goes beyond conventional attributes to include broader values such as tradition, identity, culture and/or local production (FAO, 2018a). Evidence shows that when a group of diverse actors operating in a specific territory generates new rules of interaction based on reciprocity, autonomy and an appreciation of different types of knowledge, they are able to build stable, inclusive and long-lasting market relationships both within the territory and outside of it (FAO, 2016). Some of the core activities that are part of these networks are based on the direct contact with consumers, either through farm visits, farmers' markets, direct sales or local supermarkets. Thus, where classic theories of value chains view the interactions through the lens of power struggles or transaction costs, these more recent approaches are beginning to recognize the interactions that can enhance better practices, overcome lock-in effects and contribute to more sustainable food systems. More research on how these territorial networks are organized and expand is needed in order to better be able to understand how diets can be made sustainable alongside the production and trade of food.

References

Agossou, G., Gbehounou, G., Nzamujo, G., Poisot, A.-S., Loconto, A. and Batello, C. (2016) Songhai model of integrated production in Benin. In: Loconto, A., Poisot, A.S. and Santacoloma, P. (eds) *Innovative Markets for Sustainable Agriculture: How Innovations in Market Institutions Encourage Sustainable Agriculture in Developing Countries*. Food and Agriculture Organization of the United Nations and Institut National de la Recherche Agronomique, Rome, Italy.

Andersen, M.S. (2007) An introductory note on the environmental economics of the circular economy. *Sustainability Science* 2, 133–140.

Aubry, C. and Kebir, L. (2013) Shortening food supply chains: a means for maintaining agriculture close to urban areas? The case of the French metropolitan area of Paris. *Food Policy* 41, 85–93.

Azofeifa, R. (2015) Ongoing experiences in Costa Rica: the Ecological Blue Flag Program. In: Meybeck, A. and Redfern, S. (eds) *Voluntary Standards for Sustainable Food Systems: Challenges and Opportunities*. A Workshop of the FAO/UNEP Programme on Sustainable Food Systems. Food and Agriculture Organization of the United Nations, Rome, Italy.

Barham, E. and Sylvander, B. (2011) *Labels of Origin for Food. Local Development, Global Recognition*. CAB International, Wallingford, UK.

Bowen, S. and Mutersbaugh, T. (2014) Local or localized? Exploring the contributions of Franco-Mediterranean agrifood theory to alternative food research. *Agriculture and Human Values* 31, 201–213.

Burlingame, B. and Dernini, S. (Eds) (2012) Sustainable Diets and Biodiversity: Directions and solutions for policy, research and action. *Proceedings of the International Scientific Symposium Biodiversity and Sustainable Diets United against Hunger*, 3–5 November 2010, FAO Headquarters, Rome. http://www.fao.org/docrep/016/i3004e/i3004e.pdf (accessed 28 August 2018).

Carter, C.R. and Rogers, D.S. (2008) A framework of sustainable supply chain management: moving toward new theory. *International Journal of Physical Distribution and Logistics Management* 38, 360–387.

Chiffoleau, Y. (2012) Circuits courts alimentaires, dynamiques relationnelles et lutte contre l'exclusion en agriculture. *Économie Rurale* 332, 88–101.

CIRAD-SAR (1996) *Systèmes Agroalimentaires Localisés: Organisations, Innovations et Développement Local*. Orientations et Perspectives Issues de la Consultation du CIRAD/Stratégies de Recherche dans le Domaine de la Socio-Economie de l'Alimentation et des Industries Agroalimentaires. CIRAD-SAR, Montpellier, France.

Ericksen, P., Stewart, B., Dixon, J., Barling, D., Loring, P., *et al.* (2010) The value of a food system approach. In: Ingram, J., Ericksen, P. and Liverman, D. (eds) *Food Security and Global Environmental Change*. Earthscan, London, UK.

FAO (2013) *Food Systems for Better Nutrition. The State of Food and Agriculture*. Food and Agriculture Organization of the United Nations, Rome, Italy.

FAO (2014) *Developing Sustainable Food Value Chains. Guiding Principles*. Food and Agriculture Organization of the United Nations, Rome, Italy.

FAO (2016) *Innovative Markets for Sustainable Agriculture: How Innovations in Market Institutions Encourage Sustainable Agriculture in Developing Countries*. Food and Agriculture Organization of the United Nations and Institut National de la Recherche Agronomique, Rome, Italy.

FAO (2017a) *The Future of Food and Agriculture. Trends and Challenges*. Food and Agriculture Organization of the United Nations, Rome, Italy.

FAO (2017b) *Reporte Semestral Proyecto TCP/MEX/3602 'Creación de Circuitos-Cortos de Comercialización de Productos Agropecuarios Ecológicos de la Zona Metropolitana del Valle de México (Documento de trabajo)'*. Food and Agriculture Organization of the United Nations, Mexico City, Mexico.

FAO (2018a) *Constructing Markets for Agroecology. An Analysis of Diverse Options for Marketing Products from Agroecology*. Food and Agriculture Organization of the United Nations, Rome, Italy.
FAO (2018b) *Strengthening Sustainable Food Systems through Geographical Indications: An Analysis of GI Economic Impacts*. Food and Agriculture Organization of the United Nations, Rome, Italy.
Fonte, M. (2013) Food consumption as social practice: solidarity purchasing groups in Rome, Italy. *Journal of Rural Studies* 32, 230–239.
Friedland, W.H. (2001) Reprise on commodity systems methodology. *International Journal of Sociology of Agriculture and Food* 9, 82–103.
Gereffi, G. and Korzeniewicz, M. (1994) *Commodity Chains and Global Capitalism*. Greenwood Press, Westport, Connecticut, USA.
Gobierno de la Ciudad de México (2016) Agricultura de Chinampas en la Zona Patrimonio Mundial Natural y Cultural de la Humanidad en Xochimilco, Tláhuac y Milpa Alta, Ciudad de México, México. Propuesta para la designación de sistema importante del patrimonio agrícola mundial (SIPMA) (Borrador de trabajo). Mexico City: Autoridad de la Zona Patrimonio Mundial Natural y cultural de la Humanidad en Xochimilco, Tláhuac y Milpa Alta.
Goodman, D., Dupuis, E.M. and Goodman, M.K. (2012) *Alternative Food Networks: Knowledge, Practice, and Politics*. Routledge, London, UK.
GRET (2015) Evaluation du Projet d'Appui à la Mise en Place des Indications Géographiques dans les Etats membres de l'OAPI (PAMPIG). In: Conseils, C.E. (ed.) *Rapport d'Evaluation pour l'Agence Française de Développement*. Agence Française de Développement, Paris, France.
Hinrichs, C.C. (2000) Embeddedness and local food systems: notes on two types of direct agricultural market. *Journal of Rural Studies* 16, 295–303.
Hopkins, T.K. and Wallerstein, I. (1986) Commodity chains in the world-economy prior to 1800. *Review* 10, 157–170.
Kaplinsky, R. and Morris, M. (2002) *A Handbook for Value Chain Research*. Institute of Development Studies, Brighton, UK.
Lernoud, J., Potts, J., Sampson, G., Garibay, S., Lynch, M., *et al.* (2017) *The State of Sustainable Markets – Statistics and Emerging Trends 2017*. International Trade Centre, Geneva, Switzerland.
Loconto, A. (2010) Sustainably performed: reconciling global value chain governance and performativity. *Journal of Rural Social Science* 25, 193–225.
Loconto, A. (2017) The values of value chains: Putting responsibility into action. In: Randles, S. and Laredo, P. (eds) *De-facto Responsible Innovation: Governance at Stake*. Edward Elgar, Cheltenham, UK.
Loconto, A. and Vicovaro, M. (2015) Constructing sustainable 'qualities' for local food systems in developing countries: The case of the Songhai Centre in Benin. Second International Conference on Agriculture in an Urbanizing Society: Reconnecting Agriculture and Food Chains to Societal Needs. Rome, Italy, 14–17 September 2015.
Loconto, A. and von Hagen, O. (2016) *Influencing Sustainable Sourcing Decisions in Agri-food Supply Chains*. ITC Trade Information Services Technical Paper. International Trade Centre, Geneva, Switzerland.
OECD and FAO (2016) *OECD-FAO Guidance for Responsible Agricultural Supply Chains*. OECD Publishing, Paris, France.
Ouma, S. (2015) *Assembling Export Markets : The Making and Unmaking of Global Food Connections in West Africa*. Malden, MA, John Wiley & Sons Inc., Chichester, UK.
Parker, G. (2005) *Sustainable Food? Tei-ki, Cooperatives and Food Citizenship in Japan and UK*. Working Paper in Real Estate and Planning. University of Reading, Reading, UK.
Ponte, S. and Gibbon, P. (2005) Quality standards, conventions and the governance of global value chains. *Economy and Society* 34, 1–31.
Porter, M.E. (1985) *Competitive Advantage: Creating and Sustaining Superior Performance*. Free Press, New York, New York, USA.
Porter, M.E. and Kramer, M.R. (2011) Creating shared value. *Harvard Business Review* 89, 62–77.
Potts, J., Lynch, M., Wilkings, A., Huppé, G.A., Cunningham, M. and Voora, V. (2014) *The State of Sustainability Initiatives Review 2014: Standards and the Green Economy*. IISD and IIED, Winnipeg, Canada and London, UK.
Raikes, P.L., Jensen, M.F. and Ponte, S. (2000) *Global Commodity Chain Analysis and the French Filière Approach: Comparison and Critique*. Centre for Development Research, Copenhagen, Denmark.
Renting, H., Marsden, T.K. and Banks, J. (2003) Understanding alternative food networks: exploring the role of short food supply chains in rural development. *Environment and Planning A* 35, 393–411.
Renting, H., Schermer, M. and Rossi, A. (2012) Building food democracy: exploring civic food networks and newly emerging forms of food citizenship. *International Journal of Sociology of Agriculture and Food* 19, 289–307.

SAI (2013) Sustainable Sourcing of Agricultural Raw Materials: A Practitioner's Guide. Brussels: The Sustainable Agriculture Initiative (SAI) Platform, the CSL learning platform of IMD's Global Center for Sustainability Leadership (IMD-CSL), the International Trade Centre (ITC), and the Sustainable Trade Initiative (IDH). Supporters are BSR, the Sedex Information Exchange (Sedex) and the Sustainable Food Laboratory (SFL).

Srivastava, S.K. (2007) Green supply-chain management: A state-of-the-art literature review. *International Journal of Management Reviews* 9, 53–80.

UNCTAD (2015) *Why Geographical Indications for Least Developped Countries* (LDCs)? UN Conference on Trade and Development, Geneva, Switzerland.

Vandecandelaere, E. (2016) Geographical indications: a tool for supporting sustainable food systems. In: Arfini, F., Mancini, M. C., Veneziani, M. and Donati, M. (eds) *Intellectual Property Rights for Geographical Indications: What is at Stake in the TTIP?* Cambridge Scholars Publishing, Cambridge, UK.

23 Sustainable and Healthy Gastronomy in Costa Rica: Betting on Sustainable Diets

Marcela Dumani Echandi, Patricia Sedó Masis, Randall García Viquez and Roberto Azofeifa Rodriguez

Abstract

Current consumption patterns in Costa Rica and in other countries are driven by misinformation and lack of knowledge concerning the nutritional value and sustainability of food products. Unhealthy diets are a major reason for health problems, environmental degradation and food biodiversity loss. To reverse this trend, in 2012 Costa Rica launched The National Plan on Healthy and Sustainable Gastronomy as a multi-stakeholder initiative with participation of public and private sectors. In the framework of the plan and in collaboration with international partners, Costa Rica also promotes the initiative 'Healthy and Sustainable Gastronomy' as a key driver for sustainable food systems. It is an innovative paradigm for the sustainability of natural resources based on the consumers' decision to prepare and enjoy food to be healthy. Sustainable and healthy gastronomy is alluded to by considering social, environmental and economic aspects along the entire production, marketing, service and consumption chain; and healthy in terms of the greatest concern for the nutritional situation of the population and the quality of food, whether prepared at home or offered in gastronomic establishments. In a country where nature is a key part of its brand and identity, a healthy and sustainable gastronomy is part of a new paradigm of sustainable development based on agroecology and the efficiency of agri-food systems.

Introduction

Current consumption patterns in Costa Rica and in other countries are driven by misinformation and lack of knowledge concerning the nutritional value and sustainability of food products. Unhealthy diets are a major reason for health problems, environmental degradation and food biodiversity loss. Following this trend, the future of food would face disappearance of endemic, native and local edible plants that often have high nutritional value and have traditionally provided the basis of national cuisines.

To reverse that trend, in 2012 Costa Rica launched The National Plan on Healthy and Sustainable Gastronomy as a multi-stakeholder initiative with participation of public and private sectors. While the plan has triggered important progress in this area, lack of information and awareness about linkages connecting food production and consumption is a major barrier for improving population health and invigorating the sustainable food systems.

In the framework of the plan and in collaboration with international partners, Costa Rica also promotes the initiative 'Healthy and Sustainable Gastronomy' (the initiative) as a key for sustainable food systems. Its goal is to promote, at the international level, a new healthy and sustainable cuisine as a driving force for the diversification of family farming systems, and as a new approach for the development of agriculture. The Initiative is an innovative paradigm for the sustainability of natural resources based on the consumers' decision to prepare and enjoy food to be healthy.

Food and Nutrition in Costa Rica

Food is the primary link between human beings and their environment. It is also a key component of the economic, social and cultural fabric of communities and countries. The process of food and nutrition security is a continuum among environment, production, distribution, selection, transformation and consumption. The sustainability of these factors has a direct effect on human wellbeing in terms of health, nutrition and longevity.

Costa Rica occupies 0.03% of the terrestrial surface and houses almost 4% of the species of plants and animals on the planet. It is estimated to be the most biodiverse country in the world per square kilometer and it is part of Mesoamerica, one of the World Biodiversity Centers. The abundant life zones, the presence of important forests and wetlands and the extension of its seas allow the production of a wide range of foods. An important variety of introduced species has been added to its own biodiversity, becoming part of the common food stock. Of the 12,000 species of native plants in Costa Rica, about 500 have been used as food. Unfortunately, most of the edible native species are no longer utilized due to the loss of knowledge regarding their use, acculturation and the loss of agrobiodiversity. There are exceptions in some regions such as indigenous territories where people continue using native species (Trujillo, 2014).

Although Costa Rica is one of the countries with the highest life expectancies in Latin America, and has experienced a significant reduction in malnutrition, recent data indicates an increasing prevalence of overweight and obesity in the general population, with worrying levels in children and adolescents. This is mainly due to changes in diet habits by increase in consumption of meals outside home and processed food; and reduction of vegetable and fruit consumption below recommended levels. This has been accompanied by a significant increase in the epidemiological profile of chronic non-communicable diseases linked to nutrition, and sustained micronutrient deficiency, or what is known as hidden hunger. This is compounded by the growing social inequality that poses greater vulnerability of certain sectors to food insecurity, and a high dependence of the country on the importation of staple foods.

According to several studies carried out in Costa Rica, including the National Nutrition Survey 2008–2009, the Costa Rican diet is little varied; high in consumption of food source of flour, fat and sugars, and little consumption of vegetables and fruits despite having a high availability of these products in the country. The basic dish is rice and beans, although in the last 25 years there has been a trend towards reducing the consumption of beans without an adequate substitution of a food with similar nutritional characteristics, which has impacted on the nutritional quality of basic food.

The reduction in the variety of foods observed in the Costa Rican diet in recent years has also been favored by the increase in urbanization and the dedication of the economically active population to tasks that are located far from rural areas and agriculture. A growth in the number of families that depend directly on purchasing rather than producing food has occurred, thus influencing in an increasingly important way the purchasing power of families and with it, the possibility of access to varied and nutritious foods.

A study conducted in communities in Alajuela and Heredia in 2014–2015 showed that young people have difficulty identifying crops in the field, since they have become accustomed to purchasing processed products in commercial places such as supermarkets. The loss of ancestral knowledge to make use of the naturally available food resources, as well as to cultivate traditional food species, combined with the lack of interest in cooking are aspects identified by older adults as factors affecting the continuation of food traditions in the communities studied (Sedó and Solano, 2014).

Advances and Challenges Towards Sustainable Food Systems in Costa Rica

With the goal of achieving sustainable food systems and promoting healthier food with higher nutritional value produced locally, several initiatives have been developed which integrate production and marketing of food, good environmental practices and education for consumers' health.

Maintaining the country's rich biological diversity requires good agricultural practices from the farm to the processing plant, reduction of environmental damages, adequate productivity and safe and nutritious food for people. Since the early 1990s, both the private and public sectors have been advancing towards more efficient and environmentally responsible production systems, with a significant increase in organic production. This has been stimulated by various tools, including the Program for the Promotion of Sustainable Agricultural Production which between 2004 and 2010, achieved a reduction of soil erosion in 80% of the farms and a 50% reduction in water pollution (EPYPSA, 2010). Other important efforts address climate change, especially the goal of becoming a carbon neutral country by 2021 and meeting emissions reduction targets set for 2030 (MINAE, 2015).

Efforts at the local level, such as organic fairs, farmers' markets, wholesale markets and thematic fairs are being made to improve the

Fig. 23.1. Local farmers' market. Tibás, San José, Costa Rica.

population's access to fresh and diverse foods in a way that favors both local producers and consumers economically.

With regard to sustainable consumption and production, significant efforts are being made by the public and private sectors to promote healthier diets and consumption of fresh and varied foods. These include educational programmes by the Ministries of Education and Health and the Costa Rican Social Security Fund, Dietary Guidelines, promotion of the value of food culture by the Ministry of Culture and other bodies, Inter-institutional Program for food losses and waste reduction, and an ecosystem services recognition programme, among others.

Harnessing food diversity, strengthening food culture and improving people's eating habits are part of the major challenges the country must confront along the path to inclusive and low carbon development.

Towards a Healthy and Sustainable Gastronomy

One of the objectives of the plan is to achieve a more sustainable gastronomy in Costa Rica to enhance sustainable diets and production. The gastronomy sector is part of the country's cultural environment and is an important factor in shaping the eating habits of the population and what is considered the norm. Influenced by this environment and the information they receive, consumers are tempted to adopt unsustainable diets. Availability and positioning of sustainable and healthy choices as the new norm can reverse the unfortunate current trends. This has to be accompanied by education as a way to generate awareness, knowledge and skills to disseminate new values and behaviors.

Promoting a healthy and sustainable gastronomy through the plan will shape a new national cuisine containing more local, plant-based products (especially endemic edible plants), influence consumers' eating habits and strengthen opportunities for sustainable family farming. It can inspire solutions to the challenge of food biodiversity loss due to the disappearance of endemic, native and local high nutritional value edible plants, stimulate the consumption of sustainable and more plant-based diets, and enhance the image of local gastronomies, traditional cuisine and seasonal food products. It can address soil conservation, sustainable farming, efficient food distribution and reduction of food losses and waste. Finally, it can also contribute to the national and global debates on the relevance of sustainable diets for consumer health and on the sustainability of agricultural production. As a result, a healthy and sustainable gastronomy can be the paragon for the country's food culture.

Sustainable and Healthy Gastronomy: the Concept

Sustainable and healthy gastronomy is alluded to by considering social, environmental and economic aspects along the entire production, marketing, service and consumption chain; and healthy in terms of the greatest concern for the nutritional situation of the population and the quality of food, whether prepared at home or offered in gastronomic establishments.

Costa Rican cuisine is a manifestation of the various foods of the country, with certain general characteristics as well as local nuances. It is based on the pre-Hispanic indigenous diet enriched throughout history with the broad dietary diversity of the country and ingredients and preparation techniques brought over time by different migrant groups. This has resulted in foods and beverages that are used either daily or only in a festive context through which Costa Ricans demonstrate their own identity.

Both the national population and international tourists consume Costa Rican cuisine,

Fig. 23.2. Chicasquil leaves (*Cnodoscolus chayamansa*), used in traditional Costa Rican food (a 'picadillo').

thus similar foods are found on the family table as well as in popular public places and restaurants. According to the FAO, 'sustainable consumption and production of food and agriculture is a holistic concept, driven by the consumers, which refers to the integrated implementation of sustainable patterns of consumption and production of food, respecting the capacities of natural ecosystems'. In this context, a healthy and sustainable gastronomy is expected to support the country's agro-biodiversity conservation, use biodiversity to strengthen gastronomic identity, improve the quality of its products, increase the availability of local or regional food at fair prices, and favour the livelihoods and income of families in rural areas.

A major challenge for Costa Rica is the appropriation of a healthy and sustainable gastronomy under a broad concept that transcends restaurants to also reach the homes, centers of study and work, which promotes practices, techniques and ingredients of the traditional cuisine while generating innovative proposals based on the available diversity of agricultural, aquaculture and marine food species. The plan also aims to contribute to food and nutritional security of producers and consumers. In the context of tourism, one of the country's main economic activities, having a cuisine that is also a sustainable tourism product can enhance local destinations and differentiate the country as a whole.

The plan, promoted by public and private actors over the last five years, was declared by the Government to be of public interest on March 17, 2015 (Executive Decree # 38939-S-MAG-MEIC-C). It involves efforts by public institutions (Ministries of Agriculture, Tourism, Presidency, Health, Economy, Labor and Culture), local governments, universities, producer and consumer organizations, and representatives of the private sector. The plan is creating an ideal environment to propose new projects and initiatives or revitalizing existing ones. It stimulates the creation of new ventures, the diversification of production systems, the development of human resources and the promotion of personal development. It offers opportunities for greater well-being and quality of life of the participants, whether in production, distribution, marketing and consumption. It seeks to ensure more employment with better, dignified working conditions, greater opportunities for women and youth, the integration of marginalized and at-risk people, while also helping to combat poverty, renew interest in agriculture, and reduce migration to the city (Azofeifa, 2014).

While it is true that the plan is promoting a major gastronomic change through the creative use of locally available food resources, new steps must be taken to motivate the families, especially the youth, to adopt ingredients not typically consumed, although they might be available in their communities. A significant and permanent change in the diet of communities and its impact on their quality of life requires the implementation of projects and programs through the coordinated work of institutions, organizations, producers and families. The plan requires actions by municipalities, communities, producers, managers of institutional canteens, and gastronomic businesses, among others, to develop networks and disseminate information on locally available sustainable food resources.

In a country where nature is a key part of its brand and identity, a healthy and sustainable gastronomy is part of a new paradigm of sustainable development based on agroecology and the efficiency of agri-food systems, with a view to achieving important national goals:

1. Conserve and sustainably produce the species of natural and cultivated biodiversity, relevant for food production and for the nutrition of the population.
2. Encourage the use of production practices based on technologies that favor the reproduction of agroecology cycles.
3. Promote the development of local markets for fresh food, with social and environmental added value, that invigorates local economies.
4. Facilitate spaces for the participation of family agriculture in the value chains of sustainable agri-food systems.
5. Promote good consumer practices oriented towards healthy eating, the use of seasonal foods and the reduction of food losses and waste.
6. Promote the Costa Rican food culture, strengthening the national identity and the development of a differentiated gastronomy that promotes the supply of local food and tourism.
7. Promote innovation in the use of food species and the diversification of menus based on local products from sustainable production systems.
8. Contribute to the food and nutritional security of the population through greater availability, access, consumption and biological utilization of food.

To achieve these goals, the plan works on the following deliverables: a training programme on sustainable diets; sustainability dialogues; a communication campaign in the gastronomy sector; an inventory of endemic, native and organically grown relevant species of food at the regional and national level; and gastronomy laboratories for innovation and dissemination of options for consumption of endemic, native and organic products. As a proposal for organic and sustainable family farming, the plan is an important driver of sustainable production and consumption that stimulates the creation of agri-chains and enables diverse public and private actors to work together (tourism, agriculture, agricultural trade, health and nutrition, culture, environment, biodiversity, others). In spite of the interesting progress made by so far, there are various challenges that the plan needs to address for its full implementation. The most significant are:

- Consumer education and motivation. This requires an on-going effort of information dissemination on the benefits of the production and consumption of sustainable products. Public and private investment is needed.
- Traceability systems. They help demonstrate to consumers that the food on their table is the result of sustainable production in family farms.
- Nutritional information for consumers. Research is required to evaluate the nutritional properties of menus prepared with sustainable ingredients.
- Ecolabelling and standards for sustainable products. They need to be easy and inexpensive to implement and recognized and trusted by consumers.
- Demand and supply. There is a need to establish an information system to promote value chains and enable production to satisfy demand.
- Role of other actors in the food chain. Suppliers need to adopt best practices for transparency and management of food products to guarantee quality and innocuousness.

The desired change described by the partners is:

> All Costa Rican citizens, and those who visit the country, enjoy a healthy, adequate, sustainable, innovative and quality food, with cultural identity; (the Plan) promotes consumption, harvesting and trading of locally produced food that favours family farming, protects and regenerate the ecosystem and incentivize solidarity and circular economies, contributing to just and full development in each part of the national territory.
>
> (Azofeifa, 2014)

Impacts at the International Level

The impact of the plan has taken an international scope through the Sustainable and Healthy Gastronomy Initiative promoted by Costa Rica under the umbrella of the One Planet Sustainable Food System (SFS) Programme, which is part of the One Planet Network. The initiative is motivating other countries and organizations to commit to similar proposals adapted to their own circumstances. In Costa Rica, the Ministry of Agriculture and Livestock has undertaken the initiative in collaboration with Hivos (Humanist Institute for Cooperation), IFOAM (International Federation of Organic Agriculture Movements), CACORE (Costa Rican Chamber of Restaurants), Smaackmakers (Dutch NGO) and INBio (National Biodiversity Institute). Furthermore, the Initiative contributes to several of the United Nations Sustainable Development Goals (SDGs):

- It improves food security and nutrition and promotes sustainable agriculture (SDG 2).
- It focuses on sustainable consumption towards more plant-based diets and production patterns through the gastronomy sector (SDG 12).
- Increased demand for sustainably produced food contributes to climate change mitigation and adaptation (SDG 13), and to sustainable land management and biodiversity conservation (SDG 15).
- It fosters economic growth at local, national and regional levels, and also seeks to increase diverse and local gastronomy offers within the sustainable tourism industry (SDG 8).
- It has an international coalition of partners (SDG 17) including national and international non-governmental organizations, the Government of Costa Rica in this initial phase as well as other governments in the follow-up phases.

Other countries can follow the example of Costa Rica. With an international scope, the initiative provides an opportunity to bridge efforts in emerging markets and Western countries, to foster sustainable lifestyles in food production and consumption. Part of the challenge for stimulating sustainable diets is to link complementary existing approaches and the adaptation of this new paradigm to different cultural contexts. Therefore, it is important to stimulate other countries and programmes to partner with the Initiative. Some ideas for partnerships are: the initiative as a subject for research (on the effectiveness of this methodology); the initiative as an opportunity to promote and further implement sustainable diets; and the Initiative as a component of projects aimed at reducing food loss and waste.

References

Azofeifa, R. (2014) National Plan of Sustainable and Health Gastronomy: enabling conditions for organic and sustainable family farming. Proceedings of the workshop on Knowledge and Information for Sustainable Food Systems. FAO and UNEP Programme on Sustainable Food Systems. Rome, Italy, pp. 93–96.

Chinnock, A. and Leon, M. (2013) Reported morbidity related to nutrition in Costa Rica. Available at http://www.scielo.sa.cr/pdf/rcsp/v22n2/art08v22n2.pdf (accessed 4 October 2017).

EPYPSA (2010) Evaluación final del Programa de Fomento de la Producción Agropecuaria Sostenible. Available at http://www.mag.go.cr/biblioteca_virtual/bibliotecavirtual/a00243.pdf (accessed 2 October 2017).

González, A.R. (2012) La alimentación tradicional costarricense: propuestas para su revitalización Available at https://www.ministeriodesalud.go.cr/index.php/material-educativo/tradiciones-alimentarias-y-alimentos-sub-utilizados/2503-la-alimentacion-tradicional-costarricense-propuestas-para-su-revitalizacion/file (accessed 5 October 2017).

MINAE (2015) Contribución prevista y determinada a nivel nacional de Costa Rica. Available at http://cambioclimaticocr.com/recursos/documentos/biblioteca/indc-costa-rica-version-2-0-final-es.pdf (accessed 2 October 2017).

MINSA (2009) Encuesta Nacional de Nutrición Costa Rica 2008-2009. Available at http://www.paho.org/cor/index.php?option=com_docman&view=download&category_slug=alimentacion-y-nutricion&alias=67-encuesta-nacional-de-nutricion-costa-rica-2008-2009&Itemid=222, Censo Escolar de peso y talla http://www.mep.go.cr/sites/default/files/page/adjuntos/informe-ejecutivo-censo-escolar-peso-cortofinal.pdf (accessed 2 October 2017).

Museo de Cultura Popular UNA y Escuela de Nutrición UCR (2015) Informe Final Proyecto Interuniversitario fortalecimiento de la seguridad alimentaria y nutricional en la población costarricense mediante la puesta en valor de las prácticas alimentarias tradicionales de productos vegetales subutilizados y su innovación en la preparación y consumo. Alianza Museo de Cultura Popular, Universidad Nacional/Escuela de Nutrición Universidad de Costa Rica, Fondos Concursables Consejo Nacional de Rectores, San José, Costa Rica.

Sedó, P. and Solano, M. (2014) Sembrar y comer alimentos con historia, sabor y nutrición. Reflexiones y propuestas para fortalecer la seguridad alimentaria y nutricional local. Escuela de Nutrición UCR, Museo de Cultura Popular UNA, San José, Costa Rica.

Trujillo, C.L. (2014) Plantas útiles de las fincas cacaoteras de indígenas Bribrí y Cabécar de Talamanca, Costa Rica. MSc thesis. Centro Agronómico Tropical de Investigación y Enseñanza, Turrialba, Costa Rica.

24 How Organic Food Systems Support Sustainability of Diets

Johannes Kahl, Carola Strassner, Susanne Bügel, Denis Lairon and Flavio Paoletti

Abstract

Sustainable models are needed in order to transform the current food systems. This chapter presents arguments for using organic food systems as such a model. Food systems can be recognized as coupled human and natural systems, with a set of activities and outcomes in which the boundaries of the system can be defined. This chapter takes sustainability as an inherent property of a food system. The identification of 'enabling mechanisms' from the organic food system actors' perspective may give insights to drivers and factors shaping food systems towards enhanced sustainability. Organic food systems are driven by both codified principles and value-based ethical and personal responsibilities. Organic production practices are specified in international and national standards and regulations, and are undergoing continual transformation processes. Organic farming is currently practised in 172 countries by 2.3 million producers, and consumer demand is documented by a present market size of US$80 billion, thus a lot of experience and data are already available. As organic farming provides a whole range of ecosystem services and promotes biodiversity, it may contribute to environmental sustainability. Studies have shown that consumers who regularly buy and eat organic food seem to link health and environmental sustainability through their food choices. This indicates that the organic food systems may also contribute to sustainable diets in theory and practice.

Introduction

The present global food system is facing a number of challenges (Global Panel on Agriculture and Food Systems for Nutrition, 2016). These challenges have to be addressed when transforming food systems. Food systems have major outcomes such as food security, ecosystem services and social welfare (see Eriksen *et al.*, 2008). Food systems may therefore be evaluated and assessed as to how far they address the global challenges through their outcomes (Ingram, 2011). The interlinkages between the challenges and the system outcomes have been addressed in various recently published reports. The impact of the current global food system on resources and environment has been quantified by the United Nations Environment Programme (UNEP, 2016). Special attention is given to the impact on climate change (FAO, 2015). The impact of the current global food system on food security, addressed by various reports, documents undernutrition, malnutrition and an increase in related non-communicable diseases in various countries (e.g. Gillespie and van den Bold, 2017). Indeed, sustainable models for transforming current food systems are needed to guide interventions (Global Panel, 2016; iPES Food, 2015). Addressing the goals of sustainable development, actions are needed for interventions within the entire food system including inputs and production, processing, storage, transport and retailing, consumption, policymaking and education. Special attention is drawn to (re-)connecting the different actors operating within food systems (Gillespie and van den Bold, 2017) as well as to taking actions on different (spatial) levels with a focus on alternative food systems (de Shutter, 2014). For analysis and modelling of the system's transformations, researchers need to engage in interdisciplinary work from a holistic view to provide appropriate solutions to present challenges (iPES Food, 2015; FAO, 2017). Taking a system lens to organic agriculture and food production and consumption, this chapter delivers lessons learned from the organic food system model (Kahl *et al.*, 2016).

A System Approach to Food and Farming

A food system can be defined as a set of activities and outcomes that makes up the boundaries of the system (e.g. Eriksen, 2008; Ingram, 2011; Grant, 2015). For Ostrom (2009), the challenge of taking a system approach to food is to identify and analyse relationships among multiple levels of these complex systems at different spatial and temporal scales. The UNEP (2016) report linked global challenges of food systems, a system approach and the sustainable developmental goals. Allen and Prosperi (2016) take sustainability as an inherent property of a system 'that is open to interactions with the external. It is the dynamic preservation, over time, of the intrinsic identity of the system among perpetual changes'. Vergragt *et al.* (2014) identified 'enabling mechanisms' towards sustainable food systems from the actors' perspective as well as identified the main drivers of food systems as 'our (perceived) needs or wants, driven by our values'. As we propose an organic food system as a values-based sustainable food system model, we will describe those values as an essential part (De Boer *et al.*, 2007). Such values seem to be underlying principles of both sustainable food systems (HLPE, 2014) as well as the sustainable diet definition (Burlingame & Dernini, 2012). There is a strong interlinkage between food systems and diets, indicating sustainable food systems as prerequisites for sustainable diets (Moomaw *et al.*, 2012; Meybeck and Gitz, 2017). As dietary patterns co-shape the sustainability of food systems, behavioural changes are needed for healthier and more sustainable food choices (Stehfest, 2014). Therefore, it is a challenge how to shift eating habits and food choices towards enhanced sustainability.

Organic Food Systems as Models for Sustainable Food Systems in Transition

We propose the organic food system as a potential contributor to enhance sustainable diets (Strassner *et al.*, 2015; Reaganold and Wachter, 2016). The term 'organic' agriculture was coined in the first half of the 20th century. Since then, it has been used in a number of documents at international and national level, in scientific papers and has become common use in language (e.g. Vogt, 2007). The International Federation of the Organic Agriculture Movements (IFOAM) defined the worldwide principles for organic agriculture: principle of health, ecology, fairness and care (IFOAM, 2008). These principles can be taken as the shared mission values of organic agriculture and food systems on a global level (Alrøe and Kristensen, 2004). They are to be used as a whole (Daugbjerg and Botterill, 2012), and act as the basic orientation to guide the attitude and behaviour of the organic food system actors (Esmer and Pettersson, 2007). Operation values as integrity, trust and transparency seem to be shared among organic food system actors (Hertwig *et al.*, 2017). Actors of the organic food system can be identified and characterized across different scales and on different levels and based on shared values, as well as following codified principles along the food production chain. Organic principles are codified in international and national standards and regulations (Niggli, 2015). The standards are mandatory for system actors on farms and during processing and distribution. To protect against misuse, organic food production, processing and distribution is regulated via laws and controlled with audits in more than 80 countries (Willer and Lernoud, 2017). The certification under various schemes mandated by law also give organic products a legal meaning.

Farming activities are within the frame of organic standards and regulations on different levels (e.g. Bellon and Perven, 2014). This frame gives low input in terms of fertilization and pest control and avoids synthetic materials; furthermore, there is minimal use of antibiotics in animal husbandry and a ban on genetically modified organisms (GMOs). An activity especially following from the codified principles is the certification process. Organic processing regulation in the European Union strictly limits the use of food additives (only 48 are allowed). The number is furtherly reduced (to less than 10) according to some private standards (Kahl *et al.*, 2014). In many countries with a domestic organic market, activities in public procurement and services are part of the organic food system (Caldeira *et al.*, 2017). Strassner *et al.* (2016) give an overview of sustainable HORECA including organic perspectives and case studies in different countries.

The organic food market is increasing worldwide (Willer and Lernoud, 2017). This indicates that the organic food system model is not limited to any geographical or cultural domain. Several studies have been performed with the aim to identify the drivers of the organic food choice by consumers (Padel and Foster, 2005; Hjelmar, 2011; Stolz *et al.*, 2011). Differences between regular and occasional organic food consumers were also studied (Janssen and Hamm, 2012; Pino *et al.*, 2012). Ethical motivations act as driving forces of consumers' purchasing intentions (Eden, 2009; Michaelidon and Hassan, 2008), particularly for those who consume organic food regularly (Janssen and Hamm, 2012; Pino *et al.*, 2012). 'Egoistic' motives such as health consciousness and 'altruistic' motives such as environmental concern are the main determinants of the organic food choice by consumers (Chyssohoidis and Krystallis, 2005; Torjusen *et al.*, 2004; Kriwy and Mecking, 2012; Zagata, 2012; Kesse-Guyot *et al.*, 2013; Kareklas *et al.*, 2014). The main drivers to buy organic food seems to be values based (Michaelidon and Hassan, 2008; Eden, 2009; Pino *et al.*, 2012). Ethical issues and personal responsibility seem to be a major trajectory (Chryssohoidis and Krystallis, 2005; Padel and Foster, 2005). Studies have shown that consumers who regularly buy and eat organic food seem to follow healthy and more sustainable consumption patterns based on plant foods (Eisinger-Watzl *et al.*, 2015; Baudry *et al.*, 2016a, b; Kesse-Gyuot *et al.*, 2017), and seem to link health and sustainability through their food choices (Kesse-Gyuot *et al.*, 2013) without fully understanding the association or even the causality. There are various investigations trying to understand motivations and behaviour of regular and occasional organic consumers in different geographical and cultural conditions, indicating that organic consumption activities are linked to a certain degree of responsibility as well as consciousness of the actors (Pearson *et al.*, 2007, 2011; Janssen and Hamm, 2011; Kareklas *et al.*, 2014; Hemmerling *et al.*, 2015; Bashaa *et al.*, 2015).

As the organic food system is based on a set of values and codified principles, there is an intrinsically intended outcome of the system that can be identified through different documents. Organic food systems outcomes have been assessed in a number of studies. Niggli *et al.* (2011) reviewed various reports and scientific papers on environmental, social and economic impacts of four different certification schemes including organic. Compared to the other labels, the number of studies available on organic standards was significantly higher, thus indicating a higher level of documentation and data for organic agriculture. Numerous studies have compared the environmental, social and economic impacts of organic agriculture and food production with those of other farming systems (Mondelaers *et al.*, 2009; Gomiero *et al.*, 2011; Lynch *et al.*, 2012; Tuomisto *et al.*, 2012; Reaganold and Wachter, 2016; Clark and Tilman, 2017). General conclusions are difficult to draw from the results of these studies that widely differ for many aspects (farm scale, climatic conditions, inputs, etc.). Moreover, most of these studies have been performed at farm level. Therefore, the impacts of the whole food system have not been taken into account. Organic farms generally have more plant diversity, greater faunal diversity (insects, soil fauna and microbes, birds) and often more habitat and landscape diversity (Bengsston *et al.*, 2005; Crowder *et al.*, 2010; Kennedy *et al.*, 2013; Tuck *et al.*, 2014). These effects vary with the type of organism, crop, farm size, management within the farm and management of surrounding farms, and so. (Bengtsson *et al.*, 2005; Belfrage *et al.*, 2005; Gabriel *et al.*, 2010; Tuck *et al.*, 2014). From a comparison between organic and non-organic high-input systems, yield averages are 5–34% (or 8–25% as reported in Reganold and Waechter, 2016) lower in organic systems. However, with certain crops and the adoption of best practice, the yield gap can be reduced to 13% and even less (Seufert *et al.*, 2012; Ponisio *et al.*, 2015). Moreover, the organic to conventional yield gap can be reduced to about 9% by applying diversification practices, such as multi-cropping and crop rotation (Ponisio *et al.*, 2015). When assessing the impact of the production method on the food itself, most studies compared foods delivered from organic production to those from non-organic production. These comparisons are mainly based on measurements of the content of food constituents reflecting food safety and nutrition quality issues. The overall conclusion from the meta-analysis studies available so far is that because of the farming methods and feeding regimes, organic food contains significantly less pesticide and antibiotic residues, whereas food constituents show higher levels of some secondary plant metabolites and fatty acids as well as lower levels of proteins or cadmium (Mie *et al.*, 2017). Implications on human health indicate less exposure of workers to pesticides (Reganold and Wachter, 2016), but its direct effects on human health have not yet been not investigated (Brantsæter *et al.*, 2016; Mie *et al.*, 2017). Taking organic level in the diets, organic consumers who regularly buy and consume organic food are healthier with markedly less overweight and obesity (Kesse-Guyot *et al.*, 2013; Eisinger-Watzl *et al.*, 2015) as well as following recommended healthier dietary patterns (Kesse-Guyot *et al.*, 2013; Baudry *et al.*, 2016a, 2016b). Other studies reported reduced allergy prevalence in children following an organic diet (Alfven *et al.*, 2006) or a reduced risk of having a metabolic syndrome in adults (Baudry *et al.*, 2017a, b).

Organic food and farming is also part of dietary concepts, which have been and are still taken as examples of sustainable diets.

The traditional Mediterranean diet pattern, a regional plant-based one, has long been acknowledged for its protective health effect, and more recently, for its potential sustainability (Dernini *et al.*, 2017). Until the 1950s it was a traditional/natural/organic food system of production and consumption. To take into account the tremendous changes in food production that have occurred over the years, the recently updated international recommendations for Mediterranean diet for today (Bach-Faig *et al.*, 2011) stress the consumption of 'eco-friendly foods', in practice organic foods, to keep the highest nutritional quality and avoid toxic pesticides residues or GMO crops. A recent cohort study (Seconda *et al.*, 2017) has shown that consumers combining a Mediterranean dietary pattern mainly made of organic foods exhibit the best sustainability scoring for some dedicated sustainability indicators such as diet nutritional quality, plant/animal protein ratio and body mass index, thus highlighting the synergy between the two combined dimensions.

The New Nordic Diet (NND) is a chef-driven constructed diet, building on four key principles: Nordic Identity, gastronomy, sustainability and health. Taking all four principles into account eight dietary guidelines have been suggested (Mithril *et al.*, 2012, 2013). The idea is that the food should be produced locally, be organic, mainly plant based and of high quality and high biodiversity (Mithril *et al.*, 2012, 2013; Bügel *et al.*, 2016). The first scientific studies looking at the health effects of the NND suggest that the diet may be a suitable alternative for areas that have cultural difficulties adhering to the Mediterranean diet (Adamsson *et al.*, 2014; Poulsen *et al.*, 2014; Lankinen *et al.*, 2016).

Conclusions

We have described the organic food system as a sustainable food system model in transition. Organic food system actors share specific values. It remains to be investigated how far current trends in the organic food system towards globalization, multinationals, and so on, may influence how actors follow these values further and set their activities. Furthermore, we identified mission values of food chain actors such as integrity, transparency and trust, which seem crucial towards other system actors such as consumers and policy makers. What makes the organic food system different from others is that the principles are codified and further protected through control mechanisms along the food chain from farm to the point of purchase. The organic food system offers a wide range of documentation and data available for all scales and levels. It makes the organic food system valuable for studying, analysing and using it as a 'living laboratory' on how to develop interventions to make food systems more sustainable.

References

Adamsson, V., Cederholm, T., Vessby, B. and Risérus, U. (2014) Influence of a healthy Nordic diet on serum fatty acid composition and associations with blood lipoproteins – results from the NORDIET study. *Food Nutrition Research* 58, 24114.

Alfven, T., Braun-Fahrlander, C., Brunekreef, B., von Mutius, E., Riedler, J., *et al.* (2006) Allergic diseases and atopic sensitization in children related to farming and anthroposophic lifestyle – the PARSIFAL study. *Allergy* 61(4), 414–421.

Allen, T. and Prosperi, P. (2016) Modeling sustainable food systems. *Environmental Management* 57, 956–975.

Alrøe, H.F., Kristensen, E.S. (2004) Basic principles for organic agriculture: why? And what kind of principles? *Ecology and Farming*, Special issue on Principles of Organic Agriculture.

Bach-Faig, A., Berry, E.M., Lairon, D., Reguant, J., Trichopoulou, A., *et al.* (2011) Mediterranean diet pyramid today: science and cultural updates. Mediterranean Diet Foundation Expert Group. *Public Health Nutrition* 14(12A), 2274–2284.

Bashaa, M.B., Mason, C., Shamsudinc, M.F., Hussain, I.H. and Salem, M.A. (2015) Consumers' attitude towards organic food. *Procedia Economics and Finance* 31, 444–452.

Baudry, J., Touvier, M., Allès, B., Péneau, S., Méjean, C. and Galan, P. (2016a) Typology of eaters based on conventional and organic food consumption: results from the NutriNet-Sante cohort study. *British Journal of Nutrition* 116(4), 700–709.

Baudry, J., Allès, B., Péneau, S., Touvier, M., Méjean, C., *et al.* (2016b) Dietary intakes and diet quality according to levels of organic food consumption by French adults: Cross-sectional findings from the NutriNet-Santé Cohort Study. *Public Health Nutrition* 20(4), 638–648. DOI:10.1017/S1368980016002718

Baudry, J., Lelong, H., Adriouch, S., Julia, C., Allès, B., *et al.* (2017a) Association between organic food consumption and metabolic syndrome: cross-sectional results from the NutriNet-Santé study. *European Journal of Nutrition* Aug 2. DOI: 10.1007/s00394-017-1520-1

Baudry, J., Peneau, S., Alles, B., Touvier, M., Hercberg, S. and Galan, P. (2017b) Food choice motives when purchasing in organic and conventional consumer clusters: focus on sustainable concerns (The NutriNet-Sante Cohort Study). *Nutrients* 9(2), E88.

Belfrage, K., Björklund, J. and Salomonsson L. (2005) The effects of farm size and organic farming on diversity of birds, pollinators, and plants in a Swedish landscape. *Ambio* 34, 582–588.

Bellon, S. and Perven, S. (2014) *Organic Farming, Prototype for Sustainable Agricultures*. Springer New York, USA.

Bengsston, J., Ahnström, J. and Weibull, A.C. (2005) The effects of organic agriculture on biodiversity and abundance: a meta-analysis. *Journal of Applied Ecology* 42, 261–269.

Brantsæter, A.L., Ydersbond, T.A., Hoppin, J.A., Haugen, M. and Meltzer, H.M. (2017) Organic food in the diet: exposure and health implications. *Annual Review of Public Health* 38, 295–313.

Bügel, S., Hertwig, J., Kahl, J., Lairon, D., Paoletti, F. and Strassner, C. (2016) The new Nordic diet as a prototype for regional sustainable diets. In: Meybeck, A. and Redfern, S. (eds) *Sustainable Value Chains for Sustainable Food Systems*. Food and Agriculture Organization of the United Nations, Rome, Italy.

Burlingame, B. and Dernini, S. (Eds) (2012) Sustainable Diets and Biodiversity: Directions and solutions for policy, research and action. Proceedings of the International Scientific Symposium Biodiversity and Sustainable Diets United against Hunger, 3–5 November 2010, FAO Headquarters, Rome. http://www.fao.org/docrep/016/i3004e/i3004e.pdf (accessed 28 August 2018).

Caldeira, S., Storcksdieck, S., Bakogianni, I., Gauci, C., Calleja, A. and Furtado, A. (2017) Public Procurement of food for health. Joint Publication of the Maltese Presidency and the European Union, Brussels, Belgium.

Chryssohoidis, G.M. and Krystallis, A. (2005) Organic consumers personal values research: testing and validating the list of values (LOV) scale and implementing a value-based segmentation task. *Food Quality and Preference* 16, 585–599.

Clark, M. and Tilman, D. (2017) Comparative analysis of environmental impacts of agricultural production systems, agricultural input efficiency, and food choice. *Environmental Research Letters* 12, 064016.

Crowder, D.W., Northfield, T.D., Strand, M.R. and Snyder, W.E. (2010) Organic agriculture promotes evenness and natural pest control. *Nature* 466, 109–112.

Daugbjerg, C. and Botterill, L.C. (2012) Ethical food standard schemes and global trade: paralleling the WTO? *Policy and Society* 31(4), 307–317.
de Boer, J., Hoogland, C.T. and Boersema, J.J. (2007) Towards more sustainable food choices: value priorities and motivational orientations. *Food Quality and Preference* 18, 985–996.
De Shutter, O. (2014) Report of the Special Rapporteur on the right to food. Final report: The transformative potential of the right to food. Human Rights Council Twenty-fifth session. Agenda item 3A/HRC/25/5724, January 2014.
Dernini, S., Berry, E.M., Serra-Majem, L., La Vecchia, C., Capone, R., *et al.* (2017) Med Diet 4.0: the Mediterranean diet with four sustainable benefits. *Public Health Nutrition* 20(7), 1322–1330. DOI: 10.1017/S1368980016003177.
Eden, S. (2011) Food labels as boundary objects: How consumers make sense of organic and functional foods. *Public Understanding of Science* 20, 179–194.
Eisinger-Watzl, M., Wittig, F., Heuer, T. and Hoffmann, I. (2015) Customers purchasing organic food – do they live healthier? Results of the German National Nutrition Survey II. *European Journal of Nutrition and Food Safety* 5(1), 59–71.
Ericksen, P.J. (2008) Conceptualizing food systems for global environmental change research. *Global Environmental Change* 18, 234–245.
Esmer, Y. and Pettersson, T. (2007) *Measuring and Mapping Cultures: 20 Years of Comparative Value Surveys*. Brill, Leiden, Germany.
FAO (2015) *Final Report for the International Symposium on Agroecology for Food Security and Nutrition*, 18–19 September 2014, Rome. Food and Agriculture Organization of the United Nations, Rome, Italy.
FAO (2017) *The Future of Food and Agriculture – Trends and Challenges*. Food and Agriculture Organization of the United Nations, Rome, Italy.
Gabriel, D., Sait, S.M., Kunin, W.E. and Benton T.G. (2013) Food production vs biodiversity: comparing organic and conventional agriculture. *Journal of Applied Ecology* 50(2), 355–364.
Gillespie, S. and van den Bold, M. (2017) Agriculture, Food Systems and Nutrition: Meeting the Challenge, Global Challenges. DOI: 10.1002/gch2.201600002
Global Panel on Agriculture and Food Systems for Nutrition (2016) Food systems and diets: Facing the challenges of the 21st century. Global Panel on Agriculture and Food Systems for Nutrition, London, UK.
Gomiero, T., Pimentel, D. and Paoletti M.G. (2011) Environmental impact of different agricultural management practices: conventional vs organic agriculture. *Critical Reviews Plant Science* 30, 95–124.
Grant, M. (2015) A food systems approach for food and nutrition security. *Sight and Life* 29(1), 87–90.
Hemmerling, S., Hamm, U. and Spiller, A. (2015) Consumption behaviour regarding organic food from a marketing perspective – a literature review. *Organic Agriculture* 5(4), 277–313.
Hertwig, J., Kahl, J., Strassner, C., Bügel, S. and Paoletti, F. (2017) How organic food systems may support the sustainability of dietary patterns. In: *Sustainable Diets in the Context of Sustainable Food Systems*. Food and Agriculture Organization of the United Nations, Rome, Italy.
Hjelmar, U. (2011) Consumers' purchase of organic food products. A matter of convenience and reflexive practices. *Appetite* 56, 336–344.
HLPE (2014) Food losses and waste in the context of sustainable food systems. A report by the High Level Panel of Experts on Food Security and Nutrition of the Committee on World Food Security. Food and Agriculture Organization of the United Nations, Rome, Italy.
IFOAM (2008) Principles of organic agriculture. Available at http://www.ifoam.bio/sites/default/files/poa_english_web.pdf (accessed 5 July 2017).
Ingram, J.S.I. (2011) A food systems approach to researching food security and its interactions with global environmental change. *Food Security* 3, 417–431.
iPES Food (2015) The case for a new science of for sustainable food systems. Report Nr. 01, 2015. iPES Food. Available at http://www.ipes-food.org/images/Reports/IPES_report01_1505_web_br_pages.pdf (accessed 31 May 2016)
Janssen, M. and Hamm, U. (2011) Consumer Preferences and Willingness-to-Pay For Organic Certification Logos. Report of the Certco St Project (D17). Available at http://orgprints.org/18850/1/Janssen_Hamm_2011_D17_Report_Consumer_preferences_for_organic_logos.pdf (accessed 1 October 2017).
Janssen, M. and Hamm, U. (2012) Product labelling in the market for organic food: Consumers preferences and willingness-to-pay for different organic certification logos. *Food Quality and Preference* 25, 9–22.
Kahl, J., Alborzi, F., Beck, A., Bugel, S., Busscher, N., *et al.* (2014) Organic food processing: a framework for concept, starting definitions and evaluation. *Journal of the Science of Food and Agriculture* 94(13), 2582–2594.

Kahl, J., Strassner, C., Hertwig, J., Gould, D., Bügel, S., *et al.* (2016) Learning from the organic food system as a model for sustainable food systems – the Organic Food System Program. In: Meybeck, A. and Redfern S. (eds) Sustainable Value Chains for Sustainable Food Systems. Food and Agriculture Organization of the United Nations, Rome, Italy.

Kareklas, I., Carlson, J.R. and Muehling, D.D. (2014) "I eat Organic for My Benefit and Yours:" Egoistic and Altruistic Considerations for Purchasing Organic Food and Their Implications for Advertising Strategists. *Journal of Advertising* 43(1), 18–32.

Kennedy, C.M., Lonsdorf, E., Neel, M.C., Williams, N.M., Ricketts, T.H., *et al.* (2013) A global quantitative synthesis of local and landscape effects on wild bee pollinators in agroecosystems. *Ecology Letters* 16, 584–599.

Kesse-Guyot, E., Péneau, S., Méjean, C., Szabo de Edelenyi, F., Galan, P., *et al.* (2013) Profiles of organic food consumers in a large sample of French adults: results from the Nutrinet-Sante cohort study. *PLoS One*, 8(10), e76998.

Kesse-Guyot, E., Baudry, J., Assmann, K.E., Galan, P., Hercberg, S. and Lairon, D. (2017) Prospective association between consumption frequency of organic food and body weight change, risk of overweight or obesity: results from the NutriNet-Santé Study. *British Journal of Nutrition* 117(2), 325–334. DOI:10.1017/S0007114517000058

Kriwy, P. and Mecking, R.-A. (2012) Health and environmental consciousness, costs of behaviour and the purchase of organic food. *International Journal of Consumer Studies* 36(1), 30–37.

Lankinen, M., Schwab, U., Kolehmainen, M., Paananen, J., Nygren, H., *et al.* (2016) A healthy Nordic diet alters the plasma lipidomic profile in adults with features of metabolic syndrome in a multicenter randomized dietary intervention. *The Journal of Nutrition*. DOI: 10.3945/jn.115.220459.

Lynch, D.H., Halberg, N. and Bhatta, G.D. (2012) Environmental impacts of organic agriculture in temperate regions. *CAB Review* 7, 1–17.

Meybeck, A. and Gitz, V. (2017) Sustainable diets within sustainable food systems. *Proceedings of the Nutrition Society* 76, 1–11.

Michaelidou, N. and Hassan, L.M. (2008) The role of health consciousness, food safety concern and ethical identity on attitudes and intentions towards organic food. *International Journal of Consumer Studies* 32, 163–170.

Mie, A., Andersen, H.R., Gunnarsson, S., Kahl, J., Kesse-Guyot, E., *et al.* (2017) Human health implications of organic food and organic agriculture: A comprehensive review. *Environmental Health* 16(1), 111.

Mithril, C., Dragsted, L.O., Meyer, C., Blauert, E., Holt, M.K. and Astrup, A. (2012) Guidelines for the new Nordic diet. *Public Health Nutrition* 15(10), 1941–1947.

Mithril, C., Dragsted, L.O., Meyer, C., Tetens, I., Biltoft-Jensen, A. and Astrup, A. (2013) Dietary composition and nutrient content of the new Nordic diet. *Public Health Nutrition* 16(5) 777–785.

Mondelaers, K., Aertsens, J. and Van Huylenbroeck, G. (2009) A meta-analysis of the differences in environmental impacts between organic and conventional farming. *British Food Journal* 111, 1098–1119.

Moomaw, W., Griffin, T., Kurczak, K. and Lomax, J. (2012) *The Critical Role of Global Food Consumption Patterns in Achieving Sustainable Food Systems and Food for All*. A UNEP Discussion Paper. United Nations Environment Programme, Division of Technology, Industry and Economics, Paris, France.

Niggli, U. (2015) Incorporating Agroecology Into Organic Research – An Ongoing Challenge. *Sustainable Agriculture Research* 4(3), 149–157.

Niggli, U., Jawtusch, J. and Oehen, B. (2011) *Do Standards and Certification in the Agricultural Sector Matter for Sustainability? A Review of the State of Research*. Research Institute of Organic Agriculture (FiBL), Switzerland and RESOLVE, USA.

Ostrom, E. (2009) A general framework for analyzing sustainability of social-ecological systems. *Science* 325, 419.

Padel, S. and Foster, C. (2005) Exploring the gap between attitudes and behaviour: Understanding why consumers buy or do not buy organic food. *British Food Journal* 107, 606–625.

Pearson, D., Henryks, J. and Moffitt, E. (2007) What do buyers really want when they purchase organic foods? An investigation using product attributes. *Journal of Organic Systems* 2(1), 1–9.

Pearson, D., Henryks, J. and Jones, H. (2011) Organic food: What we know (and do not know) about consumers. *Renewable Agriculture and Food Systems* 26(02), 171–177.

Pino, G., Peluso, A.M. and Guido, G. (2012) Determinants of Regular and Occasional Consumers' Intentions to Buy Organic Food. *Journal of Consumer Affairs* 46(1), 157–169.

Ponisio, L.C., M'Gonigle, L.K., Mace, K.C., Palomino, J. and de Valpine, P. (2015) Diversification practices reduce organic to conventional yield gap. *Proceedings of the Royal Society Series B* 282, 20141396. DOI: 10.1098/rspb.2014.1396

Poulsen, S.K., Due, A., Jordy, A.B., Kiens, B., Stark, K.D., *et al.* (2014) Health effect of the New Nordic Diet in adults with increased waist circumference: a 6-mo randomized controlled trial. *American Journal of Clinical Nutrition* 99(1), 35–45.

Reaganold, J.P. and Wachter, J.M. (2016) Organic agriculture in the twenty-first century. *Nature Plants* 2, 1–8.

Seconda, L., Baudry, J., Allès, B., Hamza, O., Boizot-Szantai, C., *et al.* (2017) Assessment of the Sustainability of the Mediterranean Diet combined with organic food consumption: an individual behaviour approach. *Nutrients* 9, 61. DOI: 10.339nu9010061

Seufert, V., Ramankutty, N. and Foley, J.A. (2012) Comparing the yields of organic and conventional agriculture. *Nature* 485(7397), 229–232.

Stehfest, E. (2014) Food choices for health and planet. *Nature* 515, 501–502.

Stolz, H., Stolze, M., Janssen, M. and Hamm, U. (2011) Preferences and determinants for organic, conventional and conventional-plus products. The case of occasional organic consumers. *Food Quality and Preference* 22(8), 772–779.

Strassner, C., Cavoski, I., Di Cagno, R., Kahl, J., Kesse-Guyot, E., *et al.* (2015) How the organic food system supports sustainable diets and translates these into practice. *Frontiers in Nutrition* 2, 19. DOI: 10.3389/fnut.2015.00019.

Strassner, C., Bügel, S., Hertwig, J., Kahl, J., Nuutila, J. and Paoletti, F. (2016) The role of sustainable HORECA for sustainable lifestyles – identification of challenges and future work. In: Meybeck, A. and Redfern, S. (eds) *Sustainable Value Chains for Sustainable Food Systems*. Food and Agriculture Organization of the United Nations, Rome, Italy.

Torjusen, H., Sangstad, L., O'Doherty-Jensen, K. and Kjaernes, U. (2004) *European Consumers' Conceptions of Organic Food: A Review of Available Research*. Project Report 4-2004 for National Institute for Consumer Research, Oslo, Norway.

Tuck, S., Winqvist, C., Mota, F., Ahnström, J., Turnbull, L.A. and Bengtsson, J. (2014) Land-use intensity and the effects of organic farming on biodiversity: a hierarchical meta-analysis. *Journal of Applied Ecology* 51, 746–755.

Tuomisto, H.L., Hodge, I.D., Riordan, P. and Macdonald, D.W. (2012) Does organic farming reduce environmental impacts? A meta-analysis of European research. *Journal of Environmental Management* 112, 309–320.

UNEP (2016) *Food Systems and Natural Resources*. A Report of the Working Group on Food Systems of the International Resource Panel. United Nations Environment Programme, Nairobi, Kenya.

Vergragt, P., Akenji, L. and Dewick, P. (2014) Sustainable production, consumption, and livelihoods: global and regional research perspectives. *Journal of Cleaner Production* 63, 1–12.

Vogt, G. (2007) The origins of organic farming. In: Lockeretz, W. (ed.) *Organic Farming – An International History*. CAB International, Wallingford, UK, pp. 9–29.

Willer, H. and Lernoud, J. (2017) *The World of Organic Agriculture, Statistics and Emerging Trends.* Research Institute of Organic Agriculture (FiBL), Frick, and IFOAM – Organics International, Bonn.

Zagata, L. (2012) Consumers' beliefs and behavioral intentions towards organic food. Evidence from the Czech Republic. *Appetite* 59(1), 81–89.

25 Institutional Food Procurement for Promoting Sustainable Diets

Florence Tartanac, Luana F.J. Swensson, Andrea Polo Galante and Danny Hunter

Abstract

Institutional food procurement programmes (IFPP) have been receiving increasing attention from the literature, policy makers and development agencies as an important policy instrument with the potential to deliver multiple benefits for multiple beneficiaries, including food consumers and food producers. A key characteristic of IFPP is that it has the possibility – based on its policy and regulatory frameworks – to determine not only the way food is procured, but, in particular: (i) what food will be purchased (such as local, diverse, nutritious, healthy, culturally adequate, environmentally friendly); and (ii) from whom (e.g. local and smallholder producers). It can also determine how food is received, stored, prepared and its waste managed. Considering the extent of public sector demand and how these choices are made, this chapter argues that IFPP holds considerable potential to influence both food consumption and food production patterns and to deliver multiple social, economic, environmental, nutritional and health benefits to the food system that will contribute to more sustainable diets. It provides an overview of the literature on the potential benefits of IFPP linked to the promotion of sustainable diets taking into consideration the three key pillars of sustainability as well as examples of good practices from the Brazilian food procurement programme (*Programa de Aquisição de Alimentos*), Cape Verde national school feeding programme and the municipality of Rome (Italy).

Introduction

In the last few years the use of the regular demand for food on the part of government entities (i.e. institutional demand) has been recognized as an instrument with the potential to promote sustainable and transformative development of local food systems (Morgan and Sonnino, 2008; Foodlinks, 2013; De Schutter, 2014; Fitch and Santo, 2016; Kelly and Swensson, 2017).

Various countries, regions and cities from low-income to high-income economies have been developing institutional food procurement programmes (IFPP) aimed at building a direct linkage between public demand for food and local and smallholder agriculture production. Those programmes are based on the premise that public institutions, when using their financial capacity and procurement power to award contracts, can go beyond the immediate scope of simply responding to the state's procurement needs, by addressing additional social, environmental or economic objectives that contribute to the overall public good of a state (Quinot, 2013).

Different public institutions can provide the regular and predictable demand for food for the implementation of IFPP. They include prisons, hospitals, universities, armies, social programmes and, most commonly, schools.

The IFPP – and related policy and legal frameworks – can determine not only the way food is procured, but, in particular (i) what food will be purchased (such as local, diverse, nutritious, healthy, culturally adequate, environmental-friendly) and (ii) from whom (e.g. local and smallholder producers). It will also determine how food will be received, stored, prepared and its waste managed. Another key characteristic of IFPP is that it has the potential to influence both food consumption and food production patterns through its purchasing power (Foodlinks, 2013; Bontrager *et al.*, 2014; IPES, 2016; Fitch and Santo, 2016). By changing the practices of public food procurement and creating a demand for sustainable diets, governments have the power to set a positive trend. They can send a signal about their ambitions on the future directions of the food system that has the power to incentivize those involved in the supply chain to align their values accordingly, accelerating a transition towards sustainable food consumptions and production (Foodlinks, 2013; De Schutter, 2014; IPES, 2016; UNSCN, 2017).

Considering the weight of public sector demands and depending on how those choices are made, this chapter argues that IFPP holds the potential to deliver multiple social, economic, environmental, nutritional and health dividends to the food system that can contribute to sustainable diets and its main constitutive elements.

We adopt the definition of sustainable diets as those:

> diets with low environmental impacts which contribute to food and nutrition security and to healthy life for present and future generations. Sustainable diets are protective and respectful of biodiversity and ecosystems, culturally acceptable, accessible, economically fair and affordable; nutritionally adequate, safe and healthy; while optimizing natural and human resources.
>
> (Burlingame *et al.*, 2010)

Building on the existing literature on food procurement and on the experience of different countries, this chapter aims at exploring the multiple benefits that IFPP has the potential to achieve within the food system and that can contribute to the promotion of sustainable diets.

This chapter will be organized under three sections. The first section will provide an overview of the literature on the potential benefits of IFPP linked to the promotion of sustainable diets considering four main spheres: social, economic and environmental – the three key pillars of sustainability – as well as nutrition and health. The second section will focus on country experiences and provide best practices examples. The last section will be dedicated to the concluding remarks.

Multiple Benefits and Beneficiaries

One of the key characteristics of an institutional food procurement initiative is its multifaceted nature and its potential to achieve through a single policy intervention multiple benefits and beneficiaries.

There is a range of literature and policy documents that recognizes the multiple dividends that IFPP has the potential to achieve and the contributions that public procurement of food can make to sustainability and its three key spheres: economic, social and environmental (Espejo *et al.*, 2009; Foodlinks, 2013; De Schutter, 2014, 2015; FAO, 2015; Fitch and Santo, 2016; Smith *et al.*, 2016). Specific health and nutrition dividends can also be added (Morgan and Sonnino, 2008; Global Panel, 2015; Fitch and Santo, 2016).

They also recognize that institutional procurement has the potential to benefit not only those who receive the food through the public institutions (food consumers) but also those who supply the food (food producers) and the community in general.

Nutritional and health benefits

As mentioned above, one of the primary aims of IFPPs is to link public food procurement and smallholder agriculture production. However, the dimension of IFPP must also surpass the market aspect, integrating other important elements such as dietary recommendations. As a positive relation between nutrition and public procurement, IFPP has the potential to: (i) provide health and nutrition for those who receive the food; (ii) support specific food production in order to promote healthy diets; (iii) promote 'culturally adequate' food and (iv) promote biodiversity (Burlingame *et al.*, 2010; Foodlinks, 2013; Fitch and Santo, 2016; UNSCN, 2017).

Nutrition has been dominated by the study of specific nutrients, nutrient deficiencies, and sometimes specific foods or food groups and their impact on health and nutrition. In recent years, a number of initiatives and studies have focused more directly on the question of diets and their impacts on human health, the environment, and food systems (Thompson and Amoroso, 2014). As highlighted by UNICEF (2013), inadequate food consumption is the leading cause of malnutrition. Food is the fundamental basis for good nutrition; adopting a healthy, balanced diet is essential to prevent malnutrition in all its forms, as well as a range of non-communicable diseases and adverse health conditions.

As already mentioned, sustainable diets take into consideration not only the impact on food production, but also on food consumption patterns that are often unhealthy. One aspect of healthy diet is food diversification; according to FAO (2016), 95% of the calories that people obtain from food comes from only thirty different crops, and only four of these – maize, rice, wheat and potatoes – are the ones feeding the population. Furthermore, a small amount of ingredients such as refined flours, sugar and oil originated from those crops and, despite being only few products, they represent the base of ultra-processed food. Several people around the world – including from developed and developing countries – have diets based on undiversified and unhealthy food, maintaining the cycle of food insecurity and malnutrition (Popkin, *et al.*, 2012; Garnett, 2014; Johnston *et al.*, 2014).

IFPP can play a vital role in stimulating smallholders to produce more local and biodiverse crops, increasing the number of healthy foods offered in the schools and at the local market, plus contributing to the diversification of the diet. The promotion of agricultural production diversification and consumer awareness about the importance of diversified diet are both key elements to explore agricultural biodiversity, rural development and sustainable diet.

Social and economic benefits

As stated by Mason and Lang (2017), it would be impossible to do justice to the notion of sustainable diet without facing the importance of its economics and social aspects, such as income generation and value added. Consumer choices are shaped by prices, their incomes and by affordability, but also by cultural and social factors.

From a social perspective, IFPP has the potential to contribute to food and nutrition security of both food consumers and food producers. Studies show that, in the first case, it can do this by providing quality and nutritious food in the institutional settings where consumers do not have much – if any – choice or alternative food sources. In the case of food producers, IFPP can contribute to their nutrition security through improvements from agricultural income and production (Joshi *et al.*, 2008; IPC and WFP, 2013; Gyoeri *et al.*, 2016).

One of the primary objectives of IFPP is, indeed, to support local and smallholder food producers through the provision of new, stable, predictable and fair-priced market opportunities (Morgan and Sonnino, 2008; De Schutter, 2015; FAO, 2015; Kelly and Swensson, 2017).

The rationale behind those programmes is that connecting large, predictable sources of demand for agricultural products to smallholder farmers can reduce uncertainty associated with producers' engagement with markets and investment risks in improved and diversified production. This may encourage investments, improved quality and production diversity, leading to higher and steadier incomes and ultimately, improved livelihoods (Sumberg and Sabates-Wheeler, 2010; Mitchel, 2011; De Schutter, 2014; WFP, 2016).

Another key potential of IFPP is that it can provide smallholders not only with this specific market opportunity, but it can also act as a learning path for the access to other formal and more demanding private and public markets. Although findings are preliminary and mostly qualitative based, they show that IFPP has the potential to support smallholder producers to acquire the technical and organizational skills to comply with the requirements not only of this local institutional market, but also other formal and more demanding private food markets (IPC and WFP, 2013; Kelly and Swensson, 2017).

Furthermore, IFP can also constitute an important market and income opportunity for other actors in the value chain, including small traders, small processors, and small and medium food enterprises, spreading the benefits along the community (Foodlinks, 2013; FAO and WFP, 2018).

IFPP also provides the opportunity for government to target and support specific groups of vulnerable producers, including women, indigenous peoples, members of traditional communities and youth. The Brazilian experience described in the next section is a significant example of this possibility. On these cases, institutional procurement may represent an important part in social equality.

Environmental benefits

It is well established that food and diets – including production, processing, transportation, preparation and consumption – have a considerable environmental impact (Fitch and Santo, 2016; HLPE, 2017; Mason and Lang, 2017). Indeed, food consumption has been identified as one of the key drivers of environmental pressures (Mason and Lang, 2017). The definition states clearly that sustainable diets are those with low environment impact that are protective and respectful of biodiversity and ecosystems.

Within this context, IFPP can contribute to sustainable diets not only through the provision of environmentally friendly food to its consumers, but, in particular, as a demand-driven intervention. IFPP has been recognized with great potential to positively influence water and land use, biodiversity and climate change (Fitch and Santo, 2016; Foodlinks, 2013).

Institutional procurement can target food that is produced in a specific way, and, therefore, use its purchasing power to support and promote forms of agricultural production that ensure environmental sustainability as well as biodiversity. This includes the purchase of food based on: low-impact production methods with reduced carbon inputs and greenhouse gas emissions; organic production; agroecology and biodiversity attuned practices; and enhanced animal welfare criteria (Foodlinks, 2013). There is a range of available systems that IFPP can avail to incorporate sustainability criteria and guarantee the quality of the produce. Approaches for this include, for example, a national registry of agroecology producers, organic certification and a list of criteria for environmentally sustainable food.

Furthermore, IFPP has the potential not only to influence more environmental and biodiverse types of agriculture but also promote environmental benefits in terms of reduced packaging; food waste; and lower food miles (Foodlinks, 2013). The example of the city of Rome described in the next section is an interesting example of that.

Examples of Good Practices

The Brazilian food purchase programme

The Brazilian food purchase programme *Programa de Aquisição de Alimentos* (PAA) was created in 2003 as part of the Brazilian national strategy on food and nutritional security, Zero Hunger is a multidimensional programme that combines the goal of promoting food security with the broader concerns of inclusive economic and social development. Currently, PAA has nine goals that clearly state the multifaceted nature of the programme. They include: (i) support smallholder production (family farmers and rural entrepreneurs) by promoting economic and social inclusion with sustainable surplus growth and the processing and industrialization of food products; (ii) support the consumption and valorization of food produced by smallholders; (iii) promote access to food, in the quantity, quality and regularity necessary for people with food and nutritional insecurity; (iv) promote and enhance biodiversity, organic and agro-ecological food production; and (v) encourage healthy eating habits at local and regional level. The programme also prioritizes the poorest and most vulnerable producers, which include disadvantaged social groups, such as land reform settlers, indigenous people, and women.

Data on the implementation of PAA through the National Supply Company (CONAB) shows that in 2016, 88,120 tons of food was purchased, benefiting 29,318 smallholders, including both farmers and rural entrepreneurs (CONAB, 2017). PAA also benefited 9,306,019 food consumers considered to be food- and nutrition-insecure, ensuring their access to healthy and locally produced food. Women's participation in the programme has reached 57%, an important increase from the 21% participation rate in 2009 (CONAB, 2016). Similarly, spending on biodiverse products had risen from 5.36% in 2012 to 10.99% in 2015.

Although it is a small percentage compared with overall food purchases, it illustrates the potential of IFPP as a strategic tool in promoting a market for biodiverse food, as well as in supporting its conservation and sustainable use (UNSCN, 2017).

Cape Verde national school feeding programme

The Cape Verde school feeding programme, supported by World Food Programme (WFP), has been in existence since 1979. In 2007 its operational responsibilities were transferred from the WFP to the government, and since 2010 the programme has been fully funded and implemented by the government. Under the government ownership and the United Nations joint support programme, the school feeding programme started to be reformulated and a pilot project implemented that aimed to: (i) improve the nutritional status and eating habits of students and prevent diseases; (ii) contribute to poverty reduction and social cohesion; and (iii) stimulate local agriculture production through the purchase of local products (Bigaud, 2014).

This new approach includes: (i) diversification of school menus (introducing fresh fruits, vegetables, fish and local beans); and (ii) creation of market opportunities for smallholder local producers. This approach reflects the government aim of using schools' food demand as an instrument to achieve broader social, economic, nutrition and health outcomes. For instance, the improvements in the nutritional composition of the menus by diversifying with fresh and local food – combined with food and nutrition education – is directly aimed at improving health and nutrition outcomes, including the prevention of non-communicable diseases, considered one of the priorities in the government agenda (Drake *et al.*, 2016).

The use and purchase of fresh products is also aimed at supporting domestic agriculture and decreasing the dependence on imports (estimated at 80% of food consumed in the country). Particular attention is also given to the cultural adequacy and acceptability of the menus with the replacement, for instance, of lentils by local types of beans (Drake *et al.*, 2016).

During the pilot project (2012–2014) implemented in 31 pilot schools, 48 tons of foods were purchased locally and 8942 students were enrolled in the scheme (Bigaud, 2014). Despite the encouraging results, recent studies show that the implementation of the new approach still faces important challenges, including its costs and economic sustainability (Drake *et al.*, 2016).

The municipality of Rome (Italy)

The municipality of Rome and, in particular, its school feeding programme, is an interesting example of the pursuit of social, economic, nutritional and, especially environmental outcomes through the procurement of food for schools. In 2002, Rome set a target of 70% of food in school meals being organic, combined with other requirements linked to food quality and territoriality to ensure children's health and safety, but also to promote local economic development, environmental protection, and local culture and traditions (Morgan and Sonnino, 2008). The change to organic food entailed three progressive phases.

During the first phase (2002–2004), organic products were identified that could be supplied in sufficient quantities to meet demand and that were not too challenging for the city's procurement policy and tendering procedures. During the second phase (2004–2007), more specific requirements were included in the tenders, including seasonality, variety and territoriality. The third phase incorporated targets on food losses and waste, including, the separation of waste for collection, the use of low environmental impact detergents and of non-disposable plates, glasses and cutlery. Other aspects of sustainability that were prioritized included Protected Denomination of Origin/Protected Geographial Indications (PDO/PGI) products introduced to emphasize territoriality – and as such creating marketing opportunities for local producers – and fair trade products to promote social justice and solidarity beyond its most immediate boundaries (Foodlinks, 2013; Morgan and Sonnino, 2008).

Data related to the 2007–2012 call for tenders shows that 69% of the 144,000 meals served across 550 institutions included organic food (Foodlinks, 2013). More recently, within the new municipality administration most of those criteria were confirmed in the new guidelines, which included a once-a-week 'green', locally sourced, organic, vegetarian menu, aimed at increasing the environmental sustainability of the food sourced at schools (Narducci, 2016).

Concluding Remarks

The literature and the different country experiences demonstrate the potential that institutional food procurement has to address different social, economic, environmental as well as health and nutrition outcomes that contribute to the promotion of healthy diets. The great potential of IFPP is reinforced by its possibility of influencing not only sustainable consumption, but also production patterns, with the possibility of promoting sustainable diets among its direct food consumers, the food producers and the community in general.

Nevertheless, despite the enormous opportunity that IFPP offers to drive more sustainable diets, the story of public procurement is often still 'a tale of untapped potential' (Foodlinks, 2013). Many factors can justify this fact (see, Kelly and Swensson, 2017); among them, there is multifaceted and complex nature of those programmes as well as lack of data and evaluation of its effective impact against its multiple goals and beneficiaries.

IFPPs are by their essence multifaceted and complex programmes. The different outcomes that these initiatives can reach – but also the challenges faced – go beyond sector-bound single institutions. They include, as described in this chapter, health, economics, agriculture and environment. If, on the one hand this multifaceted nature of IFPP is one of its distinct characteristics, on the other it is a source of difficulties and complexity in its implementation. As recognized by the literature and country experiences, for its successful implementation IFPP may require a coordinated and collaborative multisectoral approach that recognizes and emphasizes its cross-cutting, multifunctional nature. Such an approach, however, is not always easy to adopt. It may require effective institutional coordination, with clear institutional roles and overarching legal and policy frameworks that guide collaboration between ministries, policies, strategies and institutions, running from the ministry down to the local level where food procurement takes place (FAO, 2013, 2015; Kelly and Swensson, 2017).

Another important issue is that there are still important research gaps and lack of monitoring and evaluation systems of existing programmes to document best practices and determine the effective impact of IFPP against its multiple benefits and beneficiaries. Evaluating and measuring the impact of IFPPs, especially due to its multifaceted nature, is not an easy task.

Although the literature discussed in this chapter indicate positive and important trends on IFPP potentialities to achieve social, economic, environmental, health and nutrition outcomes, further research and impact evaluation are still necessary. They are key to continue to build the evidence base and best practices linking institutional purchasing to its multiples outcomes and the promotion of sustainable diets, and to drive its effective implementation.

References

Bigaud, N. (2014) Approvisionnement des cantines scolaires en produits locaus: d'un project-pilote à la préparation d'un programme national durable: Le cas du Cap-Vert. *Mémoire de Mission Professionnelle*. Supagro, Montpellier, France.

Bontrager Yoder, A., Liebhart, J.L., McCarty, D.J., Meinen, A., Schoeller, D., *et al.* (2014) Farm to elementary school programming increases access to fruits and vegetables and increases their consumption among those with low intake. *Journal of Nutrition, Education and Behaviour* 46(5), 341–349.

Burlingame, B., Dernini, S. (Eds) (2012) Sustainable Diets and Biodiversity: Directions and solutions for policy, research and action. *Proceedings of the International Scientific Symposium Biodiversity and Sustainable Diets United against Hunger*, 3–5 November 2010, FAO Headquarters, Rome. http://www.fao.org/docrep/016/i3004e/i3004e.pdf (accessed 28 August 2018).

CONAB (2017) *Programa de aquisição de alimentos – PAA: Resultados das ações da CONAB em 2016*. CONAB, Brasilia, Brazil.

De Schutter, O. (2014) The power of procurement: public purchasing in the service of realizing the right to food. *Briefing Note of the United Nations Special Rapporteur on the Right to Food*. Brussels, Belgium.

De Schutter, O. (2015) Institutional food purchasing as a tool for food system reform. In: *Advancing Health and Well-being in Food Systems: Strategic Opportunities for Funders*. Available at https://futureoffood.org/

wp-content/uploads/2016/09/Global-Alliance-Advancing-Health-Wellbeing-Compendium-April-2015.pdf (accessed 26 June 2018).

Drake, L., Woolnough, A., Burbano, C. and Bundy, D. (2016) *Global School Feeding Sourcebook: Lessons from 14 Countries.* Imperial College Press, London, UK.

Espejo, F., Burbano, C. and Galliano, E. (2009) *Home Grown School Feeding: A Framework to Link School Feeding with Local Agricultural Production.* World Food Programme, Rome, Italy.

FAO (2013) *Alimentación escolar y las posibilidades de compra directa de la agricultura familiar: Estudio de caso de ocho países.* Food and Agriculture Organization of the United Nations, Santiago, Chile.

FAO (2015) *Las compras públicas a la agricultura familiar y la seguridad alimentaria y nutricional en América Latina y el Caribe: Lecciones aprendidas y experiencias.* Food and Agriculture Organization of the United Nations, Santiago, Chile.

FAO (2016) Plant genetic resources. Use them or lose them. Available at http://www.fao.org/fileadmin/templates/nr/documents/CGRFA/factsheets_plant_en.pdf (accessed 23 October 2017).

FAO and WFP (2018) Home-Grown School Feeding, Resource Framework. Synopsis – March 2018. Rome. 36 pp. Available at http://www.fao.org/3/i8724en/I8724EN.pdf

Fitch, C. and R. Santo (2016) *Instituting Change: An Overview of Institutional Food Procurement and Recommendations for Improvement.* Johns Hopkins Center for a Livable Future, Baltimore, Maryland, USA.

Foodlinks (2013) Revaluing public sector food procurement in Europe: An action plan for sustainability. Available at http://www.foodlinkscommunity.net/fileadmin/documents_organicresearch/foodlinks/publications/Foodlinks_report_low.pdf (accessed 26 June 2018).

Garnett, T. (2014) *What is a Sustainable Diet: A Discussion Paper.* Food Climate Research Network, London, UK.

Global Panel (2015) *Heathy Meals in Schools: Policy Innovations Linking Agriculture, Food Systems and Nutrition. Policy Brief.* Global Panel on Agriculture and Food Systems for Nutrition, London, UK.

Gyoeri, M., Miranda, A.C. and Soares, F.V. (2016) Linking vulnerable smallholder farmers to school feeding programmes: the experience of PAA Africa. *Policy in Focus* 13(2) 12–15.

HLPE (2017) *Nutrition and Food Systems.* High Level Panel of Experts on Food Security and Nutrition of the Committee on World Food Security, Rome, Italy.

IPC and WFP (2013) *Structured Demand and Smallholder Farmers in Brazil: The Case of PAA and PNAE.* International Policy Centre for Inclusive Growth and World Food Programme, Brasilia, Brazil.

IPES (2016) From Uniformity to Diversity: A Paradigm Shift from Industrial Agriculture to Diversified Agroecological Systems. Available at http://www.ipes-food.org/images/Reports/UniformityToDiversity_FullReport.pdf (26 June 2018).

Johnston, J., Fanzo, J. and Cogill, B. (2014) Understanding sustainable diets: a descriptive analysis of the determinants and processes that influence diets and their impact on health, food security and environmental sustainability. *Advances in Nutrition* 5, 418–429.

Joshi, A., Azuma, A.M. and Feenstra, G. (2008) Do farm-to-school programs make a difference? Findings and future research needs. *Journal of Hunger and Environmental Nutrition* 3(2/3) 229–246.

Kelly, S. and Swensson, L.F.J. (2017) *Leveraging institutional food procurement for linking small farmers to markets: Findings from WFP's purchase for progress initiative and Brazil's food procurement programmes.* Food and Agriculture Organization of the United Nations, Rome, Italy.

Mason, P. and Lang, T. (2017) *Sustainable Diets: How Ecological Nutrition can Transform Consumption and the Food System.* Routledge, New York, USA.

Mitchel, A. (2011) *Structured Demand and Home-Grown School Feeding: Background and Funding Options.* Bill and Melinda Gates Foundation/Global Child and Nutrition Forum (GCNF), Nairobi, Kenya.

Morgan, K. and Sonnino, R. (2008) *The School Food Revolution: Public Food and the Challenge of Sustainable Development.* Earthscan, London, UK.

Narducci, M.G. (2016) Menu vegano, bio e a km zero per le mense di Roma: ecco il bando del M5s. Available at http://www.dire.it/16-11-2016/90388-menu-vegano-bio-km-zero-le-mense-roma-bando-del-m5s/ (accessed 23 October 2017).

Popkin, B.M., Adair, L.S. and Nq, S.W. (2012) Global nutrition transition and the pandemic of obesity in developing countries. *Nutrition Reviews* 70(1), 3–21.

Quinot, G. (2013) Promotion of social policy through public procurement in Africa. In: Quinot G. and Arrowsmith, S. (eds) *Public Procurement Regulation in Africa.* Cambridge University Press, Cambridge, UK.

Smith, J., Andersson, G., Gourlay, R., Karner, S., Mikkelsen, B. E., *et al.* (2016) Balancing competing policy demands: the case of sustainable public sector food procurement. *Journal of Cleaner Production* 112(1), 1–8.

Sumberg, J. and Sabates-Wheeler, R. (2010) Linking agricultural development to school feeding. *Food Policy* 36(3), 341–349.

Thompson, B. and Amoroso, L. (2014) Improving diets and nutrition: food-based approaches. Available at http://www.fao.org/3/a-i3030e.pdf (accessed 26 June 2018).

UNICEF (2013) *Improving Child Nutrition: The Achievable Imperative for Global Progress*. UNICEF, New York, USA.

UNSCN (2017) Schools as a system to improve nutrition: A new statement for school-based food and nutrition interventions. Available at https://www.unscn.org/uploads/web/news/document/School-Paper-EN-WEB.pdf (accessed 26 June 2018).

WFP (2016) *Purchase for Progress: Improving Livelihoods to Achieve Food Security*. World Food Programme, Rome, Italy.

26 Renewing Partnerships with Non-state Actors for Sustainable Diets through Sustainable Agriculture

Kakoli Ghosh

Abstract

Any diet that is qualified as a 'sustainable diet' should be nutritionally adequate, affordable, safe, healthy and culturally acceptable. In optimizing natural and human resources, provision of sustainable diets requires strong partnerships among the stakeholders engaged in production, delivery and disposal of food. Such partnerships have to be based on value proposition, trust and commitment. This chapter will explore the role partnerships play in developing a pathway for sustainable diets, in particular in the context of the common vision of sustainable food and agriculture principles offered by the Food and Agriculture Organization (FAO). The chapter will also focus on ways and means to strengthen sustainable diets by increasing collaboration among governments and non-state actors such as civil society, farmers' organizations, the private sector, academia and research institutions. It will discuss the current style and forms of partnerships in practice with some examples from FAO experience in coordination and strengthening of strategic partnerships, to share knowledge and resources and develop capacities among countries in support of the sustainable development goals.

Introduction

Any diet that is qualified as a 'sustainable diet' must fulfil certain criteria. It should be nutritionally adequate, affordable, safe, healthy and culturally acceptable. Furthermore, it should respect biodiversity, have low environmental impacts and optimize use of natural resources. In other words, a sustainable diet can be considered to be the direct outcome of sustainable food and agriculture practices and seamless partnerships among all the actors and stakeholders engaged in the production, delivery and disposal cycle.

The concepts that underpin sustainable agriculture practices are not new; however, the orientation within the food and agriculture sector to consciously apply them and explore possible ways of adopting integrated, systems-based approaches is a rather recent phenomenon. As pointed out by numerous studies, a paradigm shift was necessary because current agricultural systems have failed to provide an optimal balance of the economic, social and environmental concerns of society. Unsustainable farming practices have led to a rapid loss of natural resources and agro-biodiversity, hastened land degradation, water scarcity and climate change. Structural short-sightedness and policy incoherence have had a negative impact on livelihoods, and where agricultural productivity is extremely low, farmers, pastoralists and other rural dwellers struggle to survive and make a decent living.

Around 2014, the Food and Agriculture Organization (FAO) developed a Common Vision for Sustainable Food and Agriculture for an integrated approach to sustainability across agriculture, forestry and fisheries. It calls for synergies between sectors and is based on five principles:

(i) improving efficiency in use of resources; (ii) conserving, protecting and enhancing natural resources; (iii) protecting and improving rural livelihoods, equity and social well-being; (iv) enhancing the resilience of people, communities and ecosystems; and (v) promoting responsible and effective governance mechanisms. These principles provide a basis for developing national policies, strategies, programmes, regulations and initiatives for transition to sustainable agriculture and recognize that there will be trade-offs in the process. It also implies that a synergistic approach must be in place for production value chain and food-based systems (see Fig. 26.1; FAO, n.d.).

Partnership as an Action

In order to understand the drivers of sustainable food and agriculture and sustainable diets, the interactions and partnerships among and between the various stakeholders and actors who determine the sustainable food and agriculture landscape have to be recognized and their role better understood. This could be through promoting policy dialogue, development and implementation of joint and coordinated actions involving governmental bodies of different ministries, the private sector and civil society, promoting youth engagement, and raising awareness across all levels. A major effort should be in place to find ways to leverage small and big non-state actors as strategic partners and bring their knowledge, experience and skills in support of the sustainable food and agriculture for nutritious diet and safe food systems. For such partnerships to flourish and provide the positive outcome, they have to be based on value proposition, trust and commitment. In this regard, the Global Panel on Agriculture and Food Systems for Nutrition in its Report of 2017 emphasized the role of partnership and inclusion and provided important recommendations for policy makers, decision makers, professionals, business people, experts and researchers with interests in food systems and diets as well as those involved in the production, processing, trade, regulation, supply and safety of food.

Broadly, partnership interactions can be at these levels:

- Partnerships in action at global and regional level in support of policy dialogue by ensuring that partners leverage together, share lessons, identify common priorities and mechanisms for cooperation, harmonization of

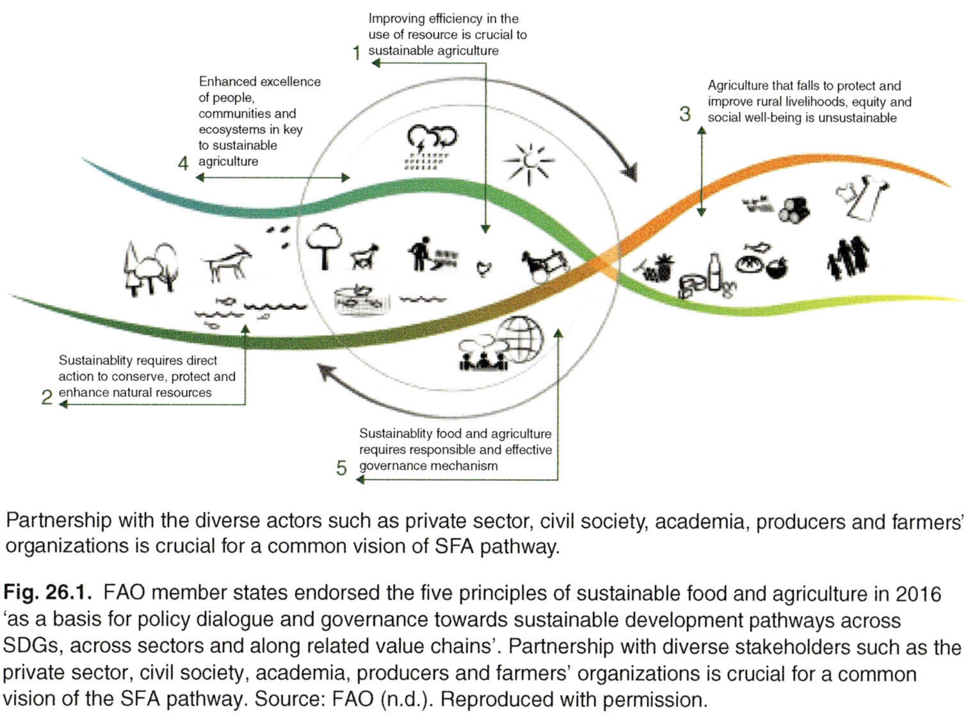

Fig. 26.1. FAO member states endorsed the five principles of sustainable food and agriculture in 2016 'as a basis for policy dialogue and governance towards sustainable development pathways across SDGs, across sectors and along related value chains'. Partnership with diverse stakeholders such as the private sector, civil society, academia, producers and farmers' organizations is crucial for a common vision of the SFA pathway. Source: FAO (n.d.). Reproduced with permission.

approaches. It should help countries participate effectively in international instruments that have a direct impact on agriculture, fisheries and forestry productivity and sustainability, develop tools and methodologies for collecting and analysing data and indicators for monitoring progress and promote an enabling environment through improved governance, mutual accountability, multi-sectoral coordination and increased investment for optimal outcomes for sustainable diets.

- Partnerships in action in support to smallholder producers at country level: At country level, principally through its country programmes, the role of various actors can be increasingly aligned to sustainable food and agriculture forming part of the basis for resource mobilization and programme design at country level. It must include policies and measures for supporting diversification of food production systems and value chains for improved access to safer, healthier and nutritious diets. The process should be transparent, and the strategies, institutions and approaches adopted must be capable of adapting to changing conditions.
- South–South cooperation/triangular cooperation: Deployment of expertise plays a key role in targeting exchange of development solutions, capacities and best practices through the deployment of long-term experts, mobilizing technical advice and the sharing of technological solutions, demonstrating new technologies, and bringing opportunities for learning by doing and hands-on training. Sharing experiences, successes and failures in improving food choices, knowledge, attitudes and practices towards healthier and safer diets can lead to action with better impact on the ground.

Alliance with Non-state Actors

The adoption of the sustainable development goals (SDGs) in 2015 is a milestone in the global community's approach to development. All UN members agreed to the 2030 Agenda that offers a vision of a fairer, more prosperous, peaceful and sustainable world in which no-one is left behind, and addresses sustainable development in all its dimensions, by paying special attention to the effects and conditions of women, youth and the most marginalized rural communities, especially those who are dependent on agriculture and natural resources for their livelihoods (FAO, 2018). The seventeen SDGs, broken down into 169 targets and 232 indicators, are linked to measurable outcomes to aid progress tracking. Partnerships at the national, regional and international levels are key mechanisms for SDG implementation, enabling collective and coherent action, catalysing financial, institutional and knowledge support and multi-stakeholder engagement in all sectors.

Food and agriculture lie at the heart of achieving the 2030 Agenda. In particular, SDG2 and SDG 12 are closely linked; SDG2 targets ending hunger and malnutrition through the promotion of good agriculture practices and farmers' livelihoods, whereas SDG12 focuses on reducing food loss and waste, improving quality and ensuring sustainable food system. This requires a transformative shift in agriculture (FAO, 2015), whereby all stakeholders place particular emphasis on sustainability in production, linkage to efficient value chains, smallholders' income and markets. Together with the national governments and its institutions, the main players who are engaged in this sector across the value chain are the private sector organizations, including the small and medium enterprises, producer and farmer organizations, civil society organizations, academia and research institutions, and regional and international intergovernmental organizations and aid agencies. To a large extent all these actors face a somewhat similar challenge as to how to increase their role and impact to better address the issues linked to food and nutrition security and the interlinkages between them. The underlying causes of change in food systems include agricultural industrialization, population growth and urbanization, climate change, globalization and technological innovations in the way food is produced, processed, retailed and marketed (FAO, 2017a). A partnership-based approach is a proven modality that can drive a collective effort for a transformative change. However, particular attention is needed to ensure that partnerships are inclusive and leverage knowledge and resources, with a particular focus on the needs of smallholders and the vulnerable sectors.

The Role of Civil Society, Farmers' and Producers' Organizations

Worldwide, civil society organizations (CSOs) have increasingly proven their ability to mobilize, campaign and launch initiatives that raise awareness and fight poverty and hunger both at the grassroots level and as a collective force at the regional and international fore. They are recognized for their work on advocacy field knowledge and information on the local context gathered through the direct presence on the ground. As partners, CSOs can enhance resource mobilization and increase participation of civil society, farmers and producers in decision-making processes. Simultaneously, engagement with CSOs is an opportunity to strengthen networking among different development actors and increase knowledge sharing through capacity development and field programmes.

Within the food and nutrition sector, civil society has a particularly crucial role to play in supporting political processes and decision making, promoting sustainability related issues at the institutional level as well as raising awareness for sustainable diets among different stakeholders, including the youth. In a local context, networks of smallholder and food producers and farmers, can be important allies to build capacities of their members to apply innovative technologies, solutions and best practices to increase the sustainability of agriculture and diets and contribute to the achievement of food and nutrition security.

The engagement of the FAO with La Via Campesina (LVC) in promoting an agroecology-based approach in the context of its contributions to food and nutrition security and climate change can be an illustrative example. The role of small-scale food producers in agroecology-related decision-making processes has been especially promoted since 2014. LVC played a proactive role in the first International Symposium on Agroecology (FAO, 2014a) where the agroecology-related technical, social, political, economic and cultural issues were discussed. During the symposium, LVC delegates and speakers from eight member countries highlighted agroecology as an alternative to the agro-industrial food system. Furthermore, they also supported organization of several regional seminars on agroecology in an effort to embed agroecology within local and regional context. These seminars brought together different actors, such as civil society representatives, policy makers, researchers and farmers, for sharing their views and knowledge exchanges on the relevance of agroecology in addressing the needs of family farmers and discussed how to mainstream agroecology through practice, research and policy.

Role of Private Sector

Private companies in the agricultural sector vary from small and medium enterprises run by local producers to multinational corporations operating across different countries and foundations supported by private companies. Every company along the value chain, large and small in size, has the potential to make a significant contribution to advance economic, social and environmental sustainable development through investing in innovative technologies and research, sharing knowledge and expertise, strengthening local agri-businesses, generating jobs and promoting sustainable development solutions and best practices at scale. At the broadest level, such alignments can expand outreach to open new possibilities for embedding sustainable food and agriculture issues in the broader context of societal choices.

In an era of challenging public sector resource constraints and demographics and technological and ecosystem transformations, the need to work with the private sector potential is perhaps greater than ever before. As the 2030 Agenda highlights, there is a need to evolve from short-term, ad-hoc partnerships with the private sector to a more transformative and systematic relationship. With regards sustainable food and agriculture and sustainable diets, strategic partnerships with the private sector can help generate significant and relevant knowledge, experience and strategic thinking and data, provided they are brought on board from the beginning. Producer organizations and cooperatives, in particular, can address the challenges of food quality and resource-use efficiency across the production chain to generate value addition and income. They can be central actors to support certification for sustainable agricultural practices and promote sustainability among consumers. Women entrepreneurs, in particular, can play pivotal roles if they are empowered through improved access to land, credit and education. All forms of collaboration are needed to achieve scale and sustained impact in the various sectors.

Despite the potential opportunities, there are not that many examples of joint actions between the corporate sector and governments that are promoting sustainable food systems taking into account smallholders and consumer concerns. A recent initiative was taken by FAO with the Italian restaurant chain Auto Grill, where the chain will offer ecological products from the Bolivian network of cooperatives and small producers to support smallholder food processing enterprises, and promote sustainable diets in Bolivia. The experience indicates that there is a need for dialogue between governments and the private sector to identify the priorities for healthy diets and nutritious food and establish the terms of engagement with the players on a case-by-case basis. It is necessary to address the complexities and the interlinkages between food systems, health and sustainability. The challenges associated in establishing the relationship should not be undermined in order to gain mutual trust, to avoid conflicts of interest, and deployment of resources. However, when handled with diligence, the advantages of a successful public–private venture can far outweigh the negative consequences.

The Role of Academia and Research Institutions

Universities and research institutions are important knowledge hubs and can support sustainable diets in multiple ways. Academia and researchers provide evidence-based solutions, nurture innovation and critical thinking, and help in developing capacities and skills essential to advance sustainable development. It can help to transform new knowledge into practical solutions on the ground, increase innovations and bring diverse and balanced perspectives to enrich policy dialogues for informed decision making. It is expected to address knowledge gaps, generate data, promote applied research and update curricula to reflect the current food systems approach to nutrition. Academia also have a very important role in bringing together different actors, including farmers, producers, civil society representatives and youth, for dialogue, knowledge exchanges and capacity building. Enhancing skills and raising awareness among students and the public on the key role they have in contributing to sustainable food and agriculture, sustainable diets, health and nutrition is an important area for joint action.

A commonly acknowledged problem in promoting healthy diets and nutrition is the dearth of appropriate data for informed decision making. Often the data produced are largely under-utilized due to a poor dissemination. There is also a lack of data harmonization that prevents comparisons across periods of time. The FAO and the World Health Organization are working together to develop GIFT platform (FAO/WHO, n.d.) – a publicly available multipurpose global database through the collation and harmonization of existing data collected within individual food consumption surveys conducted at national or sub-national level. This tool, developed in collaboration with Tufts University and other national research institutions, can be better used to build capacities, monitor food consumption and support nutritious and healthy diets.

Academia and research institutions also contributed to the international and four regional symposia organized in 2016–2017 on sustainable food systems for a healthy diet under the United Nations Decade of Action on Nutrition. The main objectives were to take stock of food security and nutrition challenges, and to share country experiences and how they influence dietary patterns in order to identify regional policy and programmatic processes for tackling nutrition issues through a food systems approach (FAO, 2016).

The Role of Indigenous Communities

The traditional food production and agriculture systems sustained by indigenous communities emphasize the critical role of sustainable agriculture practices for the ensuing healthy and sustainable diets. Most of the practices are attuned to the local condition and use local crops and production systems. Indigenous foods rely on thorough knowledge of neglected crops and breeds, agroecology and dietary patterns; however, due to over reliance on processed food and industrial food processing, most of the indigenous food systems are in decline. A good example of such systems are the globally important agriculture heritage systems (GIAHS) (FAO/GIAHS, n.d.) that have developed over millennia and reflect the indigenous knowledge systems and cultures of food producers and their place-based relationship with nature. They are referred to as land use systems and landscapes which are rich in globally significant biological diversity evolving from the co-adaptation of a community with its environment and its needs and aspirations for sustainable development. The important benefits arising from traditional food systems need special attention to raise awareness of their value and establish linkages for sustainable use of natural resources.

The Role of Parliamentarians

Parliamentarians can be important actors in establishing and enabling the political and legislative environment to achieve food and nutrition security. In most cases, improvement of food and nutrition security is principally through implementation of policies, programmes and frameworks that are anchored in appropriate legislations. Therefore, parliamentarians can position the issue of sustainable diets and nutrition at the highest level of the political and legislative agenda, thereby strengthening the political commitment and provision of budgetary allocations for the delivery. Several parliamentary alliances have been established, such as the Parliamentary Front Against Hunger in Latin America that actively support enactment of new laws in the region relating to food and nutrition security. Recognizing their critical role in food and nutrition security, the FAO has since engaged parliamentarians in Africa and Europe, by facilitating the establishment of the Parliamentary Alliance for Food Security and Nutrition (FAO, 2017b) in these regions.

Conclusion

The Framework for Action adopted by the Second International Conference on Nutrition (ICN2), held in Rome in November 2016, strengthened the premise that sustainable food systems are key to promoting healthy diets. Such food systems are reinforced by adopting SFA practices, whereby farmers (including the big land holders and smallholders), and consumers converge to adopt good agriculture practices for the production, promotion and consumption of healthy foods.

A partnership approach provides an opportunity to bring together a broad spectrum of players to work around a common theme or goal. It requires coordination, risk management and commitment. At the same time, it promotes consensus building on the path towards sustainability and catalyses coherent and collective practice change through dialogue, consultation, joint analysis and resource mobilization. It helps to improve policy processes, including policy monitoring, social outreach, message dissemination and sharing of lessons learned. Alliances can be through different modalities and on diverse topics – sustainability, resilience, trade-offs, ecosystem services, food loss – built on institutional links and complementarities. Social media tools can provide a vehicle for widening the access within the community.

References

FAO (n.d.) Sustainable food and agriculture. Available at http://www.fao.org/sustainability/en/ (accessed 29 June 2018).

FAO (2014a) Agroecology for food security and nutrition. Proceedings of the International Symposium. Available at http://www.fao.org/3/a-i4729e.pdf (accessed 25 June 2018).

FAO (2014b) Building a common vision for sustainable food and agriculture: principles and approaches. Available at http://www.fao.org/3/a-i3940e.pdf (accessed 4 July 2018).

FAO (2015) FAO and the 17 Sustainable Development Goals. Available at http://www.fao.org/3/a-i4997e.pdf (accessed 25 June 2018).

FAO (2016) International symposium on sustainable food systems for healthy diets and improved nutrition. United Nations Decade of Action on Nutrition. Available at http://www.fao.org/about/meetings/sustainable-food-systems-nutrition-symposium/en/ (accessed 29 June 2018).

FAO (2017a) Nutrition and food systems. A report by the High-Level of Experts on Food Security and Nutrition. Available at http://www.fao.org/3/a-i7846e.pdf (accessed 25 June 2018).

FAO (2017b) Meeting of the Parliamentary Front against Hunger in Latin America and the Caribbean (PFH LAC) with Members of Parliament of the Caribbean Community. Available at http://www.fao.org/americas/eventos/ver/en/c/1025205/ (accessed 25 June 2018).

FAO (2018) Transforming food and agriculture to achieve the SDGs: 20 inter connected actions to guide decision makers. Available at http://www.fao.org/3/I9900EN/i9900en.pdf (accessed 25 June 2018).

FAO/GIAHS (n.d.) Globally Important Agricultural Heritage Systems. Available at www.fao.org/giahs/en (accessed 25 June 2018).

FAO/WHO GIFT (n.d.) Global individual food consumption data tool (GIFT). Available at www.fao.org/gift-individual-food-consumption/en (accessed 25 June 2018).

27 Decalogue of Gran Canaria for Sustainable Food and Nutrition in the Community

Lluís Serra-Majem, Javier Aranceta Bartrina, Adriana Ortiz-Andrellucchi, Cristina Ruano-Rodriguez, Esther González-Padilla and Sandro Dernini

Abstract

The 'Decalogue for sustainable food and nutrition in the community: Gran Canaria Declaration 2016' aims to improve food sustainability across the globe. Public health, nutrition, consumer, social, marine and environmental sciences and tourism are important topics that have been highlighted in this decalogue, whose full implementation promotes the development of sustainable consumption and production patterns. Food sustainability is an urgent matter that depends on collaborative efforts from governments, the private and public sectors, as well as individuals. Supply and demand works both ways – a shift in the food production landscape depends on a shift in our diets. It is widely recognized that diet plays an important role in sustainable consumption, and sound science-based guidance is required as individuals, industries and policymakers address the burgeoning environmental challenges. This chapter also takes into consideration the scientific evidence that justifies the development and implementation of the ten keys for a healthier life and world. Investing in the future we want is everyone's responsibility, and a commitment of the present and future generations. The accountability of all nutrition stakeholders needs to improve if this virtuous circle between sustainable development and nutrition is to be fully realized.

Introduction

Around thirty national and international nutrition experts collaborated in production of the 'Decalogue for sustainable food and nutrition in the community: Gran Canaria Declaration 2016' (Serra-Majem *et al.*, 2017). The aim of this decalogue was to improve food sustainability across the globe. To our knowledge, this guide is a pioneer in the field, with worldwide significance. It was developed from the conclusions drawn in the 'Community Nutrition and Sustainability Expert Meeting' held at the beginning of April in the municipalities of Santa Brígida and Vega de San Mateo, Gran Canaria. It was promoted by the Spanish Academy of Nutrition and Food Sciences (AEN), the Nutrition Research Foundation (FIN), the Research Institute of Biomedical and Health Sciences University (University of Las Palmas de Gran Canaria) and the 'The Island on your Plate' project, the Spanish Society of Community Nutrition (SENC), the International Foundation of Mediterranean Diet (IFMeD), the non-governmental organization (NGO) Nutrition Without Borders, and the CIBER Physiopathology of Obesity and Nutrition.

Moreover, the document has gained the support of over 50 institutions located throughout the world, ranging from consumer associations, research institutes, scientific societies, United Nations organizations, NGOs and specialized publications, that have come together to highlight the importance of sustainable nutrition within the current food panorama. The environmental sustainability of food systems is a critical challenge for policy makers (Ridgway *et al.*, 2015).

The decalogue pointed out the idea of consuming seasonally local products to reduce the environmental footprint and the energy consumption linked to transporting goods; the value of reviving traditional recipes, buying and cooking in the company of family and friends, reducing waste and recycling adequately, or prioritizing plant-/algae-based foods and limiting the consumption of meat, processed meat and dairy products. A deeper understanding of dietary choices through integrated environmental and nutritional assessments offers a basis for better aligning environmental and health objectives of our food system at a variety of policy levels.

Other recommendations included: the importance of utilizing terrestrial and aquatic biodiversity in a sustainable manner to ensure its continuity and to maintain diet variety; take an interest in the sustainability and equity of agricultural, livestock and fishing practices; enjoy companionship and pleasure at mealtimes, always within the context of balance and moderation; and keep in mind that the Mediterranean diet represents one of the most outstanding and emblematic examples of healthy and sustainable food and nutrition, being recognized as an intangible cultural heritage by UNESCO. Box 27.1 provides a summary of this decalogue.

The Decalogue for Sustainable Food and Nutrition: Point-by-Point

1. Select and consume locally sourced foods. Choose products available at local markets

Transporting foods from remote locations to their points of consumption represents an important component of the ecological footprint of food, primarily due to energy consumption and its consequent environmental pollution. Moreover, the purchase of locally sourced products in local markets has a positive effect on the local economy, as well as in the reactivation of the rural environment and the protection of the ecosystems and landscapes. Programmes and policies that support sourcing local and regional foods for schools, hospitals, faith-based organizations and worksites may benefit institutional customers and their families, farmers/fishermen, the local community and the economy (Harris *et al*., 2012; Johns *et al*., 2013; Ishdorj *et al*., 2016).

Box 27.1. Decalogue developed based on an Expert Consensus Meeting held in Gran Canaria, Spain on the 8th and 9th of April 2016.

The ten key points for a healthier life and world are:

1. Select and consume locally sourced foods. Choose products produced in your own region and made available at local markets.
2. Preferentially consume foods that are in season. They are healthier, more economical and sustainable.
3. Revive traditional local foods and recipes. They are part of our culture and make up our identity.
4. Learn to buy and cook in the company of others. It's more fun and enriching. And we can learn from each other.
5. Plan menus and shopping lists. Try to reduce food waste and recycle adequately at home and in the community.
6. Prioritize plant-based foods. Limit the consumption of meat, processed meat and dairy products. Your health and the planet's will appreciate it.
7. Aquatic and terrestrial biodiversity is critical and we should promote it to ensure its continuity. It's everyone's responsibility.
8. Take an interest in whether the agricultural, livestock and fishing practices which provide the foods you obtain and consume are sustainable.
9. Enjoy the companionship and pleasure of mealtimes, always within the context of balance and moderation. Reduce portion sizes.
10. Enjoy the Mediterranean diet. It is one of the best examples of healthy and sustainable food and nutrition. UNESCO has declared it an Intangible Cultural Heritage of Humanity – they surely must have their reasons for doing so.

2. Preferentially consume foods that are in season

Seasonal foods respect favourable climatic conditions and facilitate foods with better organoleptic and nutritional characteristics. Moreover, foods bought in season are usually more economic and sustainable. With regard to aquatic foods, consumers should gather suitable information in order to prioritize the purchase of fish products out of the reproductive season of the donor food species/resources. As stated, in general fish and seafood have worst organoleptic and nutritional properties during their reproductive seasons. Sustainable development means improving the quality of life within carrying capacity of ecosystems. The health sector has an important role linking environmental with economic development; surveillance systems could monitor health status and the impacts on ecological and economic sustainability (Litsios, 1994).

3. Revive traditional local foods and recipes

Food education is a fundamental element for the culture and identity of people. Such education should come with measures aimed at improving the accessibility of local products to ensure feasibility and sustainability (Ouédraogo *et al.*, 2009). Therefore, the protection of gastronomic and cultural heritage in the community should be seen as a priority for the sustainability of food and nutrition in and of itself. Promoting use of local traditional food biodiversity is an essential driver of food system sustainability for peoples, and contributes to global consciousness for protecting food biodiversity and food system sustainability more broadly (Kuhnlein, 2015).

4. Learn to buy and cook in the company of others

Whether with family or with friends, learning healthy eating habits involves knowing about food and culinary techniques and having basic abilities for the purchase, and even for the production, of foods and ingredients (Ouédraogo *et al.*, 2009; Bowen and Devine, 2011).

5. Plan menus and shopping lists. Try to reduce food waste and recycle adequately at home and in the community

Planning food purchases and menus should follow the criteria for sustainability: health, environment, economy and culture. Consumers' attitudes and behaviour towards sustainable research are emerging. Consumer research has focused primarily on specific areas of sustainable food, such as organic food, local or traditional food, meat substitution and/or reduction (Pieniak *et al.*, 2016). Therefore, excess food and waste generated throughout the entire food chain (production, distribution and consumption) should be avoided (FAO, 2012; Medina, 2015). Try to adequately recycle both food waste and its packaging, putting food solidarity into practice as well. The recent review by Li *et al.* (2016) has summarized the sources, occurrence, fate and effects of plastic waste in the marine environment. Due to its resistance to degradation, most plastic debris will persist in the environment for centuries and may be transported far from its source, including great distances out to sea. Land- and ocean-based sources are major sources of plastic entering the environment, with domestic, industrial and fishing activities being the most important contributors.

6. Prioritize plant-based foods. Limit the consumption of meat, processed meat and dairy products

The production of animal foods (in particular commonly consumed red meat, processed meats and dairy products) has an environmental footprint that is greater than that of plant foods (cereals, fruits, vegetables, legume and nuts). This is especially true for greater emission of gases having a greenhouse effect, as well as increased water and energy consumption and usage of land area (Yip *et al.*, 2013; Lacirignola *et al.*, 2014; Clonan *et al.*, 2015; Machovina *et al.*, 2015). What is more, people's health will also benefit from making these recommended changes. Rising incomes and urbanization are driving a global dietary transition in which traditional diets are replaced by diets higher in refined sugars, refined fats, oils and meats. By 2050 these dietary trends, if unchecked, would be a major contributor to an estimated 80% increase in global agricultural greenhouse gas emissions from food production and to global land clearing. Diets link environmental and human health. The implementation of dietary solutions to the tightly linked diet-environment-health trilemma is a global challenge, and opportunity, of great environmental and public health importance (Belahsen, 2014; Tilman and Clark, 2014; Vanham and Bidoglio, 2014).

7. Promote aquatic and terrestrial biodiversity to ensure its continuity. It's everyone's responsibility

The loss of biodiversity over the last few decades in both vegetal and animal species may jeopardize the sustainability of the global food system, and could lead to compromised food security and the deterioration in food quality and variety. Understanding the consequences of loss of species/taxa in complex ecological communities is one of the great challenges in current biodiversity research (Brose *et al.*, 2016). The consumer, although having less responsibility for these aspects of food and nutrition, should be proactively aware of their importance and long-term significance. As consumers we should look for relevant information available in shops and markets.

8. Take an interest in whether the agricultural, livestock and fishing practices which provide the foods are sustainable

Sustainability is an important aspect of the entire food chain process (production, transformation and distribution) for both plant/algae and animal products. The environment is usually more vulnerable to intensive production than to traditional systems of production, raising livestock and fishing, and therefore sustainable and harmonious food systems should be promoted (Bruschi *et al.*, 2014). A central challenge for sustainability is how to preserve forest ecosystems and the services that they provide us while enhancing food production (Lambin and Meyfroidt, 2011). A few developing countries have managed a land use transition over the recent decades that simultaneously increased their forest cover and agricultural production. Globalization can be harnessed to increase land use efficiency rather than leading to uncontrolled land use expansion (Uhart and Milano, 2002). The same applies to the aquatic use extension, in particular with regard to deep-sea environments. The application of the ecosystem approach (https://www.cbd.int/ecosystem/) is a 'strategy for the integrated management of land, water and living resources that promotes conservation and sustainable use in an equitable way'.

9. Enjoy the companionship and pleasure of mealtimes, always within the context of balance and moderation. Reduce portion sizes

Companionship and pleasure are fundamental elements in the act of eating that provide it with a sense of identity. However, they should be combined with balance, variety and moderation to avoid that entertainment and festivity turn into excess and wastefulness. Large food portions increase total energy intake. Crucially, portion size is a modifiable determinant of dietary energy intake. In a sense, excessive food and energy consumption can be considered as food waste, with important consequences for health (Stroebele *et al.*, 2009; Vermeer *et al.*, 2011; Marteau *et al.*, 2015).

10. Enjoy the Mediterranean diet. It is one of the best examples of healthy and sustainable food and nutrition. UNESCO has declared it Intangible Cultural Heritage of Humanity – they surely must have their reasons for doing so

The Mediterranean diet, classified as a UNESCO intangible cultural heritage, embodies one of the healthiest, most traditional and most sustainable food models in the world (UNESCO, 2010). Its preservation and promotion not only have effects on the health of individuals and communities, but also impact on the health of the planet

(Burlingame and Dernini, 2011; Donini *et al.*, 2016). Community-based actions that promote healthy eating patterns based on locally available foods linked to territory (both aquatic and terrestrial), culture, equity and economy should be developed and supported. Changes in diet, reducing animal products and increasing consumption of vegetables can not only benefit human health and the overall use of land/water/resources, but can also play a decisive role in the politics of climate change mitigation. In this sense, the Mediterranean diet is presented as a sustainable cultural model respectful of the environment, whose adherence in Mediterranean countries should contribute to mitigate the global climate change (Serra-Majem *et al.*, 2011; Lake *et al.*, 2012; Sáez-Almendros *et al.*, 2013; Germani *et al.*, 2014).

Summary

This chapter highlights scientific evidence that justifies the development and implementation of the ten keys for a healthier life and world. Healthy people are vital for local development that is both economically and ecologically sound. Sustainable development is a healthy development (Litsios, 1994). Investing in the future we want is everyone's responsibility, and a commitment of the present and future generations. An educational video including the Gran Canaria's decalogue has been published and is available in English and Spanish (https://www.youtube.com/watch?v=_91m8N85cBk and https://www.youtube.com/watch?v=UC4Au7w_8xo).

Acknowledgements

Special thanks to the following partners: Spanish Academy of Nutrition and Food Sciences (AEN); Nutrition Research Foundation (FIN); Spanish Society of Community Nutrition (SENC); International Foundation of Mediterranean Diet (IFMeD); NGO Nutrition without Borders (NsF); Centre International de Hautes Etudes Agronomiques Méditerranéennes (CIHEAM); International Inter-University Center for Mediterranean Food Culture Studies (CIISCAM); Ilustre Ayuntamiento de la Vega de San Mateo; Ilustre Ayuntamiento de la Villa de Santa Brígida; Mancomunidad de Municipios de las Medianías de Gran Canaria; Forum on Mediterranean Food Cultures; UNESCO Chair of Research, Development and Planning of Local Health Systems; UNESCO Chair in Tourism Planning and Sustainable Development; and International Association of Community Nutrition (IACON).

References

Belahsen, R. (2014) Nutrition transition and food sustainability. *Proceedings of the Nutrition Society* 73(3), 385–388.

Bowen, R. and Devine, C. (2011) "Watching a person who knows how to cook, you'll learn a lot". Linked lives, cultural transmission, and the food choices of Puerto Rican girls. *Appetite* 56(2), 290–298.

Brose, U., Blanchard, J.L., Eklöf, A., Galiana, N., Hartvig, M. *et al.* (2016) Predicting the consequences of species loss using size-structured biodiversity approaches. *Biological reviews of the Cambridge Philosophical Society* 92(2), 684–697.

Bruschi, P., Mancini, M., Mattioli, E., Morganti, M., and Signorini, M.A. (2014) Traditional uses of plants in a rural community of Mozambique and possible links with Miombo degradation and harvesting sustainability. *Journal of Ethnobiology and Ethnomedicine* 10, 59.

Burlingame, B. and Dernini, S. (2011) Sustainable diets: the Mediterranean diet as an example. *Public Health Nutrition* 14(12A), 2285–2287.

Clonan, A., Wilson, P., Swift, J.A., Leibovici, D.G. and Holdsworth, M. (2015) Red and processed meat consumption and purchasing behaviours and attitudes: impacts for human health, animal welfare and environmental sustainability. *Public Health Nutrition* 18(13), 2446–2456.

Donini, L.M., Dernini, S., Lairon, D., Serra-Majem, L., Amiot, M.J., *et al.* (2016) Nutritional indicators for sustainability of healthy diet. *Frontiers in Nutrition* 3(37), 1–14.

FAO (2012) Pérdidas y desperdicio de alimentos en el mundo – Alcance, causas y prevención. Food and Agriculture Organization of the United Nations, Rome, Italy.

Germani, A., Vitiello, V., Giusti, A.M., Pinto, A., Donini, L.M. and del Balzo, V. (2014) Environmental and economic sustainability of the Mediterranean Diet. *International Journal of Food Science and Nutrition* 65(8), 1008–1012.
Harris, D., Lott, M., Lakins, V., Bowden, B. and Kimmons, J. (2012) Farm to institution: creating access to healthy local and regional foods. *Advances in Nutrition* 3(3), 343–349.
Ishdorj, A., Capps, O. Jr, Murano, P.S. (2016) Nutrient density and the cost of vegetables from elementary school lunches. *Advances in Nutrition* 7(1), 254S–260S.
Johns, T., Powell, B., Maundu, P., Eyzaguirre, P.B. (2013) Agricultural biodiversity as a link between traditional food systems and contemporary development, social integrity and ecological health. *Journal of the Science of Food and Agriculture* 93(14), 3433–3442.
Kuhnlein, H. (2015) Food system sustainability for health and well-being of indigenous peoples. *Public Health Nutrition* 13, 2415–2424.
Lacirignola, C., Capone, R., Debs, P., El Bilali, H. and Bottalico, F. (2014) Natural resources – food nexus: food-related environmental footprints in the Mediterranean countries. *Frontiers in Nutrition* 1, 23.
Lake, I., Hooper, L., Abdelhamid, A., Bentham, G., Boxall, A.B. (2012) Climate change and food security: health impacts in developed countries. *Environmental Health Perspectives* 120(11), 1520–1526.
Lambin, E. and Meyfroidt, P. (2011) Global land use change, economic globalization, and the looming land scarcity. *Proceedings of the National Academy of Sciences USA* 108(9), 3465–3472.
Li, W., Tse, H.F. and Fok, L. (2016) Plastic waste in the marine environment: A review of sources, occurrence and effects. *Science of the Total Environment* 566–567, 333–349.
Litsios, S. (1994) Sustainable development is healthy development. *World Health Forum* 15(2), 193–195.
Machovina, B., Feeley, K.J., Ripple, W.J. (2015) Biodiversity conservation: The key is reducing meat consumption. *Science of the Total Environment* 536, 419–431.
Marteau, T., Hollands, G.J., Shemilt, I. and Jebb, S.A. (2015) Downsizing: policy options to reduce portion sizes to help tackle obesity. *British Medical Journal* 351, h5863.
Medina, J. (2015) Pérdidas y desperdicios de alimentos en el contexto de los sistemas alimentarios sostenibles. Alianza contra el Hambre y la Malnutrición de España (ACHM-E).
Ouédraogo, H., Traoré, T., Zèba, A., Tiemtoré, S., Dramaix-Wilmet, M., *et al.* (2009) Development of an improved local-ingredient-based complementary food and technology transfer to rural housewives. *Food Nutrition Bulletin* 30(2), 153–160.
Pieniak, Z., Żakowska-Biemans, S., Kostyra, E. and Raats, M. (2016) Sustainable healthy eating behaviour of young adults: towards a novel methodological approach. *BMC Public Health* 16, 577.
Ridgway, E., Lawrence, M.A., Woods, J. (2015) Integrating environmental sustainability considerations into food and nutrition policies: insights from Australia's national food plan. *Frontiers in Nutrition* 2, 29.
Sáez-Almendros, S., Obrador, B., Bach-Faig, A. and Serra-Majem, L. (2013) Environmental footprints of Mediterranean versus Western dietary patterns: beyond the health benefits of the Mediterranean diet. *Environmental Health* 30(12), 118.
Serra-Majem, L., Bach-Faig, A., Miranda, G. and Clapes-Badrinas, C. (2011) Foreword: Mediterranean diet and climatic change. *Public Health Nutrition* 14(12A), 2271–2273.
Serra-Majem, L., Bartrina, J.A., Ortiz-Andrellucchi, A., Ruano-Rodriguez, C., González-Padilla, E., *et al.* (2017) Decalogue for Sustainable Food and Nutrition in the Community: Gran Canaria Declaration 2016. *Journal of Environmental Health Science and Engineering* 3(2), 1–5. DOI: 10.15436/2378-6841.17.1701
Stroebele, N., Ogden, L.G. and Hill, J.O. (2009) Do calorie-controlled portion sizes of snacks reduce energy intake? *Appetite* 52(3), 793–796.
Tilman, D. and Clark, M. (2014) Global diets link environmental sustainability and human health. *Nature* 515(7528), 518–522.
Uhart, M. and Milano, F. (2002) Multiple species production systems: reversing underdevelopment and nonsustainability in Latin America. *Annals of the New York Academy of Science* 969, 20–23.
UNESCO (2010) Representative list of the intangible cultural heritage of humanity. Available at http://www.unesco.org/culture/ich/en/RL/00394 (accessed 31 August 2016).
Vanham, D. and Bidoglio, G. (2014) The water footprint of Milan. *Water Science Technology* 69(4), 789–795.
Vermeer, W., Steenhuis, I.H., Leeuwis, F.H., Heymans, M.W. and Seidell, J.C. (2011) Small portion sizes in worksite cafeterias: do they help consumers to reduce their food intake? *International Journal of Obesity* 35(9), 1200–1207.
Yip, C., Crane, G., Karnon, J. (2013) Systematic review of reducing population meat consumption to reduce greenhouse gas emissions and obtain health benefits: effectiveness and models assessments. *International Journal of Public Health* 58(5), 683–693.

28 Ten Years to Achieve Transformational Change: the United Nations Decade of Action on Nutrition 2016–2025

Stineke Oenema

Abstract

The world has formulated an ambitious agenda foreseeing to eliminate all forms of malnutrition and achieving sustainability targets. This agenda is described in various globally agreed documents: the 2030 Agenda, the outcome documents of the Second International Conference on Nutrition (ICN2) and the nutrition targets of the World Health Assembly. The decade 2016–2025 has been proclaimed the United Nations Decade of Action on Nutrition, and offers a ten-year window of opportunity to intensify policies, programming and actions to improve nutrition. The Nutrition Decade should lead to the transformation of food systems in order to achieve the global nutrition targets, the elimination of all forms of malnutrition and accelerate the achievement of the 2030 Agenda. The promotion of sustainable diets is an entry point to start doing this. Sustainable diets serve to promote people's health and promote the demand for sustainably produced food as well as reduce the demand for products that have a high environmental footprint. The development of national food-based dietary guidelines (FBDG) that include sustainability criteria is an important step to promote sustainable diets. Apart from FBDG, the food environment and the space in which consumers make their dietary choice should be nudged in such a way that the healthier and more sustainable choice becomes the easier and obvious choice. This could be done through several forms of regulations, including taxes and subsidies. Despite the emerging level of evidence underpinning these measures and tools, still more insight and indicators are needed to be able to make the best decisions to change the food environment for the better. Investments are needed and are worth the effort considering the rate of return for investments in nutrition is 1:16. But we have to act now: the Nutrition Decade has been underway for two years, eight years to go…

Introduction

In 2016 the United Nations General Assembly declared the first ever Decade of Action on Nutrition: 2016–2025. The decade is a follow-up of the Second International Conference on Nutrition (ICN2) in 2014 and strongly refers to the sustainable development goals (SDGs). The nutrition and sustainability agendas are strongly connected in and by the Nutrition Decade and are mutually supportive. For the first time in history the world aims to *eliminate all* forms of malnutrition, leaving no-one behind. The ICN2 Framework for Action (FFA) provides the 'how' to operationalize the 'what' of the Rome Declaration (ICN2) in 2014 and the nutrition targets of the World Health Assembly in 2012 (WHO, 2012, 2014a, 2014b). Comprising sixty recommendations for action, the FFA addresses all forms of malnutrition as well as attempts to prevent the onset of malnutrition in all its forms by fostering healthy diets in a sustainable food system (UNSCN, 2017a).

 Sustainable Diets: Linking Nutrition and Food Systems (eds B. Burlingame and S. Dernini)

The Decade of Action on Nutrition further operationalizes this global framework by calling for concrete commitments and actions at country level, led by countries. At the same time, we are approaching the end of the Biodiversity Decade (2011–2020). The definition of sustainable diets explicitly references biodiversity, as does SDG 2. The nutrition and biodiversity targets reinforce each other, and sustainable diets is the focal point. Support of shorter supply chains for local produce; public and private procurement programmes for agrobiodiversity; promotion of fruits and vegetables and policy incentives and coherence for diversified production and consumption are recommended actions for more sustainable food systems and healthier diets (Wijesinha-Bettoni *et al.*, 2017).

The UN Decade of Action on Nutrition

The aim of the Nutrition Decade is to accelerate implementation of the ICN2 commitments, achieve the global nutrition and diet-related non-communicable disease (NCD) targets by 2025 and contribute to the realization of the SDGs by 2030 (Fig. 28.1) (FAO and WHO, 2017). The Food and Agriculture Organization (FAO) and the World Health Organization (WHO) in collaboration with United Nations System Standing Committee on Nutrition (UNSCN) have developed a work programme[1] for the Decade of which FAO and WHO lead the implementation, as requested by the UN General Assembly (UNGA). This implementation requires that food systems around the world be transformed and this can only happen if sufficient (massive) support is mobilized and sufficient (small and big) changes are being made, all contributing to create a tipping point that sustainably transforms the system(s). Therefore, UNGA also invited governments and other relevant stakeholders, including international and regional organizations, civil society, the private sector and academia, to actively support the implementation of the United Nations Decade of Action on Nutrition, including through voluntary contributions, as appropriate (UNGA resolution A/Res/70/259). This mobilization of support among all current and new nutrition actors at global, regional and local level is part and parcel of the decade's work programme.

Interlinkages Between the Sustainable Development Goals and Nutrition

Considering the multi-sectoral nature of nutrition and the coexistence of several forms of malnutrition (undernutrition, overweight and obesity and diet-related NCDs) the alignment and coherence of policies, programming and action is essential. Double-duty actions that have the potential to simultaneously reduce the risk or burden of the several forms of malnutrition should be considered. The Expert Group Meeting on Progress in Achieving SDG 2 (EGM), a gathering of over 100 key experts was convened in June 2017 to prepare key messages for the High Level Political Forum that took place in July 2017. They specifically looked at SDG 2 (End hunger, achieve food security and improved nutrition and promote sustainable agriculture), and developed a set of key messages[2] in which sustainable food systems was a central concept.

> Food security, adequate nutrition and sustainable agriculture will not be achieved without addressing climate change and the loss of biodiversity; using land, water and energy sustainably; tackling food loss and waste; promoting responsible consumption; tapping into the potential of agriculture to create decent jobs; expanding social protection; and achieving gender equality.
>
> (UNDESA, 2017)

Another key message emphasized the importance of diversity and diversification for achieving SDG 2. Reasoning the other way around also holds true: gradually more evidence is emerging that healthy diets are at the same time more sustainable than unhealthy ones. For example, Springman *et al.* (2016) estimated that following the WHO guidelines for healthy diets and the recommendations by the World Cancer Research Fund, greenhouse gas emissions could be reduced by 29–70% and mortality by 6–10%. Popular examples of healthy and sustainable diets are the Mediterranean diet and the new Nordic diet that both advise more consumption of fruits and vegetables, whole grains, fish and

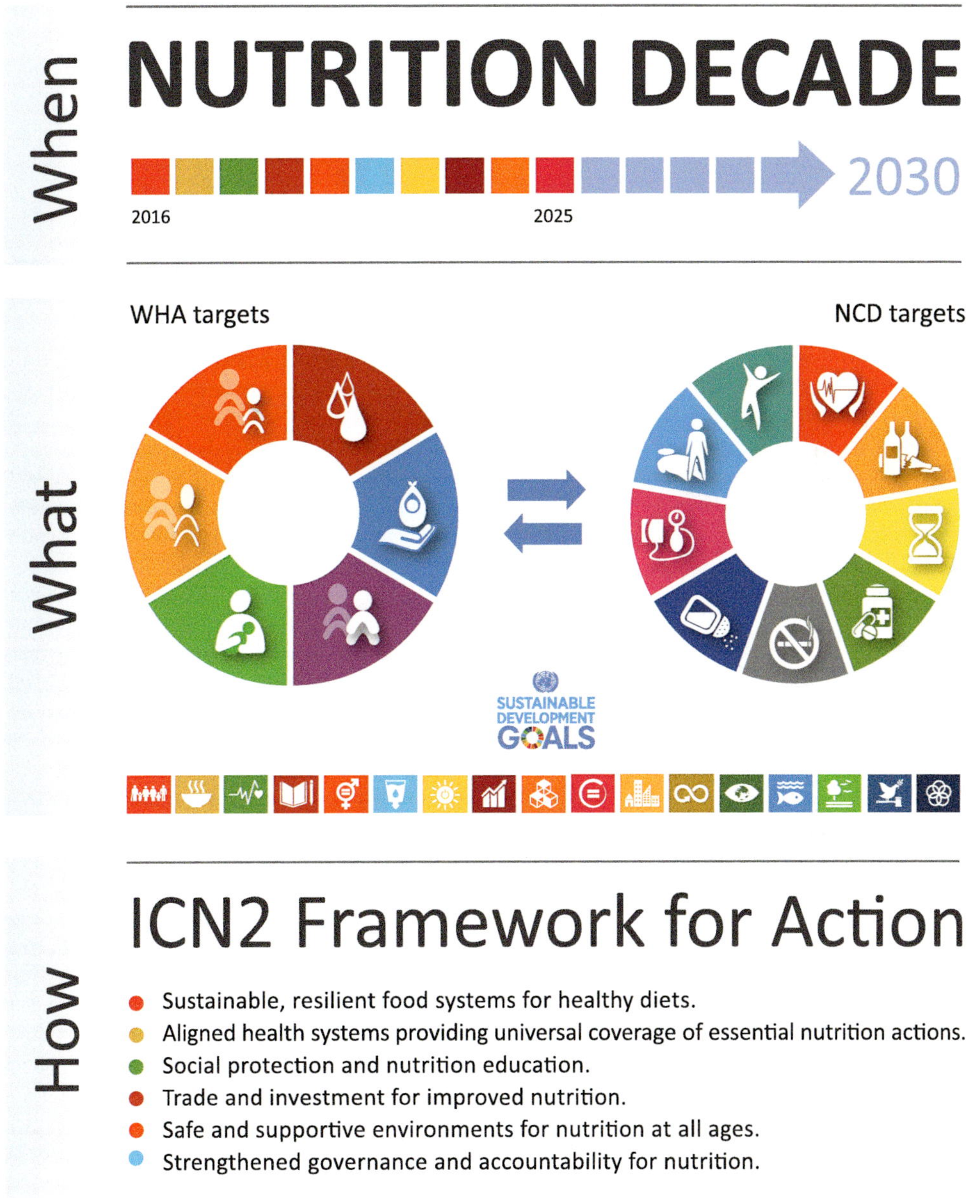

Fig. 28.1. UNSCN infographic: by 2030 end all forms of malnutrition and leave no one behind. Source: UNSCN (2017a).

non-animal proteins, and at the same time recommend less meat, processed foods and sweets, and moderate consumption of dairy.

The key messages of the EGM show the acknowledgement among experts of the need to tackle malnutrition, food insecurity *and* sustainability issues in a coherent and integrated matter. During the Nutrition Decade all actors should make use of this available knowledge, the emerging evidence to promotion of healthy and sustainable diets as an intelligent double-duty action contributing to reduction and prevention of several forms of malnutrition. Promotion of healthy and sustainable diets can even be considered a triple-duty action as it contributes to sustainable food systems and the achievement of other SDGs.

Work Programmes and Action Areas

Action during the Nutrition Decade is centred on six cross-cutting, integrative areas for impact, derived from the ICN2 FFA recommendations and relevant to related SDGs. While each of these thematic areas inform and frame action, they should not be seen as silos; in practice, policies and programmes should normally be linked to several areas at the same time. The six areas are:

- sustainable, resilient food systems for healthy diets;
- aligned health systems providing universal coverage of essential nutrition actions;
- social protection and nutrition education;
- trade and investment for improved nutrition;
- safe and supportive environments for nutrition at all ages; and
- strengthened governance and accountability for nutrition (FAO and WHO, 2017).

Sustainable food systems with healthy diets at their centre form a perfect entry point: the 'healthy diet' approach provides a key entry point to address the six action areas. For example, promotion of healthy and sustainable diets could help stimulate demand for sustainably produced food in sustainable food systems. Promotion of healthy diets should be routinely incorporated into preventative nutrition interventions in health systems. This is particularly relevant considering that malnutrition and unhealthy diets are the largest risk factor contributing to the global burden of disease. Trade and investments should look harder at the outcomes their policies have on diets. The food environment should be conducive to people choosing and consuming healthy diets – including the protection and promotion of good breastfeeding practices. Consumer education and social protection can and should be designed to protect and promote healthy diets; good examples exist of this in several countries. Good governance practices should promote sound policies that support these actions, including monitoring and evaluation activities to create feedback loops between policy and practice, as well as, as part of good governance, holding duty bearers to account (UNSCN, 2017a).

Promoting Sustainable Diets During the Nutrition Decade

There are many ways one can eat healthily, that is why countries have developed food-based dietary guidelines (FBDGs). However, research has shown that a healthy diet is not necessarily a sustainable diet (Vieux *et al.*, 2013) whereas a sustainable diet is, by definition, a healthy diet (Burlingame & Dernini, 2012), taking into account local availability, ecosystem health and food culture. To strengthen the sustainability aspects of healthy diets, FBDGs should ideally contain sustainability criteria. In this way healthy diets would indeed function as a double-edged sword: improving health and improving the sustainability of our food systems through sustainable production, transport and transformation.

At this moment, just a handful of countries (Brazil, Qatar, Sweden, Netherlands and Germany) have included sustainability criteria in their dietary guidelines. Within the context of the Nutrition Decade there is great potential for FBDGs to be further utilized for creating a supportive environment for nutrition, tying in with the six work areas (Wijesinha-Bettoni *et al.*, 2017). It is essential that more countries commit to include sustainability criteria in their dietary guidelines and do so during the Nutrition Decade.

UNSCN News 42[3] is dedicated to the Nutrition Decade and contains a set of articles providing examples and insights on what several actors think could be done in each of the six action areas. The article by Wijesinha-Betoni *et al.* (2017) describes the result of a desk review of how FBDGs can contribute to the several action areas. Among others: the agricultural sector could promote the local production of foods that are recommended in the FBDGs; the health sector should provide nutrition education based on FBDGs through institutional sessions and the alignment of health and nutrition programmes with the FBDG; social protection schemes should be aligned with FBDGs by aligning food stamps/vouchers with FBDGs and linking small scale farmers with social protection schemes such as school food programmes (Wijesinha-Bettoni *et al.*, 2017). In addition, nutrition can be promoted in institutional settings such as schools, hospitals, and so on. The UNSCN discussion paper 'Schools as a system to improve nutrition' promotes several integrated interventions of which the improvement of diet is an important one. One recommendation states that in order to improve the quality of diets, school-based interventions should be linked to national FBDG; promote dietary diversity including the utilization of traditional, neglected, and underutilized foods, while enhancing biodiversity conservation and environmental sustainability and; use strategically local procurement engaging with female and male smallholder farmers and incorporate fortified foods or nutrient supplements if the nutrient gap cannot be filled otherwise (UNSCN, 2017b). This recommendation may be valid for other institutional settings as well.

Inclusion of Nutrition in Health Systems

In order to achieve the global nutrition targets, it is essential to achieve universal health coverage (UHC) that provide essential nutrition actions. SDG 2 (food security and nutrition) and SDG 3 (health) are closely linked and the achievement of one cannot be realized without the achievement of the other. Currently, the inclusion of nutrition in the UHC package is far from ideal whereas the inclusion of preventative nutrition, health promoting actions is lacking behind with curative nutrition actions (see IFPRI [2015, Tables 4.5 and 4.6] and IFPRI [2016, p. 53] for details). Improvement of diets is a preventative action that needs far more attention and as stated above FBDGs is a useful tool to do so.

Safe and Supportive Environment

FBDGs are one way to promote and facilitate the healthy and sustainable choice and discourage the unsustainable choices. In addition to the development of FBDGs, countries should put in place other measures that encourage a supportive environment for sustainable diets. Examples of such tools are: pricing, taxation and subsidies, marketing strategies and labelling. Stakeholders, including private companies, need to be involved in planning and, where needed, strict agreements about supply, marketing and product composition (voluntary or binding) need to be made. Binding rules would need the involvement of legal experts with nutrition experts, currently not yet a default area of collaboration.

Countries in the Lead, Action Networks

Nutrition and sustainable diets are highly context specific. Cultural and individual preferences play a role, whereas the sustainable production of healthy foods relies on the local physical conditions of the environment, including the soil. The Nutrition Decade's work programme therefore clearly states that countries should take the lead. Of course, countries and their governments can be supported in their endeavours by the UN System. During the 70th session of the World Health Assembly, two countries, Brazil and Ecuador, made commitments to improve nutrition in their countries.

Brazil focuses on halting the rise of obesity among the adult population and does so by promoting healthier diets, more specifically reducing the consumption of sugar-sweetened beverages and increasing the consumption of fruits and vegetables. Ecuador explicitly looks at improving the environment favouring health and healthy diets, including the health system. Both countries include sustainability, particularly looking at sustainable production (being an essential part of food sovereignty).

Apart from commitments by governments, countries are encouraged to take part in action networks in order to stimulate progress in a specific topic related to one of the six areas of work. Norway has announced its plan to establish an action network on sustainable food from the ocean for food security and nutrition. This network is another promising step towards linking sustainability with healthy diets during the Nutrition Decade.

The Way Forward

Considering the six action areas of the Nutrition Decade it is clear that the work programme relies on both direct and indirect nutrition actions. The second action area, which calls for aligned health systems that include essential nutrition actions, basically includes the direct nutrition interventions for which there exists plenty of evidence of their effectiveness, costs and coverage. However, the other action areas call for indirect nutrition interventions, interventions that may not directly address nutrition but that certainly have an impact on nutrition (e.g. trade and investments). For these areas, emerging evidence becomes available but is not yet up to the level of the direct interventions. In order to measure progress and stimulate learning, indicators are needed to be able to measure progress of indirect interventions, along with the earlier suggested double-duty actions, many of which are described in other chapters.

Apart from indicators, more insights are needed as to how the food environment, including trade and marketing measures, taxes and subsidies impact dietary choices. How best to protect consumers from too heavy influence of (f)actors that would convince them to make unhealthy choices? How to convince and facilitate consumers to make the healthier choices? What are the costs and benefits? According to the Global Nutrition Report, the rate of return on investment is 1:16 as a mean. However sometimes the investing sector is not the sector that reaps the benefits, so decisions about investments in nutrition are preferably taken at a higher national level (e.g. the prime minister or the cabinet of the president). At international level, thanks to the Nutrition Decade, nutrition *per se* is tabled at the highest level, the UNGA. Every biennium the FAO and WHO will have to report progress on improving nutrition to the UNGA, an opportunity that should be used strategically to make way for nutrition and sustainable diets that simultaneously address both human and ecosystem well-being and health. That is in the spirit of the SDG and the Nutrition Decade. The Nutrition Decade started early 2016, we only have a little over eight years to go. Let's make use of it.

Notes

[1] The work programme is available at https://www.unscn.org/uploads/web/news/Work-Programme_UN-Decade-of-Action-on-Nutrition-20170517.pdf (accessed 28 June 2018).

[2] The full set of key messages is available at http://www.fao.org/fileadmin/templates/lon/2017_HLPF/EGM2/EGM_SDG2_Key_Messages_11_July_2017.pdf (accessed 28 June 2018).

[3] UNSCN News 42. A spotlight on the Nutrition Decade. Available at https://www.unscn.org/en/resource-center/Unscn-news?idnews=1682 (accessed 28 June 2018).

References

Burlingame, B. and Dernini, S. (Eds) (2012) Sustainable Diets and Biodiversity: Directions and solutions for policy, research and action. *Proceedings of the International Scientific Symposium Biodiversity and Sustainable Diets United against Hunger*, 3–5 November 2010, FAO Headquarters, Rome. http://www.fao.org/docrep/016/i3004e/i3004e.pdf (accessed 28 August 2018).

FAO and WHO (2017) United Nations Decade of Action on Nutrition 2016-2025: Work Programme. Available at http://www.fao.org/3/a-bs726e.pdf (accessed June 28, 2018).

IFPRI (2015) Actions and accountability to advance nutrition and sustainable development. Global Nutrition Report (GNR) Available at http://ebrary.ifpri.org/utils/getfile/collection/p15738coll2/id/129443/filename/129654.pdf (accessed 28 June 2018).

IFPRI (2016) Actions and accountability to advance nutrition and sustainable development, GNR (2016). Available at http://ebrary.ifpri.org/utils/getfile/collection/p15738coll2/id/130354/filename/130565.pdf (accessed 28 June 2018).

Spring Mann, M., Godfray, H., Rayner, M. and Scarborough, P. (2016) Analysis and valuation of the health and climate change cobenefits of dietary change. *Proceedings of the National Academy of Sciences USA* 113(15), 4146–4151. DOI: 10.5287/bodleian:XObxm2ebO

UNDESA (2017) Expert group meeting on strategies for eradicating poverty to achieve sustainable development for all. Available at https://www.un.org/development/desa/dspd/wpcontent/uploads/sites/22/2017/04/report-egm-poverty2017.pdf (accessed 28 June 2018).

UNSCN (2017a) By 2030, end all forms of malnutrition and leave no one behind. Available at https://www.unscn.org/uploads/web/news/document/NutritionPaper-EN-may2017-WEB.pdf (accessed 2 July 2018).

UNSCN (2017b) Schools as a system to improve Nutrition. Available at https://www.unscn.org/uploads/web/news/document/School-Paper-EN-WEB-nov2017.pdf (accessed 28 June 2018).

Vieux, F., Soler, L.-G., Touazi, D. and Darmon, N. (2013) High nutritional quality is not associated with low greenhouse gas emissions in self-selected diets of French adults. *The American Journal of Clinical Nutrition* 97(3), 569–583.

WHO (2012) Sixty-Fifth World Health Assembly, Geneva, 21–26 May 2012: Resolutions decisions and annexes. Available at http://www.who.int/nutrition/topics/WHA65.6_resolution_en.pdf?ua=1 (accessed 2 July 2018).

WHO (2014a) Comprehensive implementation plan on maternal, infant and young child nutrition. Available at http://apps.who.int/iris/bitstream/10665/149018/1/WHO_NMH_NHD_14.2 (accessed 2 July 2018).

WHO (2014b) Global nutrition targets 2025. Policy brief series. Available at http://apps.who.int/iris/bitstream/handle/10665/149018/WHO_NMH_NHD_14.2?sequence=1 (accessed 2 July 2018).

Wijesinha-Bettoni, R., Khosravi, A., Sherman, J., Hernandez-Garbanzo, Y., Vargas, M., Islas-Ramos, A., (2017) Implementing food-based-dietary guidelines for policies, programmes and nutrition education. UNSCN News 42. A spotlight on the Nutrition Decade. Available at https://www.unscn.org/uploads/web/news/UNSCN-News42-2017.pdf (accessed 2 July 2018).

29 Towards a Code of Conduct for Sustainable Diets

Barbara Burlingame

Abstract

In the global policy-setting arena, there are hundreds of international instruments in the form of guidelines, goals, targets, treaties, codes of conduct, declarations, action plans and recommendations covering a myriad of subjects. Some are binding; many are not. For nutrition, diets, and food systems, there are several key international instruments of relevance to the development of a code of conduct for sustainable diets. These can be viewed variously as building blocks for an international instrument for sustainable diets, as in the case of the Universal Declaration of Human Rights, The Code of Conduct for Responsible Fisheries, and the Right to Food. Others can be viewed as a model or template, as in the case of International Code of Marketing of Breast-Milk Substitutes. Still others can be mapped to the elements of sustainable diets, as is shown for Sustainable Development Goal 2. Within the UN system and its member states, codes of conduct are notoriously difficult to usher through, particularly when they require involvement of multiple sectors and disciplines. This chapter reviews the process to date to establish a rationale along with a transdisciplinary code or set of guidelines for sustainable diets. Regardless of the mechanism, the urgency for action associated with the elements of the definition or concept of sustainable diets cannot be overstated.

Introduction

International initiatives in nutrition and those addressing environmental sustainability have largely been sector specific. For many decades, the agriculture sector model for malnutrition focused on food security and dietary energy supply. *Food* was the basic unit of nutrition (FAO, 2003). The health sector model focused on diet-related chronic diseases and micronutrient deficiency diseases. Individual *nutrients* were the basic units of nutrition for the health sector. The disease model for malnutrition required pharmaceutical-types of interventions – hence, 'good' nutrients were delivered to diverse populations as supplements, fortificants and therapeutic formulations; and intakes of 'bad' food components were treated with drugs. Food-based approaches for dealing with micronutrient deficiencies were consequently undermined (Latham, 2010; Englberger, 2012). Adding the environment sector to the mix shows a well-documented litany of problems caused by the 'successes' of increased food production, from biodiversity loss to environmental degradation, including human and animal morbidity and mortality from agricultural chemical contamination of food and water (Ng and von Goetz, 2017). There is a long history of failures and unintended consequences in all sectors' approaches, including interventions of one sector undermining those of the other. Thus, a multi-sectoral, transdisciplinary approach seemed long overdue. Here, the basic unit of nutrition would not be individual nutrients, nor would it be food *per se*. It would have to be 'diet' and it would have to be addressed through an ecosystem approach to ensure sustainability (Sustainable Development Commission, 2009; Burlingame, 2014).

A cross-section of disciplines and sectors were brought together for a series of activities and initiatives under the banner of sustainable diets, one of which was the Platform for Action resolution during the Scientific Symposium on Sustainable Diets and Biodiversity. Seven actions were identified, including the request for a code of conduct for sustainable diets (Burlingame and Dernini, 2012).

Review of International Instruments

Preliminary work had already been undertaken, with several international instruments evaluated as having a relationship to sustainable diets, and as illustrations or models for the basis of a new code. These were vetted through a Technical Workshop (FAO, 2010) and further developed by a Working Group (Burlingame and Dernini, 2012).

In the intervening years, additional international instruments have been developed that speak to the same or similar issues as the original draft for sustainable diets. In this chapter, five key documents are reviewed and compared for the alignment with the notion of sustainable diets.

The Universal Declaration of Human Rights

The most basic international instrument is the Universal Declaration of Human Rights (UN, 1948). There is one mention of food, but no mention of sustainability in any form. Article 25 states unambiguously, 'Everyone has the right to a standard of living adequate for the health and well-being of himself and of his family, including food...'. This is the foundational document for many international instruments.

International Code of Marketing of Breast-Milk Substitutes (Code Breast-Milk)

One of the most powerful and enduring of the nutrition-related international instruments is the International Code of Marketing of Breast-Milk Substitutes (WHO, 1981a).

The basic aim of this code, laid out in its Article 1, is safe and adequate nutrition for infants, and protection and promotion of breast-feeding. Philosophically, there was alignment between breast-feeding and sustainable diets. Considering 'all people' instead of just infants, and substituting 'ecosystems and environmental sustainability' for women or mothers, provided appropriate text for a sustainable diets code. Equally appropriate would have been the substitution of mother with Mother Earth,[1] a term used in the reports leading up to the SDGs (UN, 2014) and in the final SDG Declaration (UN, 2015b).

Two different working groups were charged with the task of developing the draft code of conduct for sustainable diets using the Code Breast-Milk as the model (FAO, 2010, 2012). The congruence between the two, not just with the aim but also with the articles of the Code Breast-Milk, was surprisingly strong. Table 29.1 shows the text from the Code Breast-Milk in the first column, and the draft text for the preamble for the Code for Sustainable Diets in the second column. For example, the text 'Recognizing that the health of infants and young children cannot be isolated from the health and nutrition of women' is readily transformed into 'Recognizing that the health of humans cannot be isolated from the health of ecosystems'.

In addition to the text that provides a preamble to the Code Breast-Milk, there are articles addressing the roles and responsibilities of different stakeholder groups: educators, health workers, food industry personnel, the general public and mothers. The working groups on sustainable diets addressed stakeholders, too, identifying many others in an attempt to bring a multi-sectoral and multidisciplinary approach to the topic. The Code Breast-Milk remains a relevant document with stakeholders, particularly those in civil society organizations, demanding industry accountability.

The Code of Conduct for Responsible Fisheries

The Code of Conduct for Responsible Fisheries (FAO, 1995) promotes food and nutrition along with its core messages on sustainability, and as

Table 29.1. Comparison between the preamble for the Code for the Marketing of Breast-milk Substitutes and the Draft Code for Sustainable Diets.

Code for the Marketing of Breast-Milk Substitutes (WHO, 1981a)	Draft Code Sustainable Diets (FAO, 2010, 2012)
Affirming the right of every child and every pregnant and lactating woman to be adequately nourished, as a means of attaining and maintaining health	Affirming the right of every human being to be adequately nourished, as a means of attaining and maintaining health
Recognizing that infant malnutrition is part of the wider problems of lack of education, poverty, and social injustice	Recognizing that malnutrition is part of the wider problems of lack of education, poverty and social injustice
Recognizing that the health of infants and young children cannot be isolated from the health and nutrition of women...	Recognizing that the health of humans cannot be isolated from the health of ecosystems
Conscious that breast-feeding is an unequalled way of providing ideal food for the healthy growth and development of infants; and that there is an important relationship between breast-feeding and child-spacing	Conscious that food is an unequalled way of providing ideal nutrition for all ages and life stages; *and that there is an important relationship between nutrition and ecosystems*
Recognizing that the encouragement and protection of breast-feeding is an important part of the health, nutrition and other social measures required to promote healthy growth and development of infants and young children	Recognizing that the *encouragement of healthy diets, the protection of ecosystems and* the conservation and sustainable use of food biodiversity is an important part of human well-being
Considering that, when mothers do not breast-feed, or only do so partially, there is a legitimate market for infant formula and for suitable ingredients from which to prepare it; that all these products should accordingly be made accessible to those who need them through commercial or non-commercial distribution systems; and that they should not be marketed or distributed in ways that may interfere with the protection and promotion of breast-feeding	Considering that when ecosystems are not able to support sustainable diets, there is a legitimate use of *other foods, food substitutes and artificial sources of nutrients in the diet;*[a] that all these products should accordingly be made accessible to those who need them through commercial or non-commercial distribution systems; and that they should not be marketed or distributed in ways that may interfere with the protection and promotion of sustainable diets
Recognizing further that inappropriate feeding practices lead to infant malnutrition, morbidity and mortality in all countries, and that improper practices in the marketing of breast-milk substitutes and related products can contribute to these major public health problems	Recognizing further that when ecosystems are able to support sustainable diets, *then* nutrition programmes, policies and interventions supporting the use of *other foods, food substitutes and artificial sources of nutrients in the diet*[a] are inappropriate and can lead to malnutrition, and that the marketing of these can contribute to major public health problems
Appreciating that there are a number of social and economic factors affecting breast-feeding, and that, accordingly, governments should develop social support systems to protect, facilitate and encourage it, and that they should create an environment that fosters breast-feeding, provides appropriate family and community support, and protects mothers from factors that inhibit breast-feeding	Appreciating that there are a number of social and economic factors affecting sustainable diets, and that, accordingly, governments should develop social support systems to protect, facilitate and encourage them, and that they should create an environment that fosters sustainable diets, provides appropriate family and community support and protection from factors that inhibit it

Continued

Table 29.1. Continued

Code for the Marketing of Breast-Milk Substitutes (WHO, 1981a)	Draft Code Sustainable Diets (FAO, 2010, 2012)
Affirming that health-care systems, and the health professionals and other health workers serving in them, have an essential role to play in guiding infant feeding practices, encouraging and facilitating breast-feeding, and providing objective and consistent advice to mothers and families about the superior value of breast-feeding	Affirming that health-care systems, and the health professionals and other health workers serving in them, have an essential role to play in guiding sustainable diet practices, encouraging and facilitating sustainable diets, and providing objective and consistent advice to families, communities and governments about the superior value of sustainable diets
Affirming further that educational systems and other social services should be involved in the protection and promotion of breast-feeding	Affirming further that educational systems and other social services should be involved in the protection and promotion of sustainable diets
Aware that families, communities, women's organizations and other nongovernmental organizations have a special role to play in the protection and promotion of breast-feeding	Aware that families, communities, women's organizations and other nongovernmental organizations have a special role to play in the protection and promotion of sustainable diets
Affirming the need for governments, organizations of the United Nations system, nongovernmental organizations, experts in various related disciplines, consumer groups and industry to cooperate in activities aimed at the improvement of maternal, infant and young child health and nutrition	Affirming the need for governments, organizations of the United Nations system, nongovernmental organizations, experts in various related disciplines, consumer groups and industry to cooperate in activities aimed at the improvement of human and environmental health through sustainable diets

Note: italicized text indicates minor modifications from cited original text.
[a]Other foods, food substitutes and artificial sources of nutrients in the diet include ultra-processed foods, supplements, ready-to-use formulations, fortificants, infant formulas, etc.

such represents a useful model for sustainable diets. The fisheries sector has a long history of linking nutrition and environmental sustainability to its global and regional efforts. The final resolution of this code was an urgent warning as much as a guideline, reiterating 'the vital role of fisheries in world *food security*, and economic and social development, as well as the need to ensure the *sustainability of the living aquatic resources and their environment for present and future generations*'.

In its preface and introduction, we see statements such as these:

- Aquatic resources, although renewable, are not infinite and need to be properly managed, if their contribution to the *nutritional*, economic and social well-being of the growing world's population is to be sustained.
- The code recognizes the *nutritional*, economic, social, environmental and cultural importance of fisheries and the interests of all those concerned with the fishery sector.

Several articles in the code also explicitly address issues that are fundamental to sustainable diets:

- Article 6.7 The harvesting, handling, processing and distribution of fish and fishery products should be carried out in a manner which will maintain the *nutritional value*, quality and safety of the products, reduce waste and *minimize negative impacts on the environment*.
- Article 11.1.6 States and relevant organizations should sponsor research in fish technology and quality assurance and support projects to improve post-harvest handling of fish, taking into account the economic, social, *environmental* and *nutritional impact* of such projects.
- Article 11.2.15: ...ensure that their policies and practices related to the promotion of international fish trade and export production do not result in *environmental degradation* or adversely impact the *nutritional rights* and needs of people for whom fish is critical...

- Article 12.1: States should recognize that responsible fisheries requires…a sound scientific basis to assist…in making decisions…appropriate research is conducted into all aspects of fisheries including…*nutritional science*…taking into account the special needs of developing countries.

The term *nutritional rights* used in Article 11.2.15 above, is particularly poignant in the context of the five international instruments presented in this chapter. No other text is as explicit as this one in declaring nutritional rights. There is a boldness, along with urgency, related to both nutrition and environmental sustainability, making this code a worthy example to highlight in the case for a code of conduct for sustainable diets.

The Right to Food

The full title of this international instrument is The Right to Food, Voluntary Guidelines to support the progressive realization of the right to adequate food in the context of national food security (FAO, 2005). The inclusion of the terms 'voluntary', 'progressive realization', and 'national' represented concessions to a few states that would not affirm that food was a human right. As such, it presents a useful illustration of two things: the tenacity of the individuals and organizations supporting the right, and the power of a few states to over-rule or override both evidence-based and ethics-based international instruments.

There are no specific mentions of sustainable diets, but sustainability issues cover production, land use, water, and so on, and nutrition issues cover undernutrition, overweight/obesity and micronutrient intakes. Included in the articles are recommendations that states and international organizations should consider the benefits of local procurement for food assistance that could integrate the nutritional needs of those affected by food insecurity and the commercial interests of local producers.

Even as the discussions on the right to food were in their early stages, the UN Secretary General established the role of Special Rapporteur on the Right to Food. The first was appointed in the year 2000 and to date there have been three. The issues covered in the annual reports all touch on issues relevant to sustainable diets. Explicit reference to sustainable diets is found in the final report to the General Assembly (UN, 2015a) of the second rapporteur. He conveys the importance of sustainable diets as follows: States should reshape food systems for the promotion of sustainable diets, and the private sector should abstain from imposing nutrition-based interventions where local ecosystems and resources are able to support sustainable diets, and systematically ensure that such interventions prioritize local solutions.

The Sustainable Development Goals

As the Millennium Development Goal (MDG) era was coming to a close, the Sustainable Development Goals (SDGs) were taking shape as the new post-2015 development agenda (UN, 2014). Debates were many as the position of nutrition in the SDGs was the subject of much disagreement. Nutrition had been tied to the poverty goal in the MDGs, as MDG 1, and there was a strong push, particularly from economists, to keep that explicit link with the nutrition goal imbedded in the poverty goal. In the end, the SDGs found greater synergy directly linking nutrition to food and agriculture in its SDG 2 (UN, 2015b). The full title of SDG 2 is 'End hunger, achieve food security and improved nutrition, and promote sustainable agriculture'. Notwithstanding the fact that both nutrition and sustainability targets and indicators are directly and indirectly distributed throughout all the SDGs, SDG 2 is most notably the goal with the clearest relationship to sustainable diets, both in letter and in spirit. Table 29.2 shows an element by element mapping of sustainable diets to the elements within the five targets of SDG 2.

Conclusions

Several things were demonstrated by the exercise of proposing, drafting and promoting an international instrument for sustainable diets.

International instruments, i.e., intergovernmental negotiated texts, are notoriously difficult to achieve. Even when achieved, they are often

Table 29.2. Mapping elements of the definition of sustainable diets to Sustainable Development Goal (SDG) 2.

Sustainable diets	SDG 2
Low environmental impacts	Title: Promote sustainable agriculture Target 2.4 and 2.5 in their entirety
Food and nutrition security	Title: End hunger, achieve food security and improved nutrition Targets 2.1 and 2.2 in their entirety
Healthy life	Targets 2.1 and 2.2 in their entirety
Present and future generations	Title: End hunger, sustainable agriculture Targets 2.4 and 2.5: ensure sustainable production systems, resilient agricultural practices, capacity for adaptation, maintain genetic diversity, soundly managed.
Protective/respectful of biodiversity	Title: Sustainable agriculture Target 2.5 in its entirety: genetic diversity, related wild species
Protective/respectful of ecosystems	Title: Sustainable agriculture Target 2.4 its entirety: sustainable production systems, resilient agricultural practices, maintain ecosystems capacity for adaptation
Culturally acceptable	Target 2.5: traditional knowledge
Accessible	Target 2.1: ensure access to food Target 2.3: equal access to land, other productive resources and inputs, knowledge, financial services, markets Target 2.5: access to equitable sharing of benefits
Economically fair	Target 2.3: equal access to land, other productive resources and inputs, knowledge, financial services, markets
Affordable	Target 2.1: ensure access Target 2.3: double incomes, access to productive resources, non-farm employment Target 2.5: fair and equitable sharing of benefits
Nutritionally adequate	Title: Improved nutrition Targets 2.1 and 2.2 in their entirety
Safe	Target 2.1: safe food
Healthy	Title: Improved nutrition Target 2.1: nutritious and sufficient Target 2.2: end all forms of malnutrition
Optimizing natural resources	Title: Sustainable agriculture Targets 2.4 and 2.5 in their entirety
Optimizing human resources	Target 2.2: all people, infants, adolescent girls, pregnant and lactating women, older persons Target 2.3: women, indigenous peoples, family farmers, etc.

diluted versions of their original purpose. In many forums, the economic interests of a few powerful nations are directly and indirectly working to discredit, undermine, weaken or completely derail negotiations. Regardless of the technical and ethical basis for their content, the outcomes are nonetheless often still political. For example, the Code Breast-Milk was adopted by 118 countries, with a single vote against: United States of America (WHO, 1981b). And despite near-universal agreement that the right to food is a human right, the United States, along with a few other developed countries, refused to join even non-binding resolutions on the subject. It is therefore not surprising the USA has blocked an expert group's recommendation that sustainable diets should be a feature of its food-based dietary guidelines (Merrigan *et al.*, 2015). Several unsuccessful attempts were made in intergovernmental forums to further elaborate the sustainable diets code, yet these were thwarted by blocks of countries imagining adverse economic ramifications (e.g., for the livestock sectors).

There may never be a code of conduct for sustainable diets, as was originally conceived, proposed and partially drafted (Burlingame and

Dernini, 2012). Nevertheless, it is reasonable to pursue the development of best practices for groups of stakeholders, as identified in the draft code. For example, sustainable food-based dietary guidelines represent a reasonably straightforward step for most countries (see Chapter 16). In the meantime, for sustainable diets, there are numerous international instruments, as well as national policies, guidelines and goals, the achievement of which will address the basic concepts envisioned in a code of conduct for sustainable diets. Is this enough for the transformative changes required to avert disasters on both the human and environment fronts from unsustainable diets, that is, the way the planet currently produces and consumes food? It may not be.

Note

[1] Article 59 of the declaration states: '...and we reaffirm that planet Earth and its ecosystems are our common home and that "Mother Earth" is a common expression in a number of countries and regions.'

References

Burlingame, B. (2014) Grand challenges in nutrition and environmental sustainability. *Frontiers in Nutrition* 1, 3. DOI: 10.3389/fnut.2014.00003

Burlingame, B. and Dernini, S. (Eds) (2012) Sustainable Diets and Biodiversity: Directions and solutions for policy, research and action. Proceedings of the International Scientific Symposium Biodiversity and Sustainable Diets United against Hunger, 3–5 November 2010, FAO Headquarters, Rome. http://www.fao.org/docrep/016/i3004e/i3004e.pdf (accessed 28 August 2018).

Englberger, L. (2012) Revisiting the vitamin A fiasco, Going local in Micronesia. In: Burlingame, B. and Dernini, S. (eds) *Sustainable Diets: Directions and Solutions for Policy, Research and Action*. Food and Agriculture Organization of the United Nations, Rome, Italy.

FAO (1995) Code of conduct for responsible fisheries. Food and Agriculture Organization of the United Nations, Rome, Italy. Available at http://www.fao.org/3/a-v9878e.pdf (accessed February 2018).

FAO (2003) Measurement and assessment of food deprivation and undernutrition. In: *Proceedings of an International Symposium*. Food and Agriculture Organization of the United Nations, Rome, Italy.

FAO (2005) Right to food guidelines to support the progressive realization of the right to adequate food in the context of national food security. Available at http://www.fao.org/3/a-y7937e.pdf (accessed February 2018).

FAO (2010) Report of the technical workshop biodiversity in sustainable diets. Available at http://www.fao.org/ag/humannutrition/24994-064a7cf9328fbe211363424ba7796919a.pdf (accessed February 2018).

Latham, M. (2010) The great vitamin A fiasco. *World Nutrition* May 2010, 1, 1, 12-45. Available at https://hetv.org/pdf/the-great-vitamin-a-fiasco-world-nutrition-may2010.pdf (accessed February 2018).

Merrigan, K., Griffin, T., Wilde, P., Robien, K., Goldberg, J. and Dietz, W. (2015) Designing a sustainable diet. *Food Science* 350(6257), 165–166. DOI: 10.1126/science.aab2031

Ng, C.A. and von Goetz, N. (2017) The global food system as a transport pathway for hazardous chemicals: the missing link between emissions and exposure. *Environmental Health Perspectives* 125, 1–7. DOI: 10.1289/EHP168

Sustainable Development Commission (2009) Setting the Table: Advice to Government on priority elements of sustainable diets. Available at http://www.sd-commission.org.uk/data/files/publications/Setting_the_Table.pdf (accessed February 2018).

UN (1948) General Assembly Resolution 217 A: The Universal Declaration of Human Rights. Available at http://www.un.org/en/universal-declaration-human-rights/index.html (accessed February 2018).

UN (2014) General Assembly, 68th Session. Report of the Open Working Group of the General Assembly on Sustainable Development Goals (A/68/970). United Nations, New York, USA.

UN (2015a) General Assembly, Human Rights Council Twenty-fifth Session. Report of the Special Rapporteur on the Right to Food, Olivier De Schutter; Final report: The transformative potential of the right to food (A/HRC/25/57). Available at http://www.srfood.org/images/stories/pdf/officialreports/20140310_finalreport_en.pdf (accessed February 2018).

UN (2015b) General Assembly, 70th Session. Transforming our world: the 2030 Agenda for Sustainable Development (A/RES/70/1). Available at http://www.un.org/ga/search/view_doc.asp?symbol=A/RES/70/1&Lang=E (accessed February 2018).

WHO (1981a) International Code of Marketing of Breast-milk Substitutes. Available at http://www.who.int/nutrition/publications/code_english.pdf (accessed February 2018).

WHO (1981b) Thirty-fourth World Health Assembly. Available at http://apps.who.int/iris/bitstream/10665/155680/1/WHA34_1981-REC-2_eng.pdf (accessed Feb 2018).

Index